T0251559

BIOLOGICAL EFFECTS OF SURFACTANTS

BIOLOGICAL EFFECTS
OF SURFACTANTS

S. A. Ostroumov

Moscow State University
Russia

Taylor & Francis
Taylor & Francis Group
Boca Raton   London   New York

A CRC title, part of the Taylor & Francis imprint, a member of the
Taylor & Francis Group, the academic division of T&F Informa plc.

Published in 2006 by
CRC Press
Taylor & Francis Group
6000 Broken Sound Parkway NW, Suite 300
Boca Raton, FL 33487-2742

International Standard Book Number-10: 0-8493-2526-9 (Hardcover)
International Standard Book Number-13: 978-0-8493-2526-7 (Hardcover)
Library of Congress Card Number 2005053818

Library of Congress Cataloging-in-Publication Data

Ostroumov, Sergei Andreevich.
   Biological effects of surfactants / S.A. Ostroumov.
      p. cm.
   Includes bibliographical references and index.
   ISBN 0-8493-2526-9
   1. Aquatic ecology. 2. Surface active agents--Physiological effect. I. Title.

QH541.5.W3082 2005
577.6'27--dc22
                                                  2005053818

Taylor & Francis Group
is the Academic Division of Informa plc.

Visit the Taylor & Francis Web site at
http://www.taylorandfrancis.com

and the CRC Press Web site at
http://www.crcpress.com

Contents

Foreword

The book presents new results of research into the biological effects of surfactants on autotrophic and heterotrophic organisms. The author studied anionic, noniono-genic, cationic surfactants, and commercial surfactant-containing mixtures. Surfact-ants interact with membranes and pose a hazard to living organisms. Test organisms studied are the major blocks and trophic levels of aquatic ecosystems; they include bacteria, cyanobacteria, algae, flagellates, higher plants, and invertebrates. Support from various funds enabled the author to conduct his research in Russia, Ukraine, the U.S., and the U.K. and greatly expand the set of organisms studied and methods used.

The results of experiments were analyzed in relation to the assessment of eco-logical hazard to organisms and aquatic ecosystems, water self-purification studies and nature preservation priorities. These issues are important, and collection of new information for their analysis is useful.

The results of his work were reported by the author, who has become a leader in this field of biological, ecological, and environmental sciences, at representative forums in Russia and other countries, including the Joint Plenum of the Hydrobiolo-gical Society, Russian Academy of Sciences; the Scientific Council on Hydrobiology and Ichthyology, Russian Academy of Sciences; and Interdepartmental Ichthyo-logical Commission of the Russian Federation; symposia of the American Society of Limnology and Oceanography, and International Association for Theoretical and Applied Limnology; conferences in the U.K., Switzerland, Denmark, and Finland; and workshops in the U.S., Germany, the U.K., the Netherlands, and Belgium. Now the time has come to generalize the results of this valuable and useful work. The book is useful and interesting to research scientists in various fields, and also to post-graduate students and university teachers.

M.E. Vinogradov
Academician
Russian Academy of Sciences

Professor V.D. Fedorov
DSc (Biology)
Chair of the Department of
Hydrobiology, Moscow State
University

Foreword

I have been in correspondence with Dr. Ostroumov since 1995 and have worked with him to organize two sessions of the annual meetings of the American Society of Limnology and Oceanography. He serves on the editorial board for the book, *Phytoremediation: Transformation and Control*, of which I am co-editor. I am but one of several scientists from Germany, the U.K., the U.S., and several other countries that Dr. Ostroumov has attracted and helped self-organize to focus on the ecological basis of the assimilative capacity of surface waters.

This new book by Dr. Ostroumov sums up his specific contributions on the effects of surfactants and detergents on aquatic ecological systems, which are vital to understanding how this important class of compounds affects our environment. Without surfactants, including those found in common hand soap, modern societies would be hard pressed to function. Surfactants in soaps, detergents, and many other forms stop the spread of many diseases from hand-to-hand contact, are important in manufacturing for de-greasing metal surfaces and many other uses, have made possible many medical breakthroughs, and allow many other conveniences that are taken for granted.

While our ecosystems have evolved in the presence of natural surfactants, man-made surfactants have many acutely toxic and subtle effects. These priceless ecosystems can only be protected and managed through understanding of the effects on individual organisms, populations, communities, and ecosystems as a whole. As the dominance of man over the natural ecosystem will only increase in the short term, understanding of the effects of anthropogenic compounds becomes more important. The effect on waste assimilative capacity and the biological "machinery" or biogeo-chemical cycles of the planet must be known to preserve diversity and the intrinsic value of life on Earth.

Dr. Ostroumov's work is exceptional in that it serves as a point of focus for those in many disciplines to have another look at (1) the traditional sanitary engineering idea of stream, lake or estuary waste assimilative capacity used in the U.S. and elsewhere, (2) the eastern European concept of hydrobiology, and (3) Russian concept of biogeochemical cycling in surface waters. Sergei Ostroumov is pioneering the application of ecological and biological principles to redefine assimilative capacity, especially beyond the effects of bacteria and other microbes and nutrient cycling. In particular, I am excited about developing concepts to explore the hazardous waste assimilative capacity in broad ranges of aquatic and terrestrial ecosystems. I look forward to developing rigorous protocols and designing methods for the emerging field of phytoremediation (using green plants, especially the plant metabolic system to degrade hazardous wastes) based on general concepts that I see Dr. Ostroumov developing in his recent work.

I am pleased to have this opportunity to comment on the scientific leadership of Sergei A. Ostroumov. The book is highly recommended to those who are involved in studying ecology and solving environmental problems.

Steven C. McCutcheon, PhD
University of Georgia
U.S. Environmental Protection Agency
President, American Society of
Ecological Engineering

Acknowledgments

The author thanks Prof. V.D. Fedorov and many colleagues (Department of Hydrobiology, Faculty of Biology, Moscow University; Russian Academy of Sciences; American Society of Limnology and Oceanography; and other institutions) for cooperation and criticism. The scientific conferences and seminars organized by ASLO, SIL, Moscow University, Russian Academy of Sciences, U.S. EPA, SETAC, ECOTOX, WHOI, Plymouth Marine Laboratory, SUNY, Columbia University, University of Maryland, EAWAG, TNO, UFZ, Institute of Freshwater Ecology (Berlin), and some other national and international institutions were instrumental in better viewing important scientific problems. The author is grateful to Prof. N. Fisher, Prof. Ronald Weiner, Prof. J. Waterbury, Prof. John Widdows, Dr. P. Donkin, Prof. S.S. Stavskaya, Dr. G.A. Filenko, and Prof. Dr. N. Walz for providing laboratory space and facilities for doing research; Prof. R. Wetzel, Prof. Rita Colwell, Prof. G. Likens and Academicians (Members of the Russian Academy of Sciences) Prof. M.E. Vinogradov, Prof. D.S. Pavlov, Prof. A.F. Alimov, Prof. M.V. Ivanov, Prof. V.N. Bolshakov, Prof. G.V. Dobrovolsky, Prof. G.A. Zavarzin, Prof. G.G. Matishov, Prof. V.P. Skulachev, Prof. L.P. Rysin, Prof. T.I. Moiseenko, Prof. A.V.Tsyban, Prof. A.V. Yablokov, Prof. D.A. Krivolutsky, Prof. V.V. Malakhov, Prof. E.A. Kriksunov, Prof. G.S. Rozenberg, and Prof. V.M. Zakharov for advice; Prof. J. Widdows, Prof. Dr. Christian Steinberg, Prof. Dr. Norbert Walz, Prof. Dr. Henri Dumont, Prof. Curtis J. Richardson, Prof. S.M. Adams, Prof. Nico M. van Straalen, Prof. R. Newell, Dr. Rita Triebskorn, Prof. B.A. Kurlyandsky, Prof. V.S. Petrosyan, Prof. R. Weiner, Dr. Steven McCutcheon, Dr. T. Feijtel, Dr. P.J. van den Brink, Dr. E. Kristensen, Prof. A.P. Melikian, Prof. V.D. Samuilov, Prof. O.F. Filenko, Dr. N.S. Zhmur, and Dr. M. Scholten for discussions; N.E. Zourabova, Prof. David Page, Dr. M. Marcus, Dr. N.N. Kolotilova, Dr. N.V. Revkova, Dr. M.P. Kolesnikov, Dr. A.V. Smurov, Dr. A.G. Dmitrieva, Prof. G.E. Shulman, Dr. G.A. Finenko, Dr. Z.A. Romanova, and other colleagues at the Institute of the Biology of Southern Seas, National Academy of Ukraine, for help. Some mollusks were provided by the Institute of the Biology of Southern Seas (Ukraine) and PML (U.K.).

The author thanks Professor P.J. Wangersky, Dr. M. Caldwell, Mr. Allen Hill, Mr. Glenn Kempf, and Ms. E. Schuster for valuable comments and help in editing the text.

This work was in part supported by the MacArthur Foundation (Research and Writing Initiative of the Program on Global Security and Sustainability), Research Support Scheme of the Open Society Support Foundation (Grant No. 1306/1999), EERO, IBG, and individual sponsors (Mr. V.Ya. Etin, Mr. J. Ostroumoff, and others). The author thanks Prof. D. Page, Mr. J. Kessler, Prof. Dr. O. Kinne, Prof. Dr.

C. Steinberg, Prof. Dr. N. Walz, Prof. A. Hooke, Prof. M. Brody, Prof. R. Krueger, Dr. Peter Kuschk, Dr. Peter Donkin, Mr. F. Staff, and others for support.

Profound gratitude is expressed to T.A. Ostroumova for her everlasting help.

Special thanks are due to Dr. A. Schramm (University of Washington, U.S.); Dr. R.F. McMahon (The University of Texas at Arlington, U.S.); Prof. L. Mayer (University of Maine, U.S.); H.G. Dam (University of Connecticut, U.S.); Prof. N.G. Hairston, Jr. (Cornell University, U.S.); Prof. D. Pimentel (Cornell University); Prof. E. Laws (University of Hawaii, U.S.); Prof. L. Hekanson, Dr. C.M. Lindblom (Uppsala University, Sweden); Dr. B.W. Hansen (Roskilde University, Denmark); Dr. D. Hamilton (University of Western Australia, Crawley, Australia); Dr. S.C. Maberly (Center for Ecology and Hydrology, Windermere, U.K.); Dr. W. Quayle (U.K.); Prof. D.M. Paterson (St. Andrews University, Scotland, U.K.); Dr. N. Bahamon (Centro de Estudios Avanzados de Blanes, Spain); Prof. E. Prepas (University of Alberta, Canada); Dr. M.A. Belmont (Trent University, Canada); Prof. Dr. Thomas Weisse, Executive Director (Institute for Limnology, Austria); Dr. J. Boenigk (Austria); Prof. G.J. Herndl, Head of Department of Biological Oceanography, Netherlands Institute for Sea Research (NIOZ); Prof. K. Jazdzewski (Poland); Dr. Y.Z. Yacobi (Kinneret Limnological Laboratory, Israel); Dr. D. Angel (Israel); Dr. T. Moens (University of Gent, Belgium); Dr. T. Noji (Institute of Marine Research, Norway); Prof. Dr. M.A.H. Saad (Egypt); Prof. Wang Ziulin (The Ocean University of Qingdao, China), and many others for discussions and support of the idea of publishing this book.

The author thanks the translator of the book, Victor Selivanov, for his time and effort.

Abbreviations

AAL   apparent average length
ABS   alkyl benzene sulfonate
AE   alcohol ethoxylate
AES   alcohol ethoxysulfate
AHC   Avon Herbal Care
AS   alkyl sulfate
ATM   alkyl trimethyl ammonium

BA   biological activity
BAS   biologically active substance
BCF   bioconcentration factor
BCO   biochemical consumption of oxygen

CHMA   copolymer of hexene and maleic aldehyde
CL   confidence limit
CMC   critical micelle concentration
CSSD   change of the static state to the dynamic state
CTAB   cetyl trimethyl ammonium bromide
CV   coefficient of variation

DOM   dissolved organic matter
DNOC   dinitroorthocresol
DDA   dimethyl dioctadecyl ammonium bromide
DDTMA   dodecyl trimethyl ammonium bromide
DTMAC   dodecyl trimethyl ammonium chloride
DW   distilled water

EC_{50} effective concentration, i.e., the concentration at which
               the biological effect is 50% of the maximum possible effect
EER   effect on the efficiency of removal
EPA   Environmental Protection Agency

GIC   germination inhibition coefficient

HLB   hydrophilic-lipophilic balance
HPSS   high-polymer synthetic surfactant

LAS	linear alkyl benzene sulfonate
LC$_{50}$	concentration that induces the death of 50% of organisms
LTMAC	lauryl trimethyl ammonium chloride
MBA	minimum bacteriostatic activity
MCPC	multicatalytic proteinase complex
MPC	maximum permissible concentration
NOEC	no observed effect concentration
NP	nonylphenol
NPE	nonylphenolethoxylate
NST	nonspecific symptom of toxication
OD	optical density
QAC	quaternary ammonium compound
SDS	sodium dodecyl sulfate
SFG	scope for growth
SML	surface microlayer (of the sea)
STW	settled tap water
TCP	2,4,6-trichlorophenol
TDTMA	tetradecyl trimethyl ammonium bromide
TX100	nonionogenic surfactant Triton X-100

Preface

For about 20 years now, I have been doing research into anthropogenic effects on hydrobionts. In experimental studies of the impacts of synthetic surfactants on hydrobionts and other organisms, the author arrived at the conclusion that these substances are greater potential ecological hazards than previously believed. This conclusion led me to focus on the fundamental problem of the criteria and principles of assessing the hazards. Deeper insight into the problem highlighted the need for a more profound analysis of the extent to which the self-purification potential of an aquatic ecosystem can be a target of pollutants.

I attempted to answer this question in a series of papers published in 1997–2005 (Ostroumov et al. 1997, Ostroumov 2000, 2005a,b, etc.). In summing up the results of the entire work, it proved useful to return to the earlier published books (1983, 1985, 1991 coauthored by A.V. Yablokov) based upon the analysis of a broad range of material in accordance with the levels of living matter organization. That approach proved useful. In a concise form, the author's work was summed up in his book published in Russian in 2000 and in recent publications (Ostroumov 2005a,b).

This book makes use of and is based upon all those publications. In the process of translating the book from Russian into English, I realized that some terms need to be commented upon. The term "hydrobiont" that is used in the book is a synonym to "aquatic organism." The term "biogens" is a synonym to "nutrients," especially the dissolved phosphorus and nitrogen compounds that are easily available for algae and may lead to algal growth.

Moreover, I realized that an addendum should be written to cover some recent literature which was not mentioned in the book, just because it was written some years ago. This addendum is presented in the English edition of the book.

I would like to emphasize that the book is by no means a comprehensive review of literature on environmental aspects of surfactants. The book is based on my experimental work that used only a limited number of species and experimental approaches. I used only the most necessary literature to discuss my data. I tried to propose and substantiate some new elements in vision of the environmental hazards of surfactants and detergents. The book is only a contribution to the enormously broad area of research – a contribution made by one individual who is by definition limited in his resources.

I am grateful to the many people who helped in this work and participated in it directly or indirectly. Thanks are due to the coauthors and colleagues V.D. Fedorov, A.V. Yablokov, V.N. Maksimov, N.N. Kolotilova, M.P. Kolesnikov, S.V. Goryunova, T.N. Kovalyova, A.Ya. Kaplan, N.V. Kartashova, A.N. Tretyakova, N.A.

Semykina, N.E. Zourabova, E.V. Borisova, V.S. Khoroshilov, and A.E. Golovko; the staff of Papanin Institute of Biology of Inland Waters, Russian Academy of Sciences; Institute of Biology of the Southern Seas, National Academy of Sciences of Ukraine; and many others, including H. Nagel, P. Donkin, R. Weiner, and N. Fisher. Important were helpful comments made by many colleagues, to whom the author is deeply grateful. Critical remarks by and advice of Prof. V.D. Fedorov were of great help. Thanks for critical remarks are also due to T.V. Koronelli, O.F. Filenko, and A.G. Dmitrieva. Unable to express here his gratitude to all he would like to acknowledge, the author does this in Acknowledgments.

I did not even try to review the enormous amount of literature in the field; my aim was only to study some specific issues within a very vast area.

I realize that, in a work touching upon many organisms and analyzing controversial and difficult issues of hydrobiology and ecology, it is impossible to avoid blunders and errors. I do not shift the burden of these shortcomings from myself onto those to whom I express my gratitude; the responsibility for the possible shortcomings of the book is entirely mine.

The book is dedicated to my parents and my friends, to those who helped me and will help in future.

Sergei Ostroumov
Department of Hydrobiology
Faculty of Biology
M.V. Lomonosov Moscow State University
Moscow, Russia

Introduction

Where is the wisdom we have lost in knowledge?
Where is the knowledge we have lost in information?

T.S. Eliot (1888–1965)

One of the main problems in understanding the role of aquatic biota as an essential part of the biosphere (Vernadsky 1926, 1944) and protecting its biodiversity is a correct distribution of the efforts and limited means allocated for these purposes. Hence, of great importance are the concepts of the ecological hazards of anthropogenic impact (Fedorov 1977, 1980, 1987), a system of nature-preservation priorities (Zakharov 1999), and an understanding of the role of concrete groups of chemical substances that enter aquatic ecosystems.

Importance of a system of priorities in preventing pollution of water bodies and recognizing the impacts of anthropogenic substances on aquatic ecosystems and their components has been emphasized by the new environmental laws of the Russian Federation and the Concept of the Transition of the Russian Federation to Sustainable Development. The principles of sustainable development (The Concept ... 1996, Rozenberg et al. 1999) and the law require an analysis and prediction of the ecological situation to be made and ecological examination of projects associated with impacts on the aquatic environment to be conducted. Among the potentially hazardous impacts on the aquatic ecosystems is their pollution with chemical substances.

The literature shows a contradictory picture with respect to the general assessment of the extent of hazard, which synthetic surfactants pose by having their impact on hydrobionts. Some authors considered them major pollutants (Konstantinov 1979, Patin 1979, Filenko 1988, Steinberg et al. 1995), though surfactants are frequently put at the end of the list, after the traditional priority substances being discussed. On the other hand, many authors did not include them on the lists of major pollutants and did not pay them any significant attention (Wilson and Fraser 1977, Moore and Ramamoorthy 1984, Maki and Bishop 1985, Rand and Petrocelli 1985, Rosenbaum 1991, Bro-Rasmussen et al. 1994, World Resources 1994–1995, Bailey 1996). Most synthetic surfactants are considered to be substances of class 4 hazard (moderately hazardous, i.e., of the lowest rank from the point of view of the compilers of the list of substances) or substances of class 3 hazard, and the ones considered to be more hazardous are included into class 1 and class 2 hazards (Maximum Permissible Concentrations of Chemical Substances in Water of Aquatic Objects of Household and Social Amenities Water Use 1998). The tendency of assigning synthetic surfactants to class 3 or 4 (lower priority substances) is preserved in the ecological rate setting for substances that pollute fisheries waters.

As detailed as possible clarification of biological effects caused by synthetic surfactants in their impact on hydrobionts is required because these substances are contained in sewage and polluted waters discharged by practically all industries, in domestic and community sewage, in preparations for dispersion of oil in oil spills and in consequences of emergencies at water bodies. The content of synthetic surfactants in sewage waters of industrial enterprises reaches values as high as 30 g/l (Stavskaya et al. 1988).

Despite a vast literature and efforts by many scientists, (1) criteria for assessing and ranking substances by the extent of hazard for aquatic ecosystems, and (2) assessment of the potential hazard, which synthetic surfactants pose for hydrobionts remain insufficiently developed.

Aims and scope of the study. The aim of the study was to assess the potential hazard of a possible pollution of water bodies by synthetic surfactants based on revealing and characterizing the biological effects of synthetic surfactants and surfactant-containing preparations on autotrophic and heterotrophic organisms.

The major objectives were as follows:

1. To study the effects caused by synthetic surfactants in their impact on hydrobionts and other organisms (from prokaryotes to eukaryotes) and to obtain new data that can be used to analyze the regularities of the interaction of the substances with hydrobionts.
2. To use the results obtained and conclusions developed based on them to determine to what extent synthetic surfactants can be considered as a group of substances potentially hazardous for hydrobionts and aquatic ecosystems.
3. To elucidate which organisms are comparatively more tolerant to the effects of synthetic surfactants and determine what is required for remediation.

Scientific novelty. The new results of the author's research are presented below. New biological effects of the impacts of synthetic surfactants on hydrobionts and other organisms were revealed; new quantitative characteristics of the earlier known effects were established. Studies of the impacts of synthetic surfactants on autotrophic organisms established inhibition of growth of diatoms *Thalassiosira pseudonana* (Hustedt) Hasle et Heimdal, inhibition of growth of euglens, disturbance of growth and development of angiosperm plants, including inhibition of elongation of plant seedlings (*Sinapsis alba* L., *Fagopyrum esculentum* Moench, *Lepidium sativum* L., *Oryza sativa* L., etc.), and growth of aquatic macrophytes (*Pistia stratiotes* L.). Action of synthetic surfactants was found to disturb the morphogenetic processes in the rhizoderm that lead to the formation of root hairs. Studies of the bioeffects of synthetic surfactants on heterotrophic organisms were found to inhibit growth of marine bacteria (prosthecobacteria *Hyphomonas* sp.), filtration activity of marine and freshwater bivalve mollusks (*Mytilus edulis* L., *M. galloprovincialis* Lamarck, *Crassostrea gigas* Thunberg, *Unio tumidus* Philipsson, *U. pictorum* L.) and to change the behavior of annelids *Hirudo medicinalis* L. Based on the results obtained, some new approaches to a more adequate analysis of potential ecological hazard of the impact of anthropogenic stress on aquatic ecosystems were developed.

Theoretical significance of the work is in the development of a four-level concept of a level-block approach to the analysis of potential ecological hazards of anthropogenic effects of synthetic surfactants on the biota. As a result of synthetic surfactant effects, interactions between hydrobionts of various trophic levels can get unbalanced. We substantiated the importance of taking into account the biological effects of sublethal concentrations of synthetic surfactants and other aquatic pollutants, as well as the necessity to study the impacts of anthropogenic substances on a broader range of biological objects and ecosystem functions associated with water self-purification.

Practical importance. We developed approaches to a more adequate assessment of potential hazard of anthropogenic factors, including synthetic chemical substances entering aquatic ecosystems. We expanded the arsenal of methods for assessing the biological activity of substances (the morphogenetic index – the calculation of the apparent average length of seedlings, the method of assessing the impact on the formation of root hairs, the method of assessing the impact on the removal of unicellular organisms from water in the process of its biofiltration), proposed informative indices and variants of methods for assessing the biological activities of substances, which can act on hydrobionts when getting into aquatic ecosystems. The variants of methods proposed and tested are applicable for assessing and characterizing not only pollutants, but also a broader range of biologically active substances, as well as for assessing the efficiency of purification of polluted waters. We obtained data useful in selection of relatively tolerant organisms for purposes of bio- and phytoremediation and restoration of disturbed aquatic ecosystems. New opportunities of using more humane methods of biotesting without killing warm-blooded animals were expanded. The results are important for understanding potential hazards emergencies associated with massive contamination of water bodies with synthetic surfactants. The results were used in reading lecture courses for students of Moscow State University and in writing study guides. The books published by the author previously presented some of his earlier results. Those books were used in the teaching process by several universities in the Russian Federation and in Czech Republic.

Publications. Over 90 publications on the subject of the work, including 7 books.

The recent results obtained after publishing the Russian edition: see Addendum.

1 Anthropogenic Impacts and Synthetic Surfactants as Pollutants of Aquatic Ecosystems

1.1 Criteria and Priorities in Assessing the Hazardous Impacts on Aquatic Biota

The state of aquatic ecosystems reflects the general state of the biosphere. The situation in the biosphere affected by anthropogenic factors was characterized as "a slow explosion" (Fedorov 1987). The global change in the biosphere and climatic system of the Earth is a manifestation of this "slow explosion" (World Resources 1990–1991, Izrael et al. 1992). This change is due to man-made impact and disturbances in the aquatic and terrestrial ecosystems, which take part in the formation and regulation of biogeochemical and energetic fluxes in the biosphere (Fedorov 1987, 1992; Abakumov 1993; Kuznetsov 1993; Losev et al. 1993; Gorshkov 1987; Lovelock and Kump 1994; Lovelock 1995). The existing trends in increasing of anthropogenic changes in the ecosystems are unfavourable for preserving the biodiversity and form a dangerous basis for emergency and extraordinary situations (Izrael et al. 1992; Kondrasheva and Kobak 1996; Edgerton 1991; Gore 1992; Choucri 1993). The predicted events unfavourable for the aquatic and terrestrial ecosystems would occur within the lifetime of the current generation: the doubling of the concentration of CO_2 in the atmosphere as compared with the preindustrial level would occur in the mid- or second third of the 21st century (Kondrasheva and Kobak 1996; World Resources 1990–1991; Edgerton 1991; Gore 1992; Choucri 1993), i.e., within the lifetimes of people who were born not long ago. The rate of increase of the CO_2 level in the atmosphere does not slow down.

The trends of anthropogenic changes hazardous for the biodiversity of hydrobionts (aquatic organisms) were analyzed in many publications (Fedorov 1974, 1977, 1980, 1992; Ostroumov 1981, 1984, 1986a,b, 1989; Yablokov and Ostroumov 1983, 1985; Yablokov and Ostroumov 1991; Venitsianov 1992; Khublaryan 1992; Shiklomanov 1992, Yakovlev et al. 1992; Losev et al. 1993; Moiseyenko 1999).

In 1996, Russia passed the Concept of the Transition of the Russian Federation to Sustainable Development, which became a document to be taken into account by the Government in working out the programs for social and economical development, preparing the regulatory legal acts and making decisions (Decree of the President of the Russian Federation #440, 1996). The Concept was developed and raised to the rank of mandatory conceptual basis for decision making at the highest level in Russia mainly due to a new step in the development of the international community, which was the United Nations Conference on Environment and Development (Rio de Janeiro, 1992) and the program documents adopted by that conference.

The Concept noted that "civilization, using a great variety of technologies destroying the ecosystems, did not in fact suggest anything that could substitute for the regulation mechanisms of the biosphere." The importance of "the natural environmental biotic regulation mechanism" was emphasized. The ideas and suggestions put forward by experts (Venitsianov 1992; Khublaryan 1992; Fedorov 1992; Shiklomanov 1992; Yakovlev et al. 1992; Losev et al. 1993; Moiseyenko 1999) in the field of studies and preservation of aquatic ecosystems, which are water resources for this country, are in accord with this Concept.

To optimize the relations between man and the biosphere, it is necessary to minimize the harmful impacts of chemical pollution on hydrobionts. "The insalubrity of pollutants with respect to man, particular agricultural organisms (plants and animals), and the biotic component of the ecosystem or biosphere as a whole could be considered the main property defining their 'quality.' In other words, the insalubrity is considered as a property of pollutants to cause undesirable, harmful, hazardous, or disastrous changes in the living organisms" (Fedorov 1980, p. 26). Analysis of the ecological hazard caused by pollution of the environment emphasized the danger of disturbing the balance of the ecological processes and the sustainability of the ecosystems (Fedorov 1992). The hazard of the water-polluting substances and xenobiotics as well as other details of the impact of chemical substances on hydrobionts and other organisms is analyzed in Stroganov (1976a,b; 1979; 1981), Patin (1979, 1997), Abakumov (1980), Lukyanenko (1983), Alabaster and Lloyd (1984), Izrael (1984), Filenko (1988), Flerov (1989), Malakhov and Medvedeva (1991), Bezel et al. (1994), Ostroumov (2002, 2004, 2005a,b), and others.

Significant conceptual problems exist on the road leading towards progress in understanding the impacts of chemical substances on aquatic ecosystems (Ostroumov et al. 2003).

The following principal problems are yet unsolved. What is the ecological hazard of a substance? Which aspects of the impacts of chemical substances on aquatic biota are the most important? How should the priorities among the diversity of biotic distortions caused by anthropogenic substances be systemized and ranked? It is not by chance that to date the Russian Federation has no generally recognized or certified methods to determine the ecological risk caused by chemical pollution (Krivolutsky 1994). In order to find systematic and ecologically based approaches, we developed a concept of the analysis of anthropogenic impacts on living nature in accordance with the levels of organization of living systems (Yablokov and Ostroumov 1983, 1985, 1991). The concept was supported by other authors (e.g., Lavrenko 1984; Gilyarov 1985).

The main aspects of the problem include the necessity to analyze ecosystem consequences of the effect of xenobiotics on hydrobionts (Patin 1979; Fedorov 1980; Ostroumov 1984, 1986a, 2002, 2005a,b; Filenko 1988; Korte et al. 1997), the expansion and improvement of the arsenal of biotesting methods (Filenko 1988; Flerov 1989), as well as the need for more detailed studies of the biological activities of some large groups of substances not sufficiently studied before, including synthetic surfactants.

1.2 Ecological Hazard and Ecosystem Consequences of the Effect of Anthropogenic Substances on Hydrobionts

The system of assessing the environmental hazards of chemical substances in force in the European Union countries is based on three criteria: (1) acute toxicity (based on lethal concentrations (LC_{50})) for three groups of organisms (algae, daphnia, fish); (2) liability of substances to biodeterioration by microorganisms; and (3) ability of a substance to bioaccumulate (De Bruijn and Struijs 1997). A substance is considered to be low hazardous or not hazardous if it has a low toxicity (high LC_{50} values for the specified organisms), high ability of degradation (oxidation) by microorganisms, and if no bioaccumulation occurs or the bioaccumulation coefficient is smaller than 1000. While each of these criteria has its own merits, the very concept of hazard assessment based on this triad appears to be vulnerable to criticism from the hydrobiological point of view. Some of the critical comments are as follows: (1) the concept underestimates the possibility of a low value of LC_{50} for other organisms; (2) the capability of rapid degradation (oxidation) by microorganisms guarantees no ecological safety as the process of rapid oxidation of the chemical(s) is accompanied by rapid consumption of oxygen from water, which is fraught with hypoxia undesirable for the other oxygen-consuming hydrobionts; (3) bioaccumulation is not a necessary prerequisite for a negative impact to be manifested, as the substance can affect receptors of an organism, and this does not require its penetration into the tissues and cells of the organism. Thus, there is a need for further conceptual search for the approaches and priorities of estimating the hazards of substances for aquatic biota.

The criteria based on which the hazardous impacts of anthropogenic substances on the ecosystems should be assessed have not been finally elaborated yet (Stroganov 1976a,b; Abakumov 1979, 1985; Yablokov and Ostroumov 1983, 1985; Filenko 1988; Krivolutsky and Pokarzhevsky 1990; Yablokov and Ostroumov 1991; Bezel et al. 1994; Krivolutsky 1994, Korte et al. 1997, Ostroumov et al. 2003, and others).

Two groups of assessments of the states of ecological systems are distinguished (Fedorov 1980). The first group are integral indices, characterizing a result at the time of registration, such as biomass, number of species, and ratio of abundance, as well as various indices of species variety, diversity, relative abundance, domination, etc. (Fedorov 1980, p. 32). The second group are indices that can be expressed as a time derivative, i.e., as the rate of change of a function – such as productivity, respiration, and assimilation of substances (Fedorov 1980, p. 33).

The answers to the problem of how to reveal, characterize and rank anthropogenic changes in ecosystems, especially under the influence of pollution, continue to be developed. Transition of a population or an ecosystem from one dynamical regime to another can be triggered by small changes in the anthropogenic impact on the population (Bolshakov et al. 1987). Some of the most notable and assessible ecosystem changes are used as indicators of disturbances in an aquatic ecosystem, and those hydrobionts that prove capable of revealing and assessing the ecosystemic disturbances act as indicator organisms (see, e.g., Vinberg et al. 1977; Abakumov 1983; Vetrov and Chugay 1988; Abakumov and Maksimov 1988; Abakumov and Sushchenya 1991; Budayeva 1991). Several systems of bioindicator organisms and hydrobiological methods were developed.

The following methods and approaches are used to assess the state of aquatic ecosystems under increasing anthropogenic loads: systems using the Trent Biotic Index, Extended Biotic Index, the Verno and Taffy index, Chandler scores, Chatter biotic index, method of Pantle–Buck indicator organisms in Sládeček modification, system of points of the U.K. Department for Environment, Food and Rural Affairs, Moller Pillot system, Abakumov–Maksimov system (see Vinberg et al. 1977; Abakumov 1983; Abakumov and Maksimov 1988; Abakumov and Sushchenya 1991; Budayeva 1991). The Woodiwiss system emphasizes the role of organisms related to indicator taxa. Such organisms are stoneflies, as well as some Oligochaeta and Chironomidae larvae.

New approaches to assessing the anthropogenic impacts on ecosystems using benthic characteristics are likely to appear. The prospects of this are indicated by the revealed changes in Black Sea zoobenthos (Zaika 1992), changes in the structure of the White Sea microbenthos community (a decrease in the share of algophages, decreases in the Shannon index and Margalef index (Burkovsky et al. 1999) and changes in the trophic structures of zoobenthos of water bodies in Fennoscandia (Yakovlev 2000).

A concept of ecological modifications was proposed to characterize anthropogenic changes in ecosystems such as alteration of the structure and metabolism of biocenoses (Abakumov 1987a, 1991; Izrael and Abakumov 1991; Ecological modifications... 1991). The following stages were proposed for the general characteristic of the state of ecosystems (Abakumov 1987a, 1991; Ecological modifications... 1991): (1) the state of ecological wellbeing; (2) the state of anthropogenic ecological stress; (3) elements of ecological regress; (4) the state of ecological regress; (5) the state of ecological and metabolic regress. These stages of the state of ecosystems were used in a number of publications to estimate the anthropogenic effects on ecosystems (e.g., Geletin et al. 1991; Izrael and Abakumov 1991; Zamolodchikov 1993).

Situations are possible when anthropogenic effects (low pollution) can cause some ecological progress (sophistication of the biocenotic structure, increase in the number of species, complication of the trophic chain). Such situations were suggested to be designated as the state of anthropogenic excitation of the ecosystem (Abakumov 1991). In some cases the metabolic progress of biocenoses (increase in the biological activity of a biocenosis, i.e., the sum total of all processes of organic matter formation and degradation) is stimulated by progressing eutrophication of the water bodies under anthropogenic pollution (Abakumov 1991).

Analysis of unique information on the results of hydrobiological monitoring at 635 sites in 378 water objects of the USSR in 1989 showed that 35% of all water bodies investigated were in the state of ecological regress (Abakumov 1991; Izrael and Abakumov 1991).

Assessing the ecological hazards of chemical substances, it is necessary to take into account many factors including different tolerances to anthropogenic factors of the populations of the same species, which are at certain stages of development (the term "lokhos" was suggested to denote specific stages of the development of populations (Abakumov 1972, 1985)), and different tolerances to the pollutants of different units of the temporal structures of biogeocenoses (Abakumov 1984). Elementary units of biocenotic temporal structure – phalanges – are distinguished (Abakumov 1973, 1985). In this relation we note that the concept of "seasonal complexes" of organisms was suggested and is being currently developed (Fedorov et al. 1982; Smirnov 1994).

In Ostroumov (1981, 1984, 1986a,b), Yablokov and Ostroumov (1983, 1985, 1991), and Jablokov and Ostroumov (1991), anthropogenic effects were analyzed with respect to the organization levels of the living systems. The following levels were distinguished: molecular genetic level, ontogenetic level, population–species level, and biogeocenosis–biosphere level.

Several aspects of the problem were emphasized in relation to the anthropogenic effects at the level of ecosystems and biocenoses. (We note that the order of listing is arbitrary; many aspects are not subject to a simplified classification being related to the anthropogenic effects at several levels of organization of the living systems.) The aspects are as follows: (1) changes in the structures of ecosystems/bioccnoses, (2) disturbances of interspecies relations, (2.1) disturbances in trophic links and other biocenotic links, (2.2) disturbances in the balance between the species, (3) disturbances of ecological links resulting from broken information fluxes, (4) elimination of some types of biocenoses and vegetation as a whole, (5) transfer of substances by trophic chains and bioaccumulation of pollutants, (6) transport of toxic substances by migrants, (7) changes in primary productivity, and (8) biotransformation of pollutants in biological systems (this problem is also simultaneously related to the sphere of anthropogenic effects at the molecular level).

The latter issue is closely connected to self-purification in aquatic ecosystems considered in relation to the problems of anthropogenic impacts on hydrobionts in the papers by Fedorov and Ostroumov (1984), Ostroumov (1986a), Telitchenko and Ostroumov (1990), Jablokov and Ostroumov (1991), Yablokov and Ostroumov (1991), Ostroumov and Fedorov (1999), and others. The results, which additionally emphasize the importance of these issues, were obtained in studies of the actions of organotin compounds on mesocosms (Stroganov 1979; Filenko 1988) and in the analysis of the effect of some organic compounds on plankton in experimental reservoirs (Schauerte et al. 1982; Lay et al. 1985a,b; see also Korte et al. 1997). An imbalance between some groups of plankton was shown when 2,4,6-trichlorophenol (TCP) (Schauerte et al. 1982), benzene and 1,2,4-trichlorobenzene (Lay et al. 1985a,b) were introduced into the reservoirs, which emphasized the role of sublethal effects of pollutants.

The tendencies of increasing interest to such characteristics of substances as low acute toxicity were noted by Korte and co-workers: "Before now, we considered only the ecological and chemical properties of agrochemical products such as their stability to the effect of biotic and abiotic processes of transformation and degradation against the background of the production and application of these products. Now, ever greater attention would be paid to such ecotoxicological characteristics as low acute toxicity and mandatory exclusion of harmful impact on the useful organisms" (Korte et al. 1997; translated from the Russian edition).

Our experimental work revealed noticeable effects of synthetic surfactants on water filtration by bivalve mollusks (Ostroumov et al. 1997a,b; 1998; Ostroumov and Donkin 1997), which is important in view of the significant contribution of water filtration by hydrobionts to the processes of self-purification in aquatic ecosystems (e.g., Konstantinov 1979). Other hydrobionts also play a significant role in self-purification of water (e.g., Konstantinov 1979; Ostroumov 1998, 2002, 2004; Ostroumov and Fedorov 1999).

It is important to focus attention not only on the assertion that anthropogenic disturbances take place, but also on revealing the disturbed links that are especially important for maintaining a given ecosystem and preventing its further rapid degradation. A disturbance of water self-purification in an ecosystem caused by pollutants implies a threat of a positive feedback and unwinding the spiral of further disturbances and degradation of the ecosystem. A necessary stage on the way to understanding the ecosystemic effects and the ecological role of pollutants is accumulation of knowledge of the biological effects of these substances on particular species.

1.3 Biological Effects of Substances and the Need of Refining the Arsenal of Biotesting Methods

Methodological issues of biotesting are important for assessing, predicting, and preventing the consequences of pollution of the hydrosphere (Abakumov et al. 1981; Braginsky et al. 1979, 1983, 1987; Izrael 1984; Krivolutsky 1988; Filenko 1989; Flerov 1989). An important role was played by the works of N.S. Stroganov (Stroganov 1976a,b, 1979, 1981, 1982) and of his scientific school (e.g., Filenko 1985, 1986, 1988, 1989, 1990; Filenko and Lazareva 1989; Filenko et al. 1989; Artyukhova et al. 1997a,b), and also of A.G. Dmitrieva (Dmitrieva 1976; Dmitrieva et al. 1989, 1996a,b), A.I. Putintsev, E.F Isakova, V.M. Korol, M.S. Krivenko, G.D. Lebedeva, V.I. Artyukhova (Artyukhova 1996); and other faculty of the Department of Hydrobiology, Moscow State University: L.V. Ilyash (Belevich et al. 1997), L.D. Gapochka (Gapochka 1983, 1999; Gapochka et al. 1978, 1980; Gapochka and Karaush 1980), S.E. Plekhanov (Plekhanov et al. 1997), V.I. Kapkov and others. The issues of biotesting were developed in relation to issues of environmental pollution by A.G. Gusev, L.A. Lesnikov, E.A. Veselov, S.A. Patin, A.N. Krainyukova, their co-workers and many other authors. The impacts of pollutants on hydrobionts were studied by the faculty of several departments of Moscow State University: V.A. Veselovsky and T.V. Veselova (e.g., Veselova et al. 1993; Dmitrieva et al. 1989), A.O. Kasumyan (Kasumyan 1997), S.V. Kotelevtsev (Kotelevtsev et al. 1986), D.N. Matorin (Matorin 1993; Matorin et al. 1989, 1990) and of other institutes: A.I. Archakov, Yu.G. Simakov (Simakov 1986), S.A. Sokolova; scientists at the Institute of Biophysics, Siberian Branch of the Russian Academy of Sciences (e.g., Kratasyuk et al. 1996), and others.

The problems of assessing the biological activity of substances are related to many aspects of ecotoxicology (Dmitrieva 1976; Slepyan 1978; Lukyanenko 1983; Simakov 1986; Bocharov 1988; Bocharov et al. 1988; Bocharov and Prokofyev 1988; Rand and Petrocelli 1985; Maki and Bishop 1985; Juchelka and Snell 1995; Donkin et al. 1997; Ostroumov 2003a,b; and others), monitoring (e.g., Izrael 1984; Filippova et al. 1978; Pokarzhevsky 1985; Khristoforova 1989; Dmitrieva et al. 1996a; Klyuev 1996; Krivolutsky 1990; Hill et al. 1994; Kotelevtsev et al. 1994, 1997; Smaal and Widdows 1994), and self-purification of aquatic ecosystems (e.g, Gladyshev et al. 1996; Ostroumov 2004; and others). The work on biotesting of substances, analysis of the results and improvement of the methods was carried out in view of the preparation and regular update of the lists of maximum permissible concentrations and reference safe levels of impact, e.g., by M.Ya. Belousova, T.V. Avgul, N.S. Safronova, G.N. Krasovsky, Z.I. Zholdakova, T.G. Shlepnina (1987); The List of ... (1995) (compilers: S.N. Anisova, S.A. Sokolova, T.V. Mineyeva, A.T. Lebedev, O.V. Polyakova, and I.V. Semenova). Alternative methods of biotesting were developed using plant objects (Ivanov 1974, 1982, 1992; Wang 1987; Davies 1991; Davies et al. 1991; Obroucheva 1992).

It was noted that "possibly, a direct transfer of laboratory experiments on biotesting of environmental toxicity would not guarantee an error-free prediction of changes in a water body ... Therefore, ... it is useful ... to carry out biotesting not only on the organismal level, but also on the level of model ecosystems." Also, "a water body ... is a complex system and a significant difficulty is to find the main components, which determine the behavior of the system, and their interrelations" (Stroganov et al. 1983a).

In spite of the diversity of the existing methods of biotesting (Filenko 1988; Simakov 1986; Krainyukova 1988; Barenboim and Malenkov 1986; Kotelevtsev et al. 1986; Rand 1985; Rand and Petrocelli 1985; Leland and Kuwabara 1985; Maki and Bishop 1985; Nimmo 1985; Hill et al. 1994; Volkov et al. 1997), there is a pressing necessity for developing new methods of biotesting and refining the existing methods as well as intensification of the work on biotesting of synthetic chemical compounds, which is stipulated by the following.

First, the objectives of biotesting are rather diverse, and no universal method of biotesting has been found yet. "Diverse organisms – bacteria, algae, higher plants, leeches, water fleas, mollusks, fish, etc. – are used as objects for biotesting ... . Each of these objects deserves attention and has its own advantages, but none of the organisms could serve a universal object equally applicable for different goals" (Filenko 1989). A similar opinion was voiced or, in fact, reasoned by other authors (e.g., Volkov et al. 1997).

Second, work on biotesting new substances stays behind that of developing new chemical substances. According to the estimates by the National Institute of Environmental Health (U.S.) and National Toxicology Program (NTP), the level of knowledge of potential pollutants is absolutely insufficient and NTP "welcomes ... suggestions on innovation methods for testing" (Rall 1991, Telitchenko and Ostroumov 1990). The total number of known and commercially produced chemical substances significantly exceeds the number of compounds studied using the biotesting techniques. As early as in 1990, the number of unique chemical substances in the Chemical Abstract Services computer catalogue exceeded 10 million (Rall 1991, Telitchenko and Ostroumov 1990). About 100,000 compounds are in commercial use (Barenboim and Malenkov 1986). Annually about 25,000 to 30,000 new substances are synthesized, and approximately 2,000 of them become widely used. Of the more than 100,000 compounds used, not more than 10% were subject to detailed toxicological and ecotoxicological tests and tests for carcinogenicity and mutagenicity. The hygienic norms developed on this basis exist even for a smaller number of substances. According to an estimate of the National Research Council of the National Academy of Sciences (U.S.), information on potential impacts of chemical substances on the most studied biological species – man – is available only for 20% of thousands of most common chemical substances (Rall 1991, Telitchenko and Ostroumov 1990). According to the data by the Organization for Economic Cooperation and Developments (OECD), only about half of the most mass-produced chemicals were subject to adequate toxicological assessment (OECD Press Release, Paris, April 9, 1990). The Environmental Protection Agency (U.S.) makes estimates of the ecological hazards of substances, but this work lags behind the preparation of new lists of substances that are planned for such assessments, and the list of substances to be tested has more than 13,000 entries (according to the Toxic Substances Control Act of 1976 (TSCA)). According to estimates, 5–10% of new substances put forward for ecological assessments would be recognized to be hazardous (Rosenbaum 1991).

In a similar manner, determination of the biological activities of natural substances stays behind identification of new alkaloids, terpenes, flavonoids, glycosides, steroids, and other secondary metabolites in plants, invertebrates, fungal and microbial cultures.

There is a certain dissatisfaction with the existing arsenal of methods for the assessment of chemicals. Criteria and requirements that the ideal or optimal set of methods for assessing the biological activity of substances should meet include the diversity of the objects, cost efficiency, operational efficiency, etc. (Alabaster and Lloyd 1984; Barenboim and Malenkov 1986; Filenko 1988). "The results of experiments [to determine the sublethal toxicity of pollutants, S.O.] should allow us to interpret them from the point of view of viability of particular species *and ecosystems* [italicized by the author, S.O.] … ." (Alabaster and Lloyd 1984). A justified requirement put forward here and in other publications (Patin 1988a,b,c; Filenko 1988; Bolshakov 1990; Bezel et al. 1994; Krivolutsky 1994) to interpret the results from the point of view of viability and functionality of ecosystems is not met in practice (Maki and Bishop 1985) and is not even analyzed in detail theoretically except in a comparatively minor number of works (Abakumov 1980; Bezel et al. 1994; Ostroumov 2003a).

Here is a list of some important criteria to be taken into account in refining the methods for assessing the biological activities of substances (in arbitrary order, i.e., the order in the list is not related to their possible correlative importance). The list was prepared on the basis of the above-cited papers by various authors, and also of the experience of the author: (1) presentation of test organisms with different sensitivity (excessive sensitivity entails additional methodological difficulties; revealing low-sensitivity organisms is also useful as they can be used to develop purification and bioremediation systems), (2) sufficient operational efficiency, (3) cost efficiency, (4) representation of all major trophic levels and ecological groups of organisms, (5) representation of parameters important for the ecosystem – including those that characterize its capability of self-purification, (6) representation of alternative methods of biotesting requiring no mammals or vertebrates; for humanitarian reasons, such methods should be used as much as possible, and (7) convenience of statistical processing of the data.

We emphasize the importance of methods that are characterized by high operational efficiency, i.e., provide information in a short time. This property is especially important when information gathering on the biological activities and toxicities of substances lags behind their finding and synthesis of new chemicals.

Evidently, one should not expect that a single test would satisfy all requirements at once. It seems expedient to focus on a set of several tests (Filenko 1988, 1989; Kotelevtsev et al. 1986; Krainyukova 1988; Hill et al. 1994; Volkov et al. 1997). Investigators should try to refine and expand the set of tests already in their arsenals.

1.4 Substantiating the Need for Further Research into Biological Effects of Synthetic Surfactants

One of the most important and large classes of substances, whose biological effects were studied by many authors but were not characterized well enough for clear conclusions about the degree of their hazardous properties to be made, are synthetic surfactants. These surfactants are the most important components of commercial detergents.

There is no consensus of opinion in the literature about the degree of ecological hazard of synthetic surfactants. On the one hand, there are many publications on different biological effects and disturbances in the structure and function of organisms under the influence of synthetic surfactants (e.g., Ganitkevich 1975; Denisenko and Rudi 1975; Komarovsky 1975; Shevchuk et al. 1975; Yusfina and Leontyeva 1975; Mozhayev 1976; Braginsky et al. 1979, 1980, 1983; Yanysheva et al. 1982; Gapochka 1983, 1999; Gapochka et al. 1978, 1980; Gapochka and Karaush 1980; Pashchenko and Kasumyan 1984; Khanislamova et al. 1988; Parshikova 1990, 1996; Parshikova et al. 1994; Lenova and Stupina 1990; Sirenko 1991; Khristoforova et al. 1996; Davydov et al. 1997; Vives-Rego et al. 1986; Versteeg et al. 1997a,b; Ostroumov 2003a,b, and a series of our other works published from the mid-1980s). Some papers about the effects of synthetic surfactants are mentioned below in this chapter and in the references (Metelev et al. 1971; Koskova and Kozlovskaya 1979; Patin 1979; Sivak et al. 1982; Malyarevskaya and Karasina 1983; Stavskaya et al. 1988; Lewis 1991a,b; Painter 1992) and in Chapters 3, 4 and 5.

On the other hand, some of the authors do not include surfactants among the most important pollutants (Moore and Ramamoorthy 1984) and believe them to pose almost no ecological hazard for aquatic ecosystems (Fendinger et al. 1994). An experiment was described in which six volunteers received 100 mg of alkyl benzene sulfonate for four months. "Changes in their urine and body weight were analyzed but no harmful effect for their health was found" (Bakacs 1980). This experiment suggested a relative harmlessness of synthetic surfactants.

The opinion that "synthetic surfactants can be assigned to the group of substances of relatively low toxicity and are not distinguished with pronounced cumulative properties" (Shtannikov and Antonova 1978) agrees with the statement that "from the ecotoxicological point of view, modern chemical means of oil spill control pose no serious threat for marine biota as the toxicity of most preparations is lower than that of oil (LC_{50} for major dispersants is usually $102-104$ mg/l)" (Patin 1997). Oil emulsifier EPN-5 developed at the Institute of Oceanology, Russian Academy of Sciences at concentrations from 0.1 to 10 mg/l not only failed to inhibit the development of bacteria but, on the contrary, stimulated saprophytic bacteria. This preparation did not manifest any harmful action on other organisms, which also contributed to the view that synthetic surfactant-containing dispersants and emulsifiers are relatively harmless substances (Nesterova 1980). Seymour and Geyer are also certain that dispersants pose no ecological hazard and cause no damage to ecosystems (Seymour and Geyer 1992). An increase in the abundance of the saprotrophic group of microorganisms was demonstrated in the presence of dispersant DN-75 (5 mg to 10 g per liter). It was concluded that application of DN-75 is an effective means to stimulate self-purification of water bodies from oil pollution (Mochalova and Antonova 2000).

Some reputable publications on environmental pollution by harmful substances do not mention synthetic surfactants at all. Thus, synthetic surfactants are absent in the subject index of the monograph *Environmental Hazards: Toxic Waste and Hazardous Material* (Miller and Miller 1991) though the entry "pesticides" is cited on 23 pages. In the second edition of W. Rosenbaum's monograph "Environmental Politics and Policy," which purports to be comprehensive (and is on the whole rather complete and comprehensive), a detailed subject index does not refer to surfactants and detergents, although pesticides are cited both in the index and in the text on at least 15 pages (Rosenbaum 1991). Neither synthetic surfactants nor detergents were mentioned in the subject indices of other reputable publications on environmental problems including chemical pollution: a three-volume *Environmental Viewpoint* (Lazzari 1994); a solid *Global Accord* published at the Massachusetts Institute of Technology (Choucri 1993); an important book on the policy in the field of environmental protection, *Environmental Policy in the 1990s* (Vig and Kraft 1994).

Evidence of the insufficient knowledge of synthetic surfactants and relatively low attention to them is also presented by the fact that the number of publications on the ecological hazards and biological effects of these substances are much less than for the other groups of pollutants, e.g., pesticides and biocides studied in more detail (e.g., Stroganov 1979; Filenko and Parina 1983; Nimmo 1985; Ilyichev et al. 1985; Bogdashkina and Petrosyan 1988; Bocharov 1988; Bocharov et al. 1988; Bocharov and Prokofyev 1998; Widdows and Page 1993; Donkin et al. 1997), some other organic substances (Golubev et al. 1973; Klyuev 1996; Plekhanov 1997), and heavy metals (e.g., Filenko and Khobotyev 1976; Slepyan 1978; Leland and Kuwabara 1985, Beznosov et al. 1987; Chernenkova 1987; Marfenina 1988; Flerov et al. 1988; Khristoforova 1989; Malakhov and Medvedeva 1991; Artyukhova and Dmitrieva

1996; Dmitrieva et al. 1996b; Khristoforova et al. 1996; Belevich et al. 1997; Kasumyan 1997). Heavy metals were studied, e.g., by the scientists at the Moscow State University: V.N. Maksimov, O.F. Filenko, A.G. Dmitrieva, V.I. Artyukhova, L.D. Gapochka, S.E. Plekhanov, V.I. Kapkov, and others.

Our analysis of the contents of the abstracts journal published by VINITI (All-Russian Institute of Scientific and Technical Information, Moscow) showed that the monthly average number of papers on water body pollution and impacts of substances on aquatic organisms in the issues on "General Ecology. Biocenology. Hydrobiology" is 17.55 (1996) on heavy metals, 7.91 (1996) on pesticides, and 0.82 (1996) on synthetic surfactants. In 1997, there was approximately the same number of abstracts on these substances: approximately 15.25 on heavy metals, 7.5 on pesticides, and 0.75 on synthetic surfactants per month. We used 25 issues of the journal for 1995 (issues 11 and 12), 1996 (all issues except for no. 3, which was not available) and 1997 (all 12 issues). During the entire period under analysis (25 issues of the journal) the monthly average number of abstracts on heavy metals was equal to 16.08, on pesticides it was 7.52, and was 0.8 on synthetic surfactants (Table 1.1).

Table 1.1 Number of publications on surfactants, pesticides and metals as water pollutants, abstracted in *Referativny Zhurnal** (Series "General Ecology. Biocenology. Hydrobiology").

Year	Issue number	Abstracts on pesticides	Abstracts on metals	Abstracts on surfactants
1995	11	6	5	1 (patent)
	12	5	21	1
1996	1	5	9	0
	2	5	9	0
	4	7	10	1
	5	15	34	2
	6	4	12	0
	7	14	28	0
	8	4	15	1 (our paper)
	9	16	26	3
	10	2	15	0
	11	12	21	2
	12	3	14	0
1997	1	8	18	1
	2	16	20	1
	3	9	18	0
	4	5	10	0
	5	6	12	3
	6	9	5	0
	7	7	14	0

Table 1.1   (continued)

Year	Issue number	Abstracts on pesticides	Abstracts on metals	Abstracts on surfactants
1997	8	3	14	0
	9	7	13	1 (soil)
	10	4	11	1
	11	6	16	1
	12	10	32	1
For the entire period	Average per month	7.52	16.08	0.8

*Also known as *VINITI Abstracts Journal* and published by VINITI (All-Russian Scientific and Technical Information Institute, Russian Academy of Sciences, Moscow).
Note: Analyzed from No 11, 1995 (as the subject index started to publish from this number); No 3, 1996 was not available.

1.5   Ambiguity of Biological Effects Caused by Surfactants

Traditionally, candidates for consideration as being hazardous for aquatic ecosystems are the substances that exert noticeable lethal effects on hydrobionts. Exceptions are effects caused by small concentrations owing to the so-called phaseness and hormesis (in detail, see Filenko 1988, 1990).

Particular classes of surfactants (e.g., nonionogenic) are considered low-toxic or nontoxic and, respectively, attention to their ecological significance is weakened. Studies of the mutagenic and teratogenic effects caused by a nonionogenic surfactant Nonoxinol 9 belonging to the class of alkyl phenol derivatives (widely used as an intravaginal spermicide contraceptive) suggested that this surfactant does not cause pronounced mutagenic effects (Meyer et al. 1988), though one of the bacterial strains in the Ames test demonstrated an effect under the influence of this surfactant. Investigations on the molecular level showed that surfactants caused significant stimulation of some enzymes or recovery of previously disturbed enzyme activities (Witteberg and Triplett 1985; Monk et al. 1989; Saitoh et al. 1989; Fujita et al. 1987; Yamaoka et al. 1989).

Indeed, surfactants noticeably differ from "classical" pollutants in that they exhibit a rather large range of examples of their pronounced stimulatory action on many enzyme activities of hydrobionts. For instance, an anionic surfactant sodium dodecyl sulfate (SDS) stimulated the activity of tyrosinase from the skin of African clawed frog *Xenopus laevis* (Witteberg and Triplett 1985). Activation started at surfactant concentrations below the critical micelle concentration (CMC) and continued at a concentration of 30 mM or about 1%, which is a high concentration for a potential pollutant.

SDS stimulated another enzyme, ATPase, in membrane vesicles of yeast plasma membrane (Monk 1989). SDS activates the chemotrypsin-like activity of multicatalytic proteinase complex (MCPC), an enzyme complex widely presented in various tissues of animals and lower eukaryotes including yeasts (Saitoh et al. 1989). This surfactant stimulated MCPC, an enzyme involved in the division of fertilized eggs, from the eggs of the ascidians *Halocynthia roretzi*. SDS activates NADPH-dependent formation of superoxide (O_2^-) in the system of sonicated neutrophils (Fujita et al. 1987).

An important enzyme, *Pseudomonas denitrificans* nitrate reductase was found to be stimulated by the action of cationic synthetic surfactants, alkyl ammonium chlorides, on the cells. Addition of 0.5 M $C_3H_7NH_3Cl$ stimulated the enzyme by a factor of 3.9, and addition of 0.5 M of a longer-chain homolog $C_2H_5NH_3Cl$ stimulated it by a factor of 4.3 (Yamaoka et al. 1989).

The numbers indicating a comparatively low toxicity of synthetic surfactants for egg hatching in *Chironomus riparius* were published by the research center of Procter & Gamble, a manufacturer of synthetic surfactant-containing preparations (Pittinger et al. 1989). No significant inhibition of egg hatching was observed at the following concentrations of synthetic surfactants: anionic synthetic surfactant LAS (linear alkyl benzene sulfonate, ABS), 18.9 mg/l; cationic synthetic surfactant DSDMAC (distearyl dimethyl ammonium chloride), 21.5 mg/l; cationic synthetic surfactant DTMAC (dodecyl trimethyl ammonium chloride), 15.4 mg/l. However, newly hatched larvae were more susceptible than eggs and the values of LC_{50} for them (48–72 h) were smaller (Pittinger et al. 1989). Moreover, the larvae of chironomids are among the most susceptible invertebrates to ABS. Death was observed at concentrations of 0.5 mg/l (Mozhayev 1989).

The effect of nonionogenic surfactant Triton X-100 (TX100) on the cultures of *Chlorella fusca* Shihers et Krauses was studied. No inhibition was observed during the growth of the culture on 10% Bristol's medium and the action of 0.2 mM TX100 (about 120 mg/l), and an insignificant stimulation on the 5th and 14th days. A slight inhibition (about 25%) was found to occur at 0.4 and 0.8 mM TX100 (about 240 or 480 mg/l) only on the 14th day. Before this, no inhibition was observed. However, the pattern of biological effects of TX100 changed significantly if the nutrient medium based on deionized distilled water was changed to a base of natural filtered water (through a 0.45-µm membrane filter) from nine Canadian lakes with addition of 10% Bristol's medium. In this case, addition of TX100 (0.4–1.0 mM, i.e., about 240–600 mg/l) was found to cause a significant increase of growth of *Chlorella fusca*. A growth increase as compared with a medium without TX100 was 10- to 20-fold (1000–2000%) (Wong 1985). In this protocol, no toxic effect was observed even at a rather high concentration of TX100.

In a similar way, Aizdaicher and coworkers showed stimulation of the growth of marine phytoplankton (*Dunaliella tertiolecta, Platymonas* sp.) under the action of surfactant-containing preparations (detergents) (1–10 mg/l) (Aizdaicher et al. 1999). Under the action of the preparation Kristall on *Gymnodinium kovalevskii* a sufficiently high concentration (140 mg/l) is required for the 100% loss of mobility (Aizdaicher 1999), which indicates a high resistance of cells of this species.

A recent work (Ono et al. 1998) found that addition of synthetic surfactants (4–5 ppm) to the water medium, which contains young fish of *Seriola quinqueradiata* (the species is valuable in aquaculture), protected the fish from the harmful effect of toxic raphydophyte (Raphydophyceae = Chloromonadophyceae) phyto-flagellates *Chattonella marina* (Subrahmanyan) Hara & Chihara 1982 and *C. antique* (Hada) Ono 1980 (in Ono and Takano 1980). Without addition of synthetic surf-actants, the young fish perished within an hour (polyoxyethylene alkyl ethers synthe-sized from fatty acids $C_{12}-C_{18}$ were used).

This and other examples indicate that the effects of surfactants on the organ-isms are far from being straightforward and differ from the known negative effects produced by heavy metals, organometallic compounds and pesticides. It is not by chance that, as mentioned above, surfactants and surfactant-containing compounds were not included in the list of priority pollutants of aquatic environment (Moore and Ramamoorthy 1984; Maki and Bishop 1985; Seymour and Geyer 1992; Fendinger et al. 1994; Donkin 1997).

Some of the authors do not consider surfactants hazardous for living organisms at all (Wilson and Fraser 1977; see also Maki and Bishop 1985). Following a detailed analysis of the maximum diverse parameters of anionic surfactant LAS (mean length of alkyl chain, 11.8; molecular weight, 245) according to the standard protocol for assessing the hazard of the chemical, Maki and Bishop make an optimistic con-clusion. This substance, at its maximum expected use, is said "to cause no harm to aquatic life" (Maki and Bishop 1985, p. 633). Characteristically, the publication by Maki and Bishop was included as the concluding chapter "Chemical Safety Evalu-ation" into an authoritative manual, *Fundamental Basics of Aquatic Toxicology*, 666 pages of which give a thorough and detailed analysis of almost all major problems in this field of knowledge (Rand and Petrocelli 1985). A detailed review (Fendinger et al. 1994) ends with the conclusion that "combined analysis of the data on their use, biodegradation or removal during the treatment of sewage waters, concentrations in the environment and evaluation of the risk for aquatic media demonstrates the safety of LAS, AS, and AES in consumer products" (AS, alkyl sulfates; AES, alcohol ethoxysulfates).

Additional evidence that surfactants are not considered important pollutants is, e.g., as follows.

In the reviews of the state of the environment in the USSR (Review of the Back-ground State of the Environment in the USSR for 1988 and 1989; Review of the State of the Environment in the USSR for 1990) information on synthetic surfactants occu-pies much less space than heavy metals and pesticides. Though for some synthetic surfactants maximum permissible concentrations have been determined, for many nonionogenic and many cationic synthetic surfactants they are still unknown (Review of the State of the Environment in the USSR for 1990; Anisova et al. 1995). A signi-ficant proportion of synthetic surfactants was assigned by hazard level to the fourth class of pollutants (Anisova et al. 1995). Some indications that synthetic surfactants are not in fact considered high priority pollutants are presented in Table 1.2.

Additional indications that synthetic surfactants are underestimated and even ignored as aquatic pollutants can be readily found in other authoritative publications (e.g., Rosenbaum 1991; Miller and Miller 1991; Fendinger et al. 1994). Surfactants are missing from the list of parameters of the quality of water used in agriculture for irrigation (Unified Criteria of Water Quality 1982). Only anionic synthetic surf-actants are mentioned in the detailed standards for the quality of surface lotic waters

Table 1.2 Evidence of insufficient knowledge of surfactants and underestimation of their ecological hazard as environment pollutants.

Publications	Author	Year	Comment
Prophylactic Toxicology. United Nations Environment Program. International Register of Potentially Toxic Chemicals. Moscow, Vol. 1, 380 pp. (in Russian)	Izmerov, N.F. (ed.) 31 contributors	1984	Surfactant-containing preparations are mentioned only once
Collection of articles on the new ecological laws of the Russian Federation. Moscow, 372 pp. (in Russian)		1996	Surfactants are not mentioned at all, though pesticides and heavy metals are
Problems of Russian Ecology. Moscow, 348 pp. (in Russian)	Losev, K.S. et al.	1993	Surfactants are mentioned only once
Gidrobiologichesky Zhurnal, 6 issues, total over 600 pages (in Russian)	Over 70 papers by various authors	1995	No papers on the effect of surfactants within the year; some papers on other pollutants
Freshwater Plankton in Toxic Environment. Kiev: Naukova Dumka, 180 pp. (in Russian)	Braginsky, L.P., Velichko, I.M., and Shcherban, E.P.	1987	Surfactants are discussed on 11 pages only (pp. 147–157). The other text contains data on the action of metals and pesticides.
Organic Chemicals in Natural Waters. New York: Springer, 289 pp.	Moore, J. and Ramamoorthy, S.	1984	Surfactants are not included into the list of priority or essential pollutants of the aquatic environment
National Marine Pollution Program. Federal plan for ocean pollution research. Washington D.C., 205 pp.	NOAA	1988	Surfactants are not mentioned at all either in text or in Table 3 on page 30 (which gives a list of major substances toxic for the marine environment)
Fundamentals of Aquatic Toxicology, N.Y.: Hemisphere Publ. Corporation, 666 pp.	Rand, G. and Petrocelli, S. (eds)	1985	Surfactants are mentioned on two pages only. Pesticides are discussed on 55 pages, PAHs on 38 pages
Handbook of Teratology: General principles. N. Y.: Plenum Press, 476 pp.	Wilson, J. and Fraser, F.	1977	The authors do not consider synthetic surfactants hazardous to living organisms
EEC water quality objectives for chemicals hazardous to aquatic environment (List 1)	Bro-Rasmussen, F. et al.	1994	A list of 133 priority pollutants for Western Europe mentions no synthetic surfactant

from the ecological point of view; the other types of surfactants (such as noniono-genic and cationic synthetic surfactants) are not mentioned (Unified Criteria of Water Quality 1982).

In other countries, nonionogenic and cationic synthetic surfactants are frequently absent from the lists of criteria for water quality. Thus, for instance, in France, only anionic surfactants are listed as undesirable substances (nonionogenic and cationic synthetic surfactants are not mentioned). Up to 0.2 mg/l of surfactants are allowed in drinking water and up to 0.5 mg/l of surfactants are allowed in the water for non-drinking applications (termed as resource water, *La Recherche* 1990, Vol. 21, p. 600). In the U.S., synthetic surfactants are not on the list of the most important criteria for water quality assessment by the Environmental Protection Agency and the Council on Environmental Quality (Rosenbaum 1991). According to the norms established by the Public Health Service of the U.S., the recommended level of alkyl benzene sulfonates (ABS) in drinking water is 0.5 mg/l (the other synthetic surfactants are not regulated), and according to the World Health Organization the admissible level of the ABS is even greater: 1 mg/l (MacBerthouex and Rudd 1977).

Insufficient attention to nonionogenic and cationogenic synthetic surfactants (and respective gaps in the knowledge of their biological effects) is even more regrettable because these synthetic surfactants are significantly slower to decompose in the environment than anionic surfactants. Thus, according to the data by V.T. Kaplin (1979), the rate of biochemical oxidation of synthetic surfactants in water for OP-7 and OP-10 (which contain nonionogenic surfactants) is $0.006-0.007$ day^{-1}; for cationic synthetic surfactant trimethyl alkylammonium chloride, it is 0.002 day^{-1}. For comparison, the value of this coefficient for a lignin derivative lignosulfonate is 0.06 (Kaplin 1979). This means that lignosulfonate (a substance rather stable in water) decomposes 10 times faster than the above nonionogenic surfactants and 30 times faster than the above cationogenic synthetic surfactants.

It follows from the aforesaid that additional analysis is required to determine to what degrees synthetic surfactants as pollutants of water reservoirs are hazardous for the biota. Such an analysis should include (1) the degree of the biospheric pollution by these substances and (2) the character of biological effects they perform. The former issue is discussed in the next section; the latter is the main topic of Chapters 3–7.

1.6   Pollution of Aquatic Ecosystems by Synthetic Surfactants

Production of synthetic surfactants and their discharge into the aquatic environment are rapidly increasing. Synthetic surfactants are the main components of detergents and abstergents produced. Their worldwide consumption is measured in millions of tons (Berth and Jeschke 1989; Stavskaya 1990; Stavskaya et al. 1988, 1989; Lewis 1991a,b).

For instance, only in the USSR production of synthetic detergents in 1988 was 1,301,000 t; in 1989, it was 1,424,000 t; and in 1990, 1,503,000 t. By the end of that period (in 1990) the total consumption of surfactants only in the U.S. was approximately 1,134,000 t (Facts and Figures [statistical data about surfactants], 1992). Production of synthetic surfactants in the U.S. is much greater as part of the product is exported. In 1984, the total production of surfactants in the U.S. reached ~2.4

million t per year. According to estimates (Greek and Layman 1989), in 1989 the consumption of surfactants in the U.S. was ~3.3 million tons.

The use of surfactants in Western Europe was in 1987 (in thousand tons): 493 in FRG, 409 in France, 405 in Italy, 299 in Great Britain, 282 in Spain, and 167 in Benelux countries (European Market 1988). In Japan, the production of surfactants in 1986 was (in thousand tons): 619, dry detergents; 365, liquid detergents. Besides, the production of softeners was 275 thousand tons and of bleaching agents 108 thousand tons. In Brazil, the annual production of cleaning products was approximately 1.3–1.5 million tons.

In recent years, annual production and consumption of surfactants were observed to increase steadily; hence, their entry into the hydrosphere increased by several percent per year (Dean 1985). Surfactants enter the hydrosphere not only in relation with the use of detergents but also due to the use of these substances in industries, in mining, refining, and transporting of various raw materials. Consumption of synthetic surfactants (and their daily entry into the sewage waters) per person in Germany was equal to: 6.71 g of anionic surfactants, 4.07 g of nonionogenic surfactants, and 1.16 g of cationogenic synthetic surfactants (Steinberg et al. 1995). The content of surfactants in sewage waters can reach 30 g/l (Stavskaya et al. 1988).

Analysis of the composition of treated municipal effluents revealed that the contribution of anionic surfactants to the dissolved organic matter was 11–20% (Rebhun and Manka 1971). Other sources of pollution in the water (marine) medium by synthetic surfactants are dispersants, which are added to water to treat oil spills (Mochalova and Antonova 2000). Up to 75% of the compositions of the dispersants can be synthetic surfactants (Singer et al. 1990). The toxicity of dispersants was demonstrated on particular objects. For instance, Corexit 9527 was shown to be toxic to marine species *Macrocystis pyrifera*, *Haliotis rufescens*, *Holmesimysis costata*, and *Atherinops affinis* (Singer et al. 1991). Dispersants EPN-5 and DN-75 were toxic to fish, chironomids, phytoplankton, and other organisms (Nesterova 1989, see Section 10.2: "Toxicological characteristics of dispersants," p. 170).

LC_{50} (*Daphnia magna*, 48 hours, 20°C) for mixtures of three types of oil and five dispersants (Corexits 9527, 7664, 8667, 9660, and 9550; dispersant/oil volume ratio, 1:20) varied within 1.1–5.2 mg/l (Bobra et al. 1989). LC_{50} values of a mixture of Corexit 7664 and three types of oil were several times smaller than the LC_{50} of the dispersant only. Toxicity of all mixtures of Corexits and oil was higher than the toxicity of physical dispersions of oil without Corexits. Hence, it follows that under conditions of the experiments the toxicity and ecological hazard of oil pollution increased when synthetic surfactant-containing dispersants were added into the system (Bobra et al. 1989).

The discharge of synthetic surfactants into the water bodies of Russia and the former Soviet Union is significant; in some cases, it exceeds the input of pollutants of other classes. Thus, the input of synthetic surfactants with the river discharge into the Caspian Sea in 1991 was 12,200 tons, which was more than 10 times the amount of phenols, which was 1,220 tons (Review of the Ecological State of the Seas of the Russian Federation and Specific Regions of the World Ocean in 1991 1992). The transfer of detergents with the waters of the Don and Kuban rivers into the Azov Sea reached 4,000 tons (Review of the State of Pollution ... 1976, p. 109), which exceeded several times the discharge of phenols (approximately 800 t) and pesticides (approximately 1,000 t). According to other sources, the annual discharge of synthetic surfactants was equal to: approximately 3,600–4,300 tons into the Caspian Sea (with the river discharge and industrial sewage in 1986–1988), approximately 1,800 thousand tons into the Azov Sea (with the discharge of the Don and Kuban rivers in 1981–1985) (Kuksa 1994). The annual discharge of detergents with the river waters into the Dnieper–Bug estuary (liman) reached 6.11 thousand tons (Review of the State of Pollution … 1976, p. 162).

The concentrations of synthetic surfactants measured in water reservoirs reached significant levels. Synthetic surfactants easily form complexes with other compounds and are rapidly adsorbed at the interfaces, which hampers their determination by analytical methods (Gonzalez-Mazo and Gomez-Parra 1996) and can lead to underestimating the determined values compared to the real pollution of the aquatic ecosystem. Probably, it is not by chance that analysis of the most dramatic situations of chemical pollution of water reservoirs in the territory of Russia and the USSR (the cases of pollution of water reservoirs with pesticides, heavy metals, oil, and phenols at a level exceeding the maximum permissible concentration (MPC) by a factor of 30 and more) (Izrael and Abakumov 1991) presented not a single case of pollution by synthetic surfactants. This is more evidence of the incompleteness of information and contradictions surrounding the problem of environmental pollution with synthetic surfactants.

There are reports of significant levels of pollution of particular water reservoirs. Even in the territory of a national park, the concentration of synthetic surfactants in the lake (Lake Chernoye, Shatsk National Park, Ukraine) was recorded at a level of 640 µg/l (Oksiyuk 1999). In river water, a synthetic surfactant concentration of 720 µg/l (the Don River; Bryzgalo et al. 2000) or even as high as 15 MPC was observed (which, at MPC of 0.5 mg/l for communal and amenity water reservoirs, is 7 mg/l) (the Poltava River in the region of Lvov and Busk cities) (Review of the State of Environment in the USSR 1990).

Concentrations of detergents reaching 1.24 mg/l and greater were recorded in the open waters of the Black Sea (Review of the State of Pollution … 1976). In the region of Port Zhdanov, the maximum level of detergents reached 1.6–2.5 mg/l, while the monthly average level reached 0.8 mg/l. In Berdyansk Bay, the level of detergents reached 1.76 mg/l. In the Tuapse region of the Black Sea, the recorded concentrations of detergents reached 2.2–2.8 mg/l at a mean annual level of 1.38 mg/l (Review of the State of Pollution … 1976). In the regions near Odessa and Ochakov the pollution of seawater was recorded at levels of 10–32 MPC (Kuksa 1994), which is 1.0–3.2 mg/l. In the Azov Sea (Karievsky estuary and Dzherelievsky collector: the receiver of sewage waters from rice fields, 1989) the pollution of waters with surfactants was recorded at a level of 6–9 MPC (Kuksa 1994). It should be noted that pollutant concentrations were measured according to the accepted rules, a distance no less than 300–500 m from the source of pollution. This means that the real pollution between the source and sampling point is even greater than the numbers presented.

Of special interest is the surface film in the water reservoirs. "Almost all the surface of the natural reservoirs is permanently covered by a film of surfactants of natural or artificial origin" (Gladyshev 1999). An increase in the concentration of pollutants is observed in the surface microlayer (SML) of the sea (e.g., Rumbold and Snedaker 1997) including pesticides, phthalate ethers, aromatic hydrocarbons, heavy metals, and synthetic surfactants. The concentrations of surfactants measured in the SML of the Black Sea reached $50-1200$ µg/l, i.e., up to 12 MPC (Keondzhyan et al. 1990). The mean concentration of anionic detergents in the surface film $60-100$ µm thick can exceed 85 times the concentration of surfactants in the water column. Thus, in Arcachon Bay (France) the mean concentration of anionic detergents in the surface film was equal to 850 µg/l, which was 85 times greater than in the water column (Patin 1977, p. 328). It should be emphasized that the data on the levels of pollution by surfactants in the reservoirs of the USSR and Russia reflect only the level of pollution by anionic surfactants (other surfactants were not measured yet). Taking into account the real concentrations of nonionogenic and cationogenic surfactants, the cumulative rates of pollution by synthetic surfactants can be even greater.

Along with synthetic surfactants, the sewage and polluted waters also contain ecologically unsafe products of their transformation and biodegradation. For instance, alkylphenol polyethoxylates belong to an important class of synthetic surfactants. Their annual production in the U.S. only is about 140,000 tons and in Germany it is over 65,000 tons. The worldwide production of these substances exceeds 360,000 tons (Ahel et al. 1993). During the decomposition of many nonionogenic surfactants from the class of nonylphenol polyethoxylates the product of their degradation, nonylphenol (NP) enters the water medium. NP is relatively persistent and widespread in the environment. The LC_{50} of NP for *Mytilus edulis* L. was equal to 3 mg/l (96 h), 0.5 mg/l (360 h), 0.14 mg/l (850 h) (the semi-static and flowing test systems were used) (Granmo et al. 1989). Such sublethal effects like a decrease in the strength of byssus and a decrease in the ecologically important SFG (Scope For Growth) indicator (characterizes the potential for reproduction and population growth) were revealed at an even lower concentration of 0.056 mg/l (Granmo et al. 1989). This set of quantitative characteristics of the biological effects of NP indicates that the real hazard of the pollutant related to synthetic surfactants can manifest itself at the level of its concentration in water almost two orders of magnitude smaller than LC_{50} obtained in a 96-hour experiment. NP $(0.01-10$ µg/l) suppressed sedimentation and colonization of the substrate by cypris-like *Balanus amphitrite* larvae (Billinghurst et al. 1998). In the wastewater treated at sewage treatment plants, the NP were found at concentrations from 2 to 4000 µg/l, i.e. up to 4 mg/l (Giger et al. 1981; Etnier 1985, cited from Ekelund et al. 1990). The latter value (4 mg/l), according to the results presented in the paper cited above, is hazardous. Besides, one's attention is attracted by the fact that it is significantly higher than the concentration of synthetic surfactants usually measured in the environment, which is likely to be partially explained by the fact that NP is highly persistent. On the other hand, a relatively high measured concentration of NP can indirectly indicate that the real concentration of its predecessors, nonionogenic surfactants, is also high in the environment, but it is possible that some nonionogenic surfactants avoid detecting by the analytical methods owing to their greater ability to be bound to the interface surfaces than nonylphenols.

The efficiency of water purification from nonionogenic surfactants is low in the sewage treatment plants with mechanical and biological water treatment.

Approximately 60% of nonylphenol polyethoxylates entering these plants in polluted waters are passed through and enter the environment with the so-called purified waters (Ahel et al. 1993). Nonionogenic surfactants are examples of biologically hard synthetic surfactants, and no more than 40% of them are removed as a result of biological purification (Zhmur 1997).

Indications that sorption of synthetic surfactants on particles, including particles of sediments, is important were obtained in comparison of the biological effects of anionic surfactants LAS in two systems: (1) anionic surfactants were sorbed on benthic sediments and (2) anionic surfactants were completely in the aqueous phase (Bressan et al. 1989). Anionic surfactant LAS in the aqueous medium showed a biological effect on the mussels *Mytilus galloprovincialis* and other organisms at concentrations lower than 1 mg/l (see Chapter 3 for details), whereas LAS sorbed on particles had no effect on the organisms even at concentrations 3–10 times greater. However, one should not be misled by the seemingly saving effect of sorption of synthetic surfactants on detritus particles and sediments formed from them. In the course of time, the organic particles of detritus are destroyed to a certain degree and are consumed by detritophages or bacteria. Synthetic surfactants sorbed on detritus and organic sediments can either enter the organisms of detritophages or can be released into the environment, thus presenting the hazard of a new wave of pollution.

The data on the amount of synthetic surfactants transported into the reservoirs of the Russian Federation are contradictory. According to the report "On the State of the Natural Environment of the Russian Federation in 1996" the input of synthetic surfactants into the reservoirs in 1996 was 4,000 tons, while in 1992 it was equal to 8,900 tons. The latter number is evidently smaller than the above mentioned amount for the discharge into the Caspian Sea only (more than 12,000 tons in 1991).

One more variant of calculation is possible based on the mean transfer of synthetic surfactants into the sewage system per day. The amount of synthetic surfactants in the calculation of sewage water discharge into the system per one inhabitant in Russia per day is considered to be 2.5 g (Akulova and Bushtuyeva 1986). Therefore, the annual amount calculated for one million inhabitants is equal to 910 tons. Assuming that the urban population in Russia exceeds 70 million people (which is an underestimation; the real number is much greater), the total input of synthetic surfactants is greater than 63,700 tons. The efficiency of water purification in the water treatment plants is approximately 48–80%, and during the winter season it is only 20% (Boichenko and Grigoryev 1991). According to the estimates of Kostovetsky and coauthors (1975), the efficiency of purification from synthetic surfactants in aeration tanks is 47–78.3% and on biological filters, 40–48%. The content of synthetic surfactants in the urban sewage waters reached 15 mg/l (anionic surfactants; Mozhayev 1989). The content of synthetic surfactants in the water purified through biological filters was 3 mg/l and greater at an initial concentration of 5.3 mg/l synthetic surfactants in water supplied to biological filters (Kostovetsky et al. 1975). Since the time of this publication, the use of synthetic surfactants and their discharge into the urban sewage waters increased significantly.

Assuming even an overestimated efficiency of sewage water purification at 80%, we obtain that no less than 12,600 tons of synthetic surfactants enter the reservoirs, which significantly exceeds the value of 4,000 tons given above. It is likely that the real supply of synthetic surfactants is significantly greater. For instance, only from three inhabited places 4.53 tons of synthetic surfactants are discharged into the Ivan'kovskoye reservoir every day (Boichenko and Grigoryev 1991), which means that the annual amount is over 1.6 million tons. The real supply of synthetic surfactants into the Russian reservoirs should be significantly greater than this number also because not all polluted water goes to biological treatment plants. Thus, according to the data (Yakovlev et al. 1992) the annual discharge of water into the reservoirs of the Russian Federation is 76,353 million cubic meters, of which 27.146 million cubic meters is polluted water, which is equal to 36.6% (the other 45.720 million cubic meters is rated pure, and 3.487 million cubic meters is rated purified).

Synthetic surfactants pollute almost all rivers of Russia with inhabited localities, including the Moskva River (Manyakhina 1990). In recent years, the discharge of synthetic surfactants into the Moskva River continues to increase against the background of a decrease of many other pollutants (Otstavnova and Kurmakayev 1997) (Table 1.3). The mean concentrations of anionic surfactants measured in the water of the Volga River were 0.25 mg/l and in the Klyazma River they were equal to 0.33 mg/l (Mozhayev 1989).

Table 1.3   Amount of pollutants discharged into Moscow open-water reservoirs in 1992–1996 (thousand tons). Increase of surfactant discharge against the background of a decreased discharge of other pollutants (Otstavnova and Kurmakayev 1997).

Parameter	1992	1993	1994	1995	1996	1996 as compared with 1992
Synthetic surfactants	0.20	0.42	0.34	0.39	0.43	Increase (215% as compared with 1992)
Petroleum products	2.34	2.12	1.68	1.56	0.66	Decrease
Sulfates	128.2	116.1	110.5	108.3	111.4	Decrease
Ammonium nitrogen	28.88	17.99	17.72	14.17	13.55	Decrease
Chlorides	232.00	185.7	164.5	146.9	144.6	Decrease
Copper	0.095	0.059	0.054	0.059	0.046	Decrease
Suspended matter	27.67	24.01	24.61	24.03	23.13	Decrease
Total for 22 items	3102	2777	2649	2542	1305	Decrease

Synthetic surfactants were used in the elimination of the consequences of the accident at the Chernobyl atomic power station. As a result, pollution of the environment by synthetic surfactants increased in this region. Significant concentrations of cationic synthetic surfactants were found in the ecosystems of storage reservoirs (Kalenichenko 1996). In water of the aquatic farm of TPP-5 (Thermal Power Plant No 5), the MPC for fishery reservoirs was exceeded by a factor of 37–39 and more (Davydov et al. 1997). The concentrations of cationogenic surfactants in this reservoir were found to be 0.45–0.47 mg/l. Such concentrations of cationic surf-actants were observed both in the water of the bay and in fishing cribs. The water purification unit used in the incubation facility failed to decrease significantly the level of cationogenic surfactants in water (Davydov et al. 1997). Cationogenic surf-actants (from the group of quaternary ammonium compounds (QAC)) are used in the aquaculture to control fish infection. QAC are added to water at a concentration of 1–2 mg/l (Austin 1985).

In recent years, the content of anionic surfactants has grown in the Dnieper River, its tributaries, and storage reservoirs to reach in some cases 0.8 mg/l (the Kremenchug reservoir, the Vorskla River) and even more than 0.9 mg/l (the Dnieper River near Kherson; the Samara River) (Mudryi 1994).

Western data on the pollution of water reservoirs by synthetic surfactants are not readily available. The content of synthetic surfactants LAS in U.S. rivers reached 3.3 mg/l; the concentrations of anionic surfactants ABS in the rivers and estuaries of Malaysia were up to 0.54 mg/l; the level of nonionogenic surfactants alcohol ethoxy-lates in the rivers of European regions outside Russia reached 1.0 mg/l (Lewis 1991b). In FRG, the recorded concentrations of synthetic surfactants (LAS) were up to 1.6–1.7 mg/l (Steinberg et al. 1995). The mean concentrations of LAS in the waters entering the water treatment facilities of the U.S. are 3.5–4.8 mg/l. The mean concentration of LAS in the primary treatment effluent in the U.S. was equal to 2.1 mg/l, reaching 2.5 mg/l (Fendinger et al. 1997). The content of LAS in the river bottom sediments in the U.S. was up to 740 mg/kg; in Germany it reached 275 mg/kg of dry sediments (Fendinger et al. 1997).

The concentrations of synthetic surfactants found in the Aegean Sea reach 0.21 mg/l (at a depth of 0.5 m) and up to 0.35 mg/l (at a depth of 5 m) (Cosovic and Ciglenecki 1997). In the Ionian Sea, the measured concentrations of synthetic surf-actants at a depth of 0.5 m were equal to 0.18 mg/l (Cosovic and Ciglenecki 1997).

Aquatic ecosystems were found to have degradation and biotransformation products of synthetic surfactants, which possess estrogenic activity and have lower values of LC_{50} (i.e., higher toxicity) than the initial synthetic surfactants. The con-centrations of nonylphenols (NP) in the rivers of Great Britain reached 0.18 mg/l (Thiele et al. 1997).

The measured concentrations of synthetic surfactants fail to give a comprehen-sive idea of the degree of pollution of an ecosystem and cannot be compared without reservation with the concentrations of synthetic surfactants added to the experimental systems in the biotesting experiments. This is because a significant part of the surf-actants rapidly passes into a sorbed state and cannot be singled out by the standard methods, which reveal only the surfactant molecules present in aqueous solution.

Pollution of reservoirs with synthetic surfactants is to a significant degree caused by a growing use of various detergents (synthetic abstergents, foam detergents, liquid detergents), many of which contain phosphates as one of the components whose proportion can be as high as 40% of the weight (Pickup 1990). Therefore, pollution by synthetic surfactants should be considered as part of the complex pollution of the environment (Drachev 1964; Sirenko 1972; Losev et al. 1993; Mudryi 1995). According to the estimates made in Great Britain, the proportion of detergents in the total anthropogenic discharge of P is no less than 20–25%; the share of detergents coming into the water reservoirs with sewage that passed the sewage works is about 50% (Pickup 1990). The annual discharge of P with the detergents and various cleaning compositions to the sewage works of Great Britain is approximately 35,000 tons, while the transfer of P into the water reservoirs together with sewage waters purified by these facilities is approximately 56,000 tons (Pickup 1990).

Synthetic surfactants can also play a significant role as pollutants of land ecosystems. Thus, synthetic surfactants and products of their degradation can pollute soils as a result of watering and sprinkling with surfactant-containing sewage waters.

About 30–40% of the toxicity of pesticides can in some cases be provided by the additional components of pesticide preparations (Caux et al. 1986, 1988). Among them, synthetic surfactants are very important (Weinberger and Rea 1982).

Synthetic surfactants are widely used in oil extraction. They get into land ecosystems during the operation of drilling rigs and wells. Besides, open-cut mining of some minerals can include covering the surface of the soil with a layer of foam based on synthetic surfactants.

1.7 Synthetic Surfactants and Self-Purification of Water Including its Filtration by Mollusks

The overall scale of water filtration in natural ecosystems is high. Also high is the role of the filtrating organisms (filter feeders) (along with other hydrobionts [Konstantinov 1979; Kokin 1981; Koronelli 1982, 1996]) in the processes involved in self-purification of reservoirs (Bogorov 1969; Vinberg 1973, 1980; Sushchenya 1965; Ivanova 1976b; Kondratyev 1977; Konstantinov 1977, 1979; Alimov 1981; Gilyarov 1987; Zaika 1992; Alekseyenko and Aleksandrova 1995; Wotton et al. 1998; Newell and Ott 1999; Ostroumov 2005). "The well-being of the ecosystem in a reservoir is determined not only by means of indicator organisms and species diversity of hydrobionts but also by the preservation of useful biological processes: self-purification, photosynthesis, reproduction of commercially useful hydrobionts" (Stroganov et al. 1983a). Self-purification of reservoirs is a necessary prerequisite for determining critical (ecologically admissible) loads on the water reservoirs (Moiseyenko 1999) and assessing the assimilation capacity (Izrael and Tsyban 1989, 1992) of aquatic ecosystems. Preservation of the self-purification potential of reservoirs is especially relevant for Russia, where 27.7% of the cases of testing the utility and drinking water supply were found not to correspond to the required chemical criteria of water quality. The situation in three regions in the basin of the Volga River (Kaluga, Nizhny Novgorod, and Saratov regions) and in Kalmykia and Mordovia was even more grave – the inconsistency was found in more than 40% of water sources (Elpiner 1999).

Therefore, the pressing problem is to what extent the filtration activity of hydrobionts under the influence of anthropogenic factors including chemical pollution with synthetic surfactants can be suppressed (Ostroumov 1986a, 1998, 2005a,b). Judging by the available publications, this problem is not yet considered high priority in the development of the control and regulation systems for water quality in the fishery reservoirs of the Russian Federation. This system is based on establishing maximum permissible concentrations of pollutants in water by means of performing experiments with hydrobionts according to a certain protocol (Methodological Recommendations ... 1986; see also a new edition, Methodological Recommendations ... 1998). This useful protocol includes many important hydrobionts and well-grounded methods of assessing the effects of substances on hydrobionts. However, bivalve mollusks and filtration of water by them are absent from the list of recommended objects and methods of biotesting (Methodological Recommendations ... 1986, 1998). Filtration of water by hydrobionts, including by bivalve mollusks, as an object of possible effect and inhibition by pollutants, is not mentioned in Section 3 ("Effects of pollutants on the processes of self-purification") in the document (Methodological Recommendations ... 1986) nor is it mentioned in the corresponding section of the new edition (1998).

The rate of filtration of water by mollusks can be inhibited by such substances as heavy metals (Stuijfzand et al. 1995), polyaromatic compounds, organotin compounds, polychlorbiphenyls (Smaal and Widdows 1994), pesticides and other substances (Mitin 1984; Donkin et al. 1997; Ryzhikova and Ryabukhina 1998). It is the filtration activity of mollusks that is the most vulnerable function among the complex of processes, which are taken into account in the comprehensive parameter SFG (Scope For Growth) that characterizes the state and potential of reproduction and growth of the population (Smaal and Widdows 1994).

We studied the problem of the effects of synthetic surfactants on the activity of mollusks, both marine – mussels *Mytilus edulis, M. galloprovincialis,* oysters *Crassostrea gigas* – and freshwater. The results of this work are presented in Chapters 3, 4, 5, and 6 as well as in Ostroumov et al. (1997a,b), Ostroumov and Donkin (1997), Ostroumov (2000a–d, 2003a,b). It seems worthwhile to analyze new facts on the action of substances on hydrobionts in relation to the fundamental problems of hydrobiology, taking into account that "... accumulation ... of pollutants ... in amounts exceeding the ability of the biosphere to degrade them disturbs the evolutionally-developed natural systems and links in the biosphere, undermines the self-regulatory ability of the natural complexes" (Ostroumov 1986b).

Taking the aforesaid into account, the following problems appear to be important in the analysis of our experimental results: (1) Can the new data help better assess the hazard for self-purification of the systems (Zak 1960; Drachev 1964; Vinberg 1973; Bronfman et al. 1976; Konstantinov 1977, 1979; Braginsky et al. 1980; Self-purification ... 1980; Sinelnikov 1980; Vavilin 1983; Skurlatov 1988; Shtamm 1988; Bogdashkina and Petrosyan 1988; Polikarpov and Egorov 1986; Koronelli 1996; Ostroumov and Donkin 1997; Mill et al. 1980; McCutcheon 1997) under conditions of increasing chemical pollution (Guskov et al. 1986; World Resources 1994; Mudryi 1995)? Does the hazard that self-purification is the target of possible impacts exist, and what is the degree of this hazard? (2) Can the new results be used in the development of biotechnological approaches to water purification as well as to bioremediation of polluted ecosystems (McCutcheon et al. 1995; Medina and McCutcheon 1996; Varfolomeyev et al. 1997) taking into account the presence of synthetic surfactants in the complex pollution of the environment?

In the 1980s, analysis of the literature was made in Malyarevskaya and Karasina (1983), Braginsky et al. (1983), Stavskaya et al. (1988, 1989), Stavskaya (1990),

Ostroumov (1991), Sivak et al. (1982), Bock and Stache (1982), Ramade (1987). In this work, emphasis is made on later studies.

We use the term "contaminant" along with the term "pollutant," which is close in sense and widely used in international and scientific literature. The author does not claim the full coverage of vast literature in this field and confines himself only to references illustrating the main ideas. The term "biological activity" and some other terms are used in the same sense as in Ostroumov (1986a).

Test organisms used in this work included representatives of different groups of organisms belonging both to autotrophs and heterotrophs. The choice of particular organisms is substantiated in Chapter 2.

We believe that the data on the biological effects of synthetic surfactants and the degree of sensitivity or tolerance of organisms to them (our new experimental results in this field are described in the next chapters) can be of interest from several points of view, including the following.

(1) It is necessary to have a more complete idea of the potential hazards of the possible consequences of various types of environmental pollution, including those occurring when mass quantities of synthetic surfactants get into the environment due to the breach of the technological and nature-conserving regulations as well as a result of emergency and extraordinary situations.

(2) Revealing a comparatively high tolerance can also be of interest in search for and development of systems of biological purification and treatment of polluted waters, sediments, soils or other components of the ecosystems as well as in the development of approaches to remediation of polluted sites and ecosystems.

An additional substantiation of the importance of investigating this field based on the analysis of vast literature (more than 800 references) is given in large reviews (Yablokov and Ostroumov 1983, 1985; Ostroumov 1986b; and others). After those reviews were published, our view of and approaches to the assessment of the biological activity of substances and ecological hazard caused by environmental pollution were supported (Lavrenko 1984; Sokolov 1987; Stugren 1987; Symonides 1987; Gusev 1988; Dubinin 1988; Pokarzhevsky and Semenova 1988; Stavskaya 1988; Romanenko and Romanenko 1992; and others).

The scope of the problems of this work did not include the analysis of the mechanisms of the effects of synthetic surfactants on the organisms (the mechanisms are related to a greater degree to biological membranes and were the subject of earlier works and publications by the author). Problems of the transfer of the data and results of laboratory experiments to natural ecosystems are also beyond the framework of this book.

We hope that the diversity of the organisms used in studies of the effects of synthetic surfactants would contribute to the accumulation of a wide range of material for fundamental generalizations and substantiated conclusions. Below, we present new data on the effects of surfactants of freshwater and marine prokaryotes and eukaryotes, including bacteria, cyanobacteria, algae, flagellates, higher plants and invertebrates.

As synthetic surfactants are divided into three large classes: anionic, non-ionogenic and cationic (or cationogenic), these classes of substances are discussed separately in Chapters 3, 4 and 5.

2 Organisms and Methods

2.1 Organisms: Substantiation of Choice and Aspects of Methods Used

The subjects of investigation were typical representatives of the main trophic levels and large taxa from prokaryotes to eukaryotes including cyanobacteria (*Synechococcus* Näg.; *Stratonostoc linckia* (Roth) Elenk., *f. muscorum* (Ag.) Elenk. = *Nostoc muscorum* Ag., and others), marine bacteria (*Hyphomonas* (ex Pongratz 1957) Moore, Weiner and Gebers 1984), green algae (*Scenedesmus quadricauda* Bréb., *Chlorella vilgaris* Beijer, *Bracteacoccus minor* (Chodat) Petrova, and others) diatomic algae (*Thalassiosira pseudonana* Hasle et Heimdal), euglena (*Euglena* Ehr.; *Euglena gracilis* Klebs), mollusks (*Unio tumidus* Philipsson s. lato, *U. pictorum* (L.) s. lato, *Crassiana crassa* (Philipsson) s. lato, *Anodonta cygnea* (L.) s. lato, *Mytilus edulis* L., *M. galloprovincialis* Lamarck, *Crassostrea gigas* Thunberg, *Limnaea stagnalis* (L.), *Mercenaria mercenaria*), annelids (*Hirudo medicinalis* L.), macrophytes (*Pistia stratiotes* L., *Elodea canadensis* Michaux), seedlings of angiosperm plants (*Sinapis alba* L., *Fagopyrum esculentum* Moench, *Lepidium sativum* L., *Oryza sativa* L., *Camelina sativa* (L.) Crantz, *Triticum aestivum* L., and others). These objects were of theoretical and practical interest due to the details of their ecology, their role in the ecosystems, and the possibility of their use as biological resources. Diverse biological material made it possible to obtain broader and more substantiated conclusions on the possible role of synthetic surfactants as pollutants.

Below we describe the justification of the choice of the objects (organisms) and methodological aspects of their use. The nomenclature of cyanobacteria is given according to Gollerbakh et al. (1953); of marine phytoplankton, according to Tomas (1997). The nomenclature of vascular plants of Russia and adjacent countries (territories of the former USSR) is given according to Cherepanov (1995). The nomenclature of invertebrates is according to Zatsepin and Rittikh (1975), and Zatsepin et al. (1978).

2.1.1 Prokaryotes

2.1.1.1 Cyanobacteria (Cyanophycota)

This essential group of phototrophic prokaryotes uses water as the donor of electrons and, thus, produces oxygen in the light. The genus *Synechococcus* (class

Chroococcophyceae, order Chroococcales) is one of the four main genera of marine cyanobacteria. This genus also includes the species that occur in freshwaters and in terrestrial habitats. *Synechococcus* develops in eutrophic waters and can reach large concentrations of cells in seawater. It is an important component of phytoplankton and is involved in biogeochemical fluxes of elements through marine ecosystems. It is capable of nitrogen fixation and, thus, is one of the main suppliers of nitrogen to seawater (e.g., Kondratyeva et al. 1989; South and Whittick 1987). Cyanobacteria can reach a share of 60% of all chlorophyll in marine ecosystems in the upper 50 m and about 20% and more of the total primary production (e.g., Sieburth 1979). In this study, along with the other species of cyanobacteria, we studied cyanobacteria of the combined genus *Synechococcus*. These unicellular cyanobacteria (less than 3 μm in size) are widespread in the open regions of the seas. They were also found in the seas of the Arctic Basin (Mishustina et al. 1994).

The *Synechococcus* strains used were from the collection of the Woods Hole Oceanographic Institution (U.S.). The WH7805 strain (GC contents in the DNA, 59.7 mol.%) are immotile pink cells. The strain was isolated by L. Brand (cruise 48 of the "Oceanus"), sample dated June 30, 1987. The WH8103 strain (GC contents in the DNA, 58.9 mol.%) are motile yellowish cells; the strain was isolated by J. Waterbury (cruise 92 of the "Oceanus") from the sample dated March 17, 1981. The strains were maintained in the laboratory of J. Waterbury on medium SN (see [18, 20] in the paper by Waterbury and Ostroumov 1994). The cells were cultured at a temperature of 22°C and at a permanent illumination of 20 (micro Einstein) $m^{-2} s^{-1}$. Synthetic-surfactant solutions added to the cultures were sterilized by filtration through Acro-disc sterile filters (Gelman Sciences) with pore diameter of 0.45 μm. The absorption spectra were recorded using a Shimatzu UV 3101PC spectrophotometer to charac-terize the culture. Along with the native spectra, the spectra of samples with addition of sucrose to the dish (1.5 g per 3.5 ml of cell suspension) were measured, which allowed us to decrease the light scattering.

We also used the strains *Anabaena* sp. CALU 811, *Cylindrospermum* sp. CALU 306, *Synechococcus* sp. CALU 742 from the collection of the Laboratory of Micro-biology, Biological Research Institute, Leningrad University. The cultures were grown in 50-ml conical flasks on liquid medium (20 ml each) of the following composition (g/l): KNO_3, 1; K_2HPO_4, 0.2; $MgSO_4$, 0.2; $NaHCO_3$, 0.2; $CaCl_2$, 0.05; trace elements, 1 ml (solution in medium no. 6). Growth conditions: 25°C, 2000 lux. Xenobiotic NS was added to the medium at concentrations of 0.1, 0.5 and 1 mg/l. The effect of the xenobiotic on the culture growth was judged by the amount of cell biomass, which was determined by drying aliquots to a constant weight at 105°C.

Strain 33 of *Nostoc muscorum* Ag. was isolated from calcareous soil in the Kirov Region. Strain 235 was isolated from soils contaminated with oil (Almetyevsk, Tatarstan Republic). The cultures were grown in a medium containing per 1 liter (in g): KNO_3, 1.0; K_2HPO_4, 0.2; $MgSO_4 \cdot 7H_2O$, 0.2; $CaCl_2$, 0.15; $NaHCO_3$, 0.2; and 1 ml of a solution of trace elements. The trace-element solution contained (in g/l): $ZnSO_4 \cdot 7H_2O$, 0.22; $MnSO_4$, 1.81; $CuSO_4 \cdot 5H_2O$, 0.79; $(NH_4)_2Mo_7O_4 \cdot 4H_2O$, 1.0; $FeSO_4 \cdot 7H_2O$, 9.3; $CaCl_2$, 1.2; $Co(NO_3)_2H_2O$, 0.08; EDTA, 10.0; H_3Bo_3, 1.989. Distilled water was used. The inoculate was added into each flask by 1 ml. The inoculate of *Nostoc muscorum* cyanobacteria was preliminarily homogenized by an electromechanical homogenizer (5,000 rpm for 1 min). The growth flasks contained 50 ml of the medium each. Each variant was represented by two repeats. The cultures were incubated at an illumination of 3000 lux and at room temperature.

2.1.1.2 Marine heterotrophic bacteria *Hyphomonas* (ex Pongratz 1957) Moore, Weiner and Gebers 1984, 71VP

Gram-negative pleiomorphic bacteria, chemoorganotrophs, require the presence of amino acids as the source of carbon (Moore and Weiner 1989). They are included in the group of budding and/or appendaged bacteria (*Bergey's Manual of Determinative Bacteriology*, 1989, Vol. 3) and are related to the first organisms that inhabit hard surfaces in marine water and form a biological film. The film then becomes a basis for colonization of this biotope by other periphyton organisms. They are widespread in biological films at hard surfaces in various marine ecosystems, in biofouling on the surfaces of hydrotechnical constructions and ships. Thus, these bacteria are of practical importance. They have a specific lifecycle, which includes budding of the daughter cells from the end of the hypha. The maternal cell attaches to a hard substrate, while the budded daughter cell is transported away by the water flow and precipitates on a substrate in another place, attaches to the surface, elongates, and buds off a new generation of daughter cells (Moore and Weiner 1989). The effect of synthetic surfactants on these bacteria was barely studied; particular effects of non-ionogenic and cationic surfactants were not studied in detail.

The *Hyphomonas* bacteria were grown on S-1 medium proposed by the author. The composition of the medium is as follows: 1 l of NaCl solution (22 g/l) was added to 1 l of Marine Broth 2216 medium (0.5% peptone, 0.1% yeast extract; on sterile seawater; Difco Laboratories, Detroit). This medium is more advantageous as compared with the media used earlier as it is more economic and convenient for the experiments, where bacterial growth is recorded in the optical density measurements. Addition of TX100 was made before that of the inoculate. The inoculate (5%, v/v) was a 1-day culture grown on the same medium (S-1). The optical density (600 nm; optical path, 10 mm) after the inoculation was about 0.05. Cultivation and incubation were carried out in a thermostatted room at 25°C without mixing. The density of the culture in all experiments was measured spectrophotometrically at a wavelength of 600 nm. Additions of TX100 were sterilized by passing them through a Sterile Acrodisc bacterial filter (Gelman Sciences), 0.2 μm. The experiment included two repeats unless indicated otherwise.

2.1.1.3 Other objects

The bacteria *Rhodospirillum rubrum* were kindly provided by the group of Prof. V.D. Samuilov (Moscow State University, Department of Cell Physiology).

2.1.2   Eukaryotes

2.1.2.1   Diatomic algae

Thalassiosira pseudonana Hasle et Heimdal (=*Cyclotella nana* Guillard clone 3H in Guillard & Ryther) (class Bacillariophyceae, order Biddulphiales, suborder Coscino-discineae, family Thalassiosiraceae (9 genera)) (according to the other classification: class Centrophyceae, order Coscinodiscales, family Thalassiosiraceae, 11 genera, predominantly in marine plankton). The species of the genus *Thalassiosira* Cleve (about 80 modern and fossil species) are widely represented in all geographical zones, in the plankton of seas and saline reservoirs. It is a characteristic representative of the typical and mass species of marine diatoms, one of the dominating groups in marine plankton, very important as a feed resource for many species of commercial fish. The diatoms make a significant contribution to the global processes of atmospheric carbon fixation and oxygen evolution, participate in the processes of self-purification of reservoirs, and are used in assessing the sanitary state of waters.

The algae were grown on medium f/2 (Giullard and Ryther 1962) without $FeCl_3$ and EDTA. In order to prepare the medium, water was preliminarily filtered through a polycarbonate filter (Nucleopore, 0.2 μm). The cells were counted in a Fischer hemacytometer after preliminary fixation by addition of Lugol's solution (50 ml in 1 ml of algal culture). The initial density of the culture was $3 \cdot 10^4$ cells/ml in all variants. The culture in the stationary growth phase was used for inoculation. The illumination regime in the inoculation was as follows: light, 14 h; darkness, 10 h; intensity of illumination was 254 (micro Einstein) m^{-2} s^{-1}. Temperature, 17°C.

2.1.2.2   Green algae

Five classes including the Protococcophyceae, which is sometimes considered as an order. The Protococcophyceae are ecologically diverse, are present in plankton, benthos, neuston, and periphyton (the epiphyte and epizoan forms), and are common in land habitats and in soil. They are present in many types of reservoirs including fish ponds, some types of precipitation tanks, biological ponds, and filtration fields of urban water treatment facilities. Thus, they take an active part in self-purification of water by the ecosystems. Most of the Protococcophyceae are euryhaline and eurythermal organisms. The species of the general *Scenedesmus* and *Chlorella* became classical plant-cell objects and models, which were used to study many aspects of biochemistry and physiology. The Protococcophyceae are actively studied to be used for intensifying the purification of polluted waters and producing protein and vitaminized fodder. By the role they play in the natural ecosystems and biogeo-chemical processes of the biosphere they can be second (not always) only to the diatoms (Gollerbakh 1977; Matvienko 1977; Kondratyeva et al. 1989; South and Whittick 1987).

Bracteacoccus minor (Chodat) Petrova. Strain 200 was obtained from the Komarov Botanical Institute, Russian Academy of Sciences, St. Petersburg (no. 867-1 in their collection). Strain 219 was isolated from volcanic ash collected on the ash plateau free of vegetation in the vicinity of Tyatya Volcano (Kunashir Island). The algae were grown in a medium containing in 1 l (in g): KNO_3, 1.0; K_2HPO_4, 0.2; $MgSO_4 \cdot 7H_2O$, 0.2; $CaCl_2$, 0.15; $NaHCO_3$, 0.2; and also 1 ml of trace element solution. The solution of trace elements contained (in g/l): $ZnSO_4 \cdot 7H_2O$, 0.22; $MnSO_4$, 1.81; $CuSO_4 \cdot 5H_2O$, 0.079; $(NH_4)_2Mo_7O_{24} \cdot 4H_2O$, 1.0; $FeSO_4 \cdot 7H_2O$, 9.3; $CaCl_2$, 1.2; $Co(NO_3)_2H_2O$, 0.08; EDTA, 10.0; H_3BO_3, 1.989. Distilled water was used.

The growth flasks contained 50 ml of medium each. Each variant was made in two repeats. The algal cultures were incubated at an illumination of 3000 lux at room temperature.

Scenedesmus quadricauda Breb. The cultures were grown on Uspensky nutrient medium no. 1 in Luminostat at a temperature of 24–25°C and illumination of 2000 lux. The sources of illumination were fluorescent lamps LB-40. The cultures were grown in 250-ml Ehrlenmeyer flasks. The number of cells was determined by direct calculation in a Goryaev chamber. In the experiments with sodium dodecyl sulfate (SDS), the initial density of cells was 3.16 million/ml (experiment 1) and 2.47 million/ml (experiment 2).

Pulverized podzolic soil from the Kirov Region was used in the experiments with soil cultures (the collection site was the experimental field of the Kirov Agricultural Institute). The soil had the following agrochemical characteristic: $pH_{sal} = 4.6$; P_2O_5, 37.3 mg/100 g; K_2O, 3.2 mg/100 g; humus, 1.2%. Soil (30 g each) was placed in Petri dishes, where 10 ml each of an aqueous solution of TDTMA (in distilled water) was added at concentrations of 0.1 and 0.05 mg/ml. Distilled water (10 ml each) was added to control dishes. Soil samples were incubated in the light at a room temperature and 70% humidity relative to complete water capacity (watering with distilled water by weight).

Abundance of the algae in the soil cultures was determined by the generally recognized method of the direct count of cells in the soil suspension using a light microscope. The author thanks Prof. E.A. Shtina for consultations and assistance in this part of work.

We also used other algae in the experiments. They were grown on standard media mentioned in respective chapters of this work and in papers we published.

2.1.2.3 Euglenas

Euglenophyta (about 1000 species, half of them found in Russia and republics of the former Soviet Union) generally inhabit internal continental reservoirs. They possess all major types of nutrition: autotrophic, saprophytic, holozoic (peculiar of higher animals); are capable of mixotrophy. They are involved in self-purification of aquatic objects, the water of which contains many organic substances. The species of the *Euglena* genus are capable of mass development in reservoirs that can lead to water blooming. This is a favorite object for cultivation in laboratories with the objective to study the effect of various factors. They are promising for use in purification of polluted waters and for cultivating in the photoautotrophic life support systems (Safonova 1977; Kondratyeva et al. 1989).

The culture *Euglena gracilis* Klebs var. Z. Pringsheim was grown photoorgano-trophically in 100-ml flasks at a temperature of 26°C and illumination of 1,500–2,000 lux. The medium of the following composition was used (in g/l): NaCl, 0.1; $MgSO_4 \cdot 7H_2O$, 0.4; KH_2PO_4, 0.4; $CaCl_2 \cdot 6H_2O$, 0.05; glucose, 10.0; *L*-glutamic acid, 2.0; $(NH_4)_2SO_4$, 1.0; vitamin B_1 (0.2% solution), 0.2; vitamin B_{12} (0.01% solution), 0.2, solutions I and II, 1 ml each per 1 l of medium. In order to prepare solution I, 695 mg of $FeSO_4 \cdot 7H_2O$ and 930 mg of Na_2EDTA was taken and dissolved in warm bidistilled water, pH was adjusted by NaOH, and then water was added up to 100 ml. Solution II was prepared by taking (in g per liter of bidistilled water): $ZnSO_4 \cdot 7H_2O$, 10.0; $MnSO_4 \cdot 4H_2O$, 2.2; H_3BO_4, 12.2; $Co(NO_3)_2 \cdot 6H_2O$, 1.0; $NaMoO_4 \cdot 2H_2O$, 1.2; $CuSO_4 \cdot 5H_2O$, 0.001. The strain *E. gracilis* obtained from the algal collection of Göttingen University (Germany), no. 1224-5/25, was used.

Bidistilled water was used to prepare solutions and media. The volume of the inoculate, which was sampled in the mid-logarithmic phase of growth, was 5 ml at the onset of the cultivation.

2.1.2.4   Plant seedlings

Seedlings are recommended as one of the priority objects for biotesting in the field of water quality studies (Unified methods for water quality studies. Part 3. Methods of Biological Analysis of Waters, ed. by M. Gubachek, Moscow, 1975). They are used in the arsenal of methods of the U.S. Environmental Protection Agency (U.S. EPA 1982) and other U.S. agencies (U.S. Food and Drug Administration 1987), and European agencies (European Organization for Economic Cooperation and Development 1984). The method is highly economic and efficient from the point of view of the information versus biotesting cost ratio. Plant seedlings can be used in laboratories (industrial enterprises, chemical institutions), where more susceptible organisms do not survive. Therefore, the method is free of one disadvantage of highly-sensitive test objects – they are not always capable of living under conditions of inplant laboratories, where the air can be polluted with chemicals.

Seedlings serve as an alternative to animal testing, which is important from the humanitarian point of view and in the view of official recommendations by the International Union of Toxicology (IUTOX). In 1985, the IUTOX Executive Committee published an official statement that "alternative methods of testing, which do not require the use of animals, should be in common use (after their comprehensive scientific testing)" (see Telitchenko and Ostroumov 1990). High economic efficiency of this biotest is important for Russia under the current conditions of science financing. Biotesting on plant seedlings was performed by several authors – including in Russia, in the laboratories by V.B. Ivanov (Ivanov 1974, 1982, 1983, 1986, 1992), N.V. Obroucheva (Obroucheva 1992) and in the West (e.g., Wang 1987; Wang and Williams 1990; Davies 1991; Davies et al. 1991). The method was successfully used in the laboratory headed by Professor Ivanov to assess a broad class of biologically active substances (BAS), including compounds important for pharmacology. (The author is deeply grateful to V.B. Ivanov and all colleagues at the laboratory for numerous consultations and discussions of the results). In studies carried out under the supervision of Professor V.N. Maksimov, the method was used to assess the toxic impacts of metals. The method was also successfully used at the Department of Soil Science, Moscow State University, to assess the toxicity of various biological preparations. This method is one of the main tools in investigating allelopathic substances. We wrote in detail about this trend of BAS studies in Chapter 3 of an earlier book (Ostroumov 1986). The method was used in the Central Botanical Garden, Ukrainian Academy of Sciences (A.M. Grozdinsky, E.A. Golovko and other specialists; the author is grateful to them for the seeds of cress). The method was also used at the Institute of Hydrobiology (Kiev) for water quality assessment (Sirenko and Kozitskaya 1988). Though this method was recommended by the U.S. Environmental Protection Agency (U.S. Environmental Protection Agency 1982), it was comparatively rarely used in the U.S. The variant of the method we used was more advanced methodologically compared to the works by Wang and by Davis and co-authors in the sense that they did not use information about the effect of chemicals on the ratio of germinated and nongerminated seeds. Introduction of the integral morphogenetic index, which unites information on the effect of a tested chemical or polluted water on both processes – germination of seeds and elongation of a seedling – was a methodological improvement.

Various plant test objects were used (Ostroumov 1990, Table 1). Traditional techniques and some less traditional variants and approaches were evaluated. A list of some major effects, based on the biological activity of substances – various surfactants and some pesticides (Maksimov et al. 1988), is given in Ostroumov (1990) (Table 2 therein). The following details of some evaluated variants of methods should be noted (Goryunova and Ostroumov 1986; Nagel et al. 1988; Ostroumov and Maksimov 1988; and others).

1. The group of methods for estimating the biological activity of substances and pollution of aqueous medium by their effect on seeds (at 100% germination) and further growth of seedlings. In order to estimate the biological activity (BA) of a substance or aqueous medium at 100% germination of seeds, a Petri dish (10 cm in diameter) with the seeds of test objects put on filter paper was filled with a corresponding water solution (usually 7, 10 or 15 ml). Controls were filled with distilled water (DW) or settled tap water (STW). Incubation was carried out in the dark at room temperature or at 26–28°C. After a time interval t_1 the length of the seedlings (hypocotyl + root or only root) was measured. Further measurements were made at times t_2, t_3, and so on. The measurement results were processed statistically using nonparametric methods (see below). Calculations of the mean rate of elongation and percentage of inhibition were used (Ivanov 1974). This group of methods was used for white mustard *Sinapis alba*, buckwheat *Fagopyrum esculentum*, cucumber *Cucumis sativis*, watercress *Lepidium sativum* and other objects. In the summary table (Ostroumov 1990, Table 1), the methods are denoted as 1, 4, 5, 7, 8.

2. The group of methods for estimating the biological activity of substances and pollution of aqueous medium by their effect on the elongation of preincubated seedlings. When the effect of a tested substance at the initial stage of germination was to be avoided, and the objective was to study its effect on the elongation of the seedlings, the experiment was carried out as follows (see Ivanov 1974). The seeds were first preincubated in DW or STW. Then seedlings (of predominantly fixed length) were picked out and placed in Petri dishes with tested solution at different concentrations. All dishes contained different volumes of solutions (usually 7, 10 or 15 ml) and an equal number of seedlings. Control seedlings were transferred into new Petri dishes with the water used to prepare the test solutions. The length of the seedlings at the beginning of incubation (t_1) was measured. Then the incubation was carried out in the dark and the length of seedlings was measured (t_2, t_3, etc.). Elongation of the seedlings in the test solutions was compared with that of the control seedlings. The results were processed statistically. The method has the following restrictions: preincubation in DW or STW may slightly smooth the effect and decrease the sensitivity of the method as compared to the methods of group 1. The methods of biotesting on seedlings was evaluated in a large cycle of works by V.B. Ivanov (e.g., Ivanov 1974) and in our studies on buckwheat, rice, and other test objects. In the summary table (Ostroumov 1990, Table 1) the methods are denoted as 1b, 5b, 5c, 6.

   3. The group of methods for estimating the biological activity of substances and pollution of aqueous medium by their effect on the degree of germination. Some (but not all) tested substances significantly decrease the proportion of germinating seeds. It is easiest to estimate this proportion when the germination of control seeds is 100%. If part of the controls does not germinate, the following relation is used to estimate the effect of the substance tested:

$$E = [(M_0 - M_c)/(N - M_c)] \cdot 100\%,$$

where N, M_c, and M_0 are the numbers of seeds taken for testing at each concentration, of control seeds that failed to germinate, and of seeds that failed to germinate at a tested concentration of the substance, respectively. The biological meaning of this relation is that it gives an algorithm to reveal, to a certain degree, what proportion of the seeds would not germinate due to the effect of the substance tested. This approach was tested in experiments to study the effect of surfactants on *F. esculentum* and *Allium cepa*. It can be used in any experiment where not all control seeds germinate.

   4. The group of methods for estimating the biological activity of substances and pollution of aqueous medium by their effect on the mean length of seedlings. Some substances can slow down elongation of seedlings without greatly decreasing germination, whereas others can inhibit both processes. Therefore, it is of interest to develop a method of biotesting and processing of the results, which would take into account and integrate the effect of substances both on the growth rate and germination of seeds. We proposed and tested the following method. Seeds were put into Petri dishes with the tested solution. Then the length of seedlings was measured and the number of seeds that did not germinate was recorded. Further processing and calculation of the mean length of the seedlings took account of the seeds that failed to germinate as seedlings with conventional zero length. The value obtained upon averaging was called the apparent average length (AAL) of the seedlings. The parameter thus calculated combined information on the effect of a substance both on the length of seedlings and germination of seeds. This approach was tested on *F. esculentum* and *Oryza sativa*. In the summary table (Ostroumov 1990, Table 1), the corresponding methods are denoted as 1a and 5a.

Other variants of the methods tested are based on the hypocotyl orientation disorders in seedlings affected by biologically active substances (BAS). The orientation of hypocotyls in those experiments was registered by eye. At a certain stage of development the overwhelming majority of *Camelina sativa* hypocotyls are oriented vertically. An example of such a study is given in the chapter on nonionogenic surfactants. Processing of the results of experiments with seedlings: after the initial results are obtained, they need to be statistically processed. The "Statgraphics" package was used. After calculating the mean length (or apparent average length) of seedlings, in some experiments it is reasonable to calculate the rate of elongation (V) and the percentage of inhibition (I) using the relations

$$V = [x(t_2) - x(t_1)]/(t_2 - t_1),$$

$$I = (1 - x_{exp}/x_{contr}) \cdot 100\% = [(x_{contr} - x_{exp})/x_{contr}] \cdot 100\%$$

where $x(t_1)$ and $x(t_2)$ are the mean lengths of seedlings at times t_1 and t_2; x_{exp} is the mean length of seedlings in the variant where BAS of tested (polluted) aqueous medium are active; x_{contr} is the mean length of seedlings in the control.

The Student's t-test was used to assess the statistical significance of the difference between x_{exp} and x_{contr}. Nonparametric criteria were also used, such as Wilcoxon's test and Kolmogorov–Smirnov test. The use of these criteria is provided by the "Statgraphics" package. However, it can compare only samples of the same volume using Wilcoxon's and Kolmogorov–Smirnov tests, though in practice samplings can have different volumes; this is a drawback of this package. It is eliminated in the "Statis" statistical package developed by A.P. Kulaichev at the Biological Faculty of the Moscow State University.

5. Methods of experiments on the effect on the rhizoderm cells. Seeds of buckwheat *F. esculentum*, white mustard *S. alba* or soft wheat *Triticum aestivum* were put in Petri dishes on filter paper. Solutions of nonionogenic surfactant Triton X-100 (Schuchardt) (7–15 ml) in distilled water were added into the dishes, and incubation was carried out in the dark. The same volumes of distilled water were added into control dishes, and the incubation was performed likewise.

Experiments with *F. esculentum* cultivar Shatilovskaya-5 were performed in two variants.

Variant 1. A total of 17–20 seeds was put in a Petri dish and 10 ml of test solution was added. Incubation was carried out at 27°C. When the concentration of surfactants increased, the number of germinated seeds decreased, which led to reduction in the total number of seedlings. The number of seedlings that failed to fix was registered in 45 h.

Variant 2. Seeds were soaked not in a surfactant solution (as in variant 1) but in distilled water and then incubated. In 21 h, medium-length seedlings were transferred into new Petri dishes with Triton X-100 solutions. Ten seedlings were put in each dish. The number of seedlings that failed to fix was registered 43 h after the onset of soaking in distilled water. In experiments with *S. alba* VNIIMK, 15 seeds were put in a dish and 7 ml of test solution was added. Incubation was carried out at 18°C.

In experiments with *T. aestivum*, 3 or 4 seeds or seedlings of winter wheat (cultivar Zarya) were put in a Petri dish, and 7–15 ml of Triton X-100 solution in distilled water or an equal volume of distilled water was added. In one of the variants of the experiment, seedlings were put on perforated disks, and the roots were completely immersed in water through the holes. Incubation was done at 27°C in the dark.

2.1.2.5   Mollusks

Marine and freshwater mollusks were used. The role of mollusks is important both in the fishing industry and as a mariculture component. The total catch of marine bivalve mollusks exceeds that of all other groups of invertebrates taken together. In the monetary respect, the role of invertebrates (including mollusks) is more significant than in the weight aspect (Moiseyev 1985). Bivalve mollusks are of great importance as part of biofouling. Some bivalve mollusks became dangerous intruders (e.g., *Dreissena polymorpha*). Bivalve mollusks are included on the list of species of the Red Books of Russia and other republics of the former Soviet Union. (See also The IUCN Invertebrate… 1985.) The nomenclature used is according to Zatsepin and Rittikh (1975) and Zatsepin et al. (1978).

2.1.2.5.1   Freshwater mollusks

The following mollusks were used in the study: *Unio pictorum* (L.) s. lato (common pearl clam), *U. tumidus* Philipsson s. lato (common cline shaped pearl clam), *Crassiana crassa* (Philipsson) s. lato (=*U. crassus*) ((thick) oval pearl clam), *Anodonta cygnea* (L.) s. lato (common anodonta). Family Unionidae, order Actinodontida, subclass Schizodonta (=Palaeheterodonta), class Bivalvia.

The organisms were collected in the Upper Moskva River on the stone sandy silted bottom at a depth of 40–60 cm.

The rate of filtration was determined by the decrease of the optical density of the incubation medium as a result of the decrease in the number of algal, cyanobacterial or *Saccharomyces cerevisiae* cells preliminarily added and removed by filtration. The author thanks N.N. Kolotilova and E.A. Kuznetsov for assistance.

If not indicated otherwise, a typical experiment was as follows. Eight mollusks of *U. pictorum* were placed into beakers with 1.5 l of settled tap water (STW), i.e., tap water that had been kept at room temperature for at least 1–2 days. In variant A (control, no surfactant), 8 mollusks of 20.8 to 30.4 g in weight (mean weight, 24.3 g; wet weight with the shell) were put in a beaker. In variant B (with surfactant), 8 mollusks of 21.4 to 36.7 g in weight (mean weight, 26.1 g) were in a beaker. In both variants, a suspension of *S. cerevisiae* cells was preliminarily added into the water (SAF-Moment, S.I. Lesaffre, 59703 Marcq, France). The final concentration (dry weight) was 263.1 mg/l. Besides, an additional control was performed (variant C). In variant C, the beakers contained STW with a suspension of *S. cerevisiae* cells without mollusks. In variant C, no surfactant was added. The beakers in all three variants were incubated at a temperature of 17°C. Aliquots were taken and the optical density was measured at 500 nm (Hitachi spectrophotometer 200-20, optical path 10 mm).

The experiments with *U. tumidus* were performed in a similar manner. In some experiments, *Scenedesmus quadricauda*, *Synechocystis* sp. 6803 and *S. cerevisiae* (the strain from the Collection of the Department of Microbiology, Moscow State University), kindly provided by N.N. Kolotilova, were used as a suspension of plankton cells filtered by mollusks.

The culture of *S. cerevisiae* from the Collection of the Department of Microbiology, Moscow State University was grown for 96 h in shaken glass flasks at 26–28°C and constant intensive aeration (180–200 rpm). The culture growth medium contained (g/l): $(NH_4)_2SO_4$, 5.0; KH_2PO_4, 1.2; KCl, 0.15; $MgSO_4 \cdot 7H_2O$, 0.2; $CaCl_2$, 0.05; yeast autolysate, 50 ml; initial pH was 6.0.

Medium (150 ml) was introduced into 0.75-liter shaken flasks and sterilized. Before inoculation, sterile solution of glucose was added into the flasks (2%, w/w). The inoculate was added in terms of 3–4 ml per flask. The inoculate was grown in test tubes on agar slants for 2 days at 30°C. The biomass was washed off with sterile water and used for inoculation. The optical density at 500 nm was measured on a Hitachi 200-20 spectrophotometer at an optical path length of 10 mm.

Some experiments used *Lymnaea stagnalis* (L.) (=*Limnaea stagnalis*); family Lymnaeidae, order Basommatophora, subclass Pulmonata (pulmonary mollusks).

The organisms were collected from the eutrophicated pond next to the Upper Moskva River. Plastic beakers were used to keep the mollusks during the experiments. They were made from standard plastic bottles for soft drinks by cutting off the upper part of the bottle. The advantage of these beakers is their form: (1) owing to the height/square of the base ratio they occupy comparatively small space on laboratory tables; (2) due to a specific shape, the bottom has grooves where the sedimentary material (pellets) is concentrated. This feature makes sampling for subsequent analysis easy. Weighing of the phytomass and pellet samples as well as their elemental analysis was done by M.P. Kolesnikov. The methods of analysis are described in Ostroumov and Kolesnikov (2000, *Doklady RAS*, **373 (2)**: 278–280). From 5 to 10 animals were placed in the beakers. Unless indicated otherwise, the total biomass of mollusks per beaker was 10–14 g in most cases (in the experiment with TDTMA the wet weight with the shell was 11.72 g on average). The animals were fed with leaves of yellow water lily *Nuphar lutea* (L.) Smith taken from the Moskva River in June at a depth of 1.5 m. In some experiments, leaves of *Taraxacum officinale* Wigg were used. The central vein from each leaf was removed and divided into two parts. One of them was used as food for the mollusks in the beakers with surfactant TDTMA; the other was used in control beakers. The volume of water (STW) in each beaker was 1 liter. Every day during the experiment, the total volume of water was replaced with a new solution of TDTMA of a concentration given in the text above.

2.1.2.5.2 Marine mollusks

(a) *Mytilus edulis*. Subclass Filibranchia (=Pteromorphia). Order Arcoida (=Neotaxodonta) (according to one of the classification systems, Order Dysodonta). Family Mytilidae. Edible mussel (*Mytilus edulis*) is one of the most widespread species of bivalve mollusks. It inhabits the Pacific, Atlantic, and Arctic Oceans. It is a euryhaline species and is common both near the coasts and in river estuaries. The population density reaches several thousands, and the biomass can significantly exceed 1 kg/m². A eurythermal species, can survive significant fluctuations of temperature. Since the Stone Age, they have provided food for man. The annual catch exceeds 600,000 tons (e.g., in 1980 the catch was 628,000 tons) of mussels (*M. edulis* and other species of this order), which is approximately 19.5% of the total world catch of bottom-dwelling mollusks (excluding cephalopods) (Moiseyev 1985). The resources of mussels near the Black Sea coast of Russia and Ukraine were earlier estimated as about 65 million tons (now decreased significantly). In the White Sea, only near the Karelian and Pomor coasts the resource is close to 3,000 tons. The mussel mariculture yields approximately 8 tons (in some cases, significantly more, up to 150 tons) of high quality meat from one hectare or more than 1 ton of mussels from 100 m² of special strand nets ("busho") placed in the water, where mussel young are transferred. More than 80% of the world production of mussels is from mariculture. Mussel meat contains approximately 10% proteins, 1% fats, 0.5% carbohydrates (Zatsepin and Filatova 1968). The catch of mussels (mainly cultivated) is equal to (thousand tons): 80–120 in the Netherlands, 65–95 in Spain, 50–70 in France, 45–90 in Denmark, 11–23 in Germany, 7–11 in Great Britain. This is a source of valuable chemical substances. Mussels are used for monitoring seawater pollution (Burdin 1985) because they accumulate pollutants (Beznosov et al. 1987; Donkin 1994). Owing to active water filtration (Alimov 1981) they are one of the edificators of aquatic ecosystems (Zaika et al. 1990). Larvae of mussels (sterroblastulas, conchostomes, trochophores, veligers) are important components of zooplankton and food resources of many fish species.

The animals for experiments were collected from the sandy bottom of the Exmouth estuary in Southern England and were kept in tanks with automatic imitation of tide and ebb. The mollusks were manually cleaned from fouling with barnacle crawfish.

The experiments were carried out in 2-liter glass beakers with magnetic mixers. The beakers were incubated in a thermostatted room at a temperature of 16°C. Seawater was taken at a distance of 15 km from the coastline of Plymouth and filtered through nitrocellulose filters of the WCN type with 0.45 μm pore size (Whatman and Maidstone, England).

Usually, 16 animals were used in the experiments, 8 of which were subjected to a xenobiotic effect and the other 8 were used as controls. Surfactants were added to experimental beakers 1.5 h before the onset of the experiment. The surfactant concentration indicated in the tables and mentioned in the text in the discussion of the data is in all cases the initial concentration at the time of adding the xenobiotic into the beaker. In addition to 8 beakers, in which 16 animals were kept in pairs, a ninth beaker with the same volume of water was also used. The same volume of algal suspension was simultaneously added to all nine beakers.

The rate of filtration was determined by the depletion in the concentration of cells of marine phytoflagellates *Isochrysis galbana* Parke (strain CCAP 927/1) (Class Haptophyceae Christensen 1962, renamed to Prymnesiophyceae Hibberd 1976, part of Chrysophyseae; Order Isochrysidales Pascher 1910, Family Isochrysidaceae Pascher 1910). The strain was obtained from the NERC Culture Collection of Algae and Protozoa, Dunstaffnage Marine Laboratory, P.O. Box 3, Oban Argyll, PA34 4AD, Scotland, U.K.). The algae were grown at constant aeration with an air flow in 20-liter spherical glass beakers at constant illumination.

The composition of the medium for growing algae was as follows (g per liter of filtered seawater): EDTA disodium salt, 4.5; NaNO$_3$, 100; NaH$_2$PO$_4$·12H$_2$O, 20; MnCl$_2$·4H$_2$O, 0.36; FeCl$_3$, 1.3. A vitamin solution was added at a concentration of 0.1 ml per 1 l of medium: in 200 ml of distilled water, aneurine-HCl (vitamin B$_1$, thiamine), 0.2 g; cyancobalamin (vitamin B$_{12}$), 0.01 g. A trace element solution was also added (Ostroumov et al. 1997).

The concentration of algae was measured using a Coulter counter (Coulter Electronics, model Industrial D).

The following relation was used to calculate the filtration rate (CR) in l/h (Donkin et al. 1997):

$$CR = V(\ln C_1 - \ln C_2)/(t_2 - t_1),$$

where V is the volume of water in the beaker (2 l), C_1 is the concentration of cells at the onset of a time interval, C_2 is the concentration of cells in the end of a time interval, and $t_2 - t_1$ is the duration of the time interval (in hours). A similar relation was recommended in Filenko (1988). The author thanks Dr. P. Donkin and Mr. F. Staff for assistance.

The Student's t-test was used to assess the statistical significance of the difference between the quantitative values in the experiment and control. The standard error was calculated using the commonly accepted relation. Microsoft Excel 2000 was used to perform statistical analysis of the data.

(b) *Mytilus galloprovincialis* Lam

This species was used to study the effect of SDS, TDTMA, and some mixed preparations (in 1999). Two-month old mollusks were obtained in August from an aquaculture enterprise at the Institute of the Biology of Southern Seas (IBSS), National Academy of Sciences, Ukraine (Kazachiya Bay). (The author thanks A.V. Pirkova, V.I. Kholodov, and other staff of IBSS.) The larvae of mollusks used in the experiment were obtained from group breeding (8 female and 5 male species, 1.5–2 years old). Spawning was stimulated by temperature variations (cooling to 18°C followed by sharp warming to 22°C). The larvae were grown in seawater taken from a 60-m depth and filtered through a 20-μm filter (synthetic fiber net). Water temperature during growing was 18–25°C. The larvae were fed with a mixture of algae (*Thalassiosira* sp., *Monochrysis lutheri, Dunaliella viridis*) grown on Provasoli medium (Grodzinsky A.M. and Grozdinsky D.M. 1973) at a 24-h illumination of 10,000–12,000 lux.

Approximately 500 animals were placed in the beakers. The total biomass of mollusks in a beaker was 0.5 g (wet weight with the shell). Several variants of the method were used. To feed mussels, some experiments used algae *Pavlova lutheri* (Droop) Green 1975 (=*Monochrysis lutheri* Droop 1953) (Class Haptophyceae Christensen 1962, renamed to Prymnesiophyceae Hibberd 1976, part of

Chrysophyseae; Order Pavlovales Green 1976; Family Pavlovaceae Green 1976). If not indicated otherwise, the mollusks were placed in a beaker, 25 ml of seawater each was added, and 25 ml each of *Pavlova lutheri* (=*Monochrys lutheri*) algal suspension (12 mln cells per ml) so that the final concentration of the algae was 6 million cells per ml. The time of adding the algal suspension was taken as the onset of incubation. The final volume of the incubation medium in each beaker was 50 ml.

The algae for the experiment using *Dunaliella viridis* (Chlorophyceae, Order Volvocales Oltmanns 1904; Family Dunaliellaceae Christinsen 1967) (e.g., in studies of the effect of SDS) were grown in Provasoli medium (Grodzinsky A.M. and Grozdinsky D.M. 1973). Unless indicated otherwise, 30 ml of algal suspension was introduced into the beakers at an initial concentration of cells approximately equal to 3×10^6 cells/ml. Simultaneously, the optical density (OD) was measured in three beakers, two of which served as controls. Mussels were put into an experimental beaker (0.5 g of wet weight with shells, a total of approximately 500 mollusks) and then the algal suspension and surfactant were added. Control 1: the beaker contained mussels, algae, no surfactant. Control 2: the beaker contained algae, surfactant, no mussels. Control 3: the beaker contained only algae without mussels and without surfactant. The optical density was measured at 658 nm (optical path 10 mm) using a LOMO spectrophotometer SF-26 (Russia).

Some experiments used *Saccharomyces cerevisiae* (SAF-Moment, S.I. Lesaffre, 59703 Marcq, France). If not indicated otherwise, incubation was carried out at 25.8°C. The optical density of the medium was measured at 650 nm (optical path 10 mm) using a SF-26 spectrophotometer (LOMO, Russia). In some experiments, the time of visual differences (degree of turbidity of the medium) between the control and experiment was recorded. The advantage of the methods using *S. cerevisiae* is the possibility of their field use and also where laboratory equipment is limited.

(c) *Crassostrea gigas* Thunberg

Subclass Filibranchia (=Pteromorphia). Order Pectinida (=Monomyaria). Family Ostreidae. One-year old mollusks were obtained in August from an aquaculture enterprise at the Institute of the Biology of Southern Seas, National Academy of Sciences, Ukraine and the State Oceanarium of Ukraine (Kazachiya Bay). The author thanks A.V. Pirkova, V.I. Kholodov and other staff at IBSS. Larvae were settled in laboratory conditions. Then the mollusks were transferred to cages and grown in the open sea for a year. The filtration rate was determined by the depletion in the optical density of the medium at 550 nm as a result of the removal of *S. cerevisae* cells preliminarily added to the seawater and then removed by filtration. The concentration of *S. cerevisae* cells (SAF-Moment, S.I. Lesaffre, 59703 Marcq, France) was usually 100 mg/l (dry weight). The temperature is given in the tables. The optical density of the medium was measured at 650 nm (optical path 10 mm) using an SF-26 spectrophotometer (LOMO, Russia). The mean weight of the oysters used and the volume of the incubation medium is given in the notes to the tables.

(d) *Mercenaria mercenaria* (Linne)

Family Veneridae (over 400 species). Suborder Heterodonta. Order Eulamellibranchia. This mollusk (up to 10–12 cm in length) is the main commercial species among the Veneridae. Most venerids are edible, many of their species are the objects of commercial fishing, and some of the species are cultivated. The Veneridae are harvested in Japan (more than 0.1 million tons a year), U.S., China, France, Spain, and other countries.

Our studies used 4-day old larvae. The larvae were obtained from an aquaculture enterprise (hatchery) in Bluepoints Co. Inc. (Mr. Chris Pranis, P.O. Box 8, Atlantic Avenue, West Sayville, Long Island, NY 11796; (516)589-0123). Phytoflagellate *Isochrysis galbana* was used for feeding. The salinity of water used in the experiment was 26 ppt. The larvae were incubated in six-well tissue culture plates (flat bottom, Multiwell; Becton Dickinson and Co.). The volume of wells was 10 cm. The observation of the wells was performed using a Nikon binocular dissecting microscope. The author thanks Prof. N. Fischer (SUNY) for the opportunity to use his laboratory for this study. This research was supported in part by USIA and the International Programs, SUNY.

2.1.2.6 Annelids. Medical leech *Hirudo medicinalis*

Type Annelides. Class Leeches Hirudinea, Subclass Euhirudinea. Order Gnathobdellidae (Rhynchobdellinae) Gnathobdelliformes (according to other classification systems Gnathobdellea; Arhynchobdellea). Family Hirudinidae (Gnathobdellidae). This leech produces valuable medical substances, e.g., hirudine used for curing thrombophlebitis and hypertension. Approximately 12,000 kg of leeches are used annually, mainly for medical purposes (*IUNC Invertebrate Red Data Book* 1983, p. 207). The species is included into the Red Books of the International Union of Nature Conservation and into the Red Books of many countries (Yablokov and Ostroumov 1983, 1985) including Russia and other republics of the former Soviet Union. In 1848, the Russian government established an increased tax for the export of leeches due to the threat of exhausting their resources and banned their harvesting in the reservoirs from May to July. The species is a model for studying leeches important from an ecological point of view; they are active and energetic predators and, thus, the regulators of aquatic ecosystems. As a class, leeches (approximately 400 species) inhabit all types of aquatic ecosystems (freshwater and marine) and live by feeding all classes of vertebrates, mollusks, crustaceans, water insects, worms, etc. (Lukin 1968).

Young species of *H. medicinalis* obtained from the biological plant of the Main Administration of Pharmacies of Moscow were used for biotesting on leeches. The species were 6 months old; their weight was ~80–200 mg (wet weight); they were fed last 40 days before the experiment. The animals were placed, three species each, in Petri dishes 10 cm in diameter with tested solution (20–30 ml) or settled tap water (STW) in the same amount. The effect of surfactants at each concentration (1, 5, 10, 25, 50, and 150 mg/l) was studied on nine species, i.e. three Petri dishes were taken for each concentration. The temperature during the incubation was 20°C. All surfactant solutions were prepared on STW. The same water was in the controls.

For testing, all animals were placed for a few minutes into pure STW. This initial preliminary incubation was performed in Petri dishes. Three species were placed in each Petri dish, and 20–30 ml of STW was added. The animals became quiet very quickly and achieved a static state. Then, the water was removed from all dishes and solutions of surfactants or STW (for controls) were added. The time of exposure was counted starting from this moment. In 10 min, the state of the test animals was recorded. In the first 50 min, the animals in each Petri dish were observed every 2 min (the number of observations for each concentration was $9 \times 25 = 225$). In particular, the quiescent state was recorded, as well as locomotor activity ("dynamic states"), characteristic postures associated with intoxication (Lapkina and Flerov 1979; Guidelines on the bioassaying of sewage waters using medical leech ... 1986). The author thanks B.A. Flerov and L.N. Lapkina for advice and assistance.

2.1.2.7   Other objects and methods

The culture of marine phytoflagellates *Olisthodiscus luteus* N. Carter 1937 (Order Chromophyta, Class Raphidophyceae Chadefaud ex Silva 1980 = Class Chloromonadophyceae Papenfuss 1955) was kindly provided by L.B. Ilyash (Department of Hydrobiology, Moscow State University). This species is common in salt marshes in Europe and Japan (Tomas 1997).

The electrographic reaction of the *Cyprinus caprio* olfactory bulb was characterized by A.Ya. Kaplan on an installation he developed and made using his own method (Kaplan 1987, 1988). The method enabled us to characterize quantitatively the functional activity of the olfactory bulb and the signal from the receptors at the stimulation by aqueous solutions containing low concentrations of adequate chemical stimulants (natural substances, e.g., amino acids), and disturbances of normal functional activity under the influence of chemicals.

2.2   Chemical Substances Used

2.2.1   Anionic surfactants

2.2.1.1   Sodium dodecyl sulfate (SDS)

Sodium dodecyl sulfate (SDS, sodium lauryl sulfate, chemical formula $C_{12}H_{25}SO_4Na$, molecular weight 288.5) is one of the widely used primary alkyl sulfates. Properties: dissolves in water, chloroform, methanol, butanol; does not dissolve in diethyl ether, benzene, and dioxane (up to 40°C); CMC (critical micelle concentration) is 8.1 mmol/l, HLB (hydrophilic-lipophilic balance) is 42.0. Widely used as a frother, emulsifier, solubilizer, wetting agent, and dispersant. $LC_{50} = 2.7$ g/kg (white rats, intraperitoneally). An analog is manufactured in Germany by BASF under the name Waschrohstoff 818 Teig.

2.2.1.2   Sulfonol

Sulfonol manufactured in Sumgait in 1984 was used. It contains: sodium alkyl benzene sulfonates, 45% (C_{12}–C_{18}); sodium sulfate, 10%; nonsulfonated hydrocarbons, up to 3%; the balance, H_2O. For the preparation with 80% of surfactant, CMC = 1.88 g/l; HLB = 11.0–11.7. Applications: basic composition for cleaning wet wool, dyeing fabric, cleaning metal surfaces, washing workpieces, as a foaming agent, primary emulsifier in emulsion polymerization, wetting agent. It is considered to be low toxic. For the preparation with 80% of surfactant, LC_{50} = 5.45 g/kg (white rats, intraperitoneally). Analogs are manufactured in Germany, Belgium, the U.S., and Great Britain under other names (Abramzon and Gaevoy 1979, p. 284). Similar substances are used as flotation agents in rock mining industry, as plasticizers of concretes and cements, corrosion inhibitors, and in many other fields. The content of sulfonol in the detergent Lotos is 20% (19–22%).

2.2.2 Nonionogenic surfactants

2.2.2.1 Triton X-100

Triton X-100 (TX100, oxyethylated alkyl phenol, polyoxyethylene isoctylphenyl ether, molecular weight 624.9) is one of the widely used representatives of polyethylene glycol monoalkyl phenyl ethers (alkylaryl polyethers). A colorless transparent viscous fluid. Properties: dissolves in hard and soft water, ethanol, benzene. CMC, 0.24–0.9 mmol/l; aggregation number, 140; mean micellar weight, 90,000 (the fractions with different boiling temperature have different micellar weights). Substances of this class (oxyethylated alkyl phenols) are widely used as emulsifiers, solubilizers, wetting agents, dispersants, components of washing and degreasing compositions; are used in oil refining industry, petroleum industry, and many other branches (Abramson and Gaevoy 1979). TX100 is widely used in biomedical research, as well as in some preparations used in genetic engineering. LC_{50} of some technical preparations containing TX100 and its homologs is approximately 1.6 g/kg (white rats, intraperitoneally). Analogs are manufactured in the U.S. and Great Britain under different names. Substances of this class include NP-4, Triton X-207, Triton N-101, and others.

2.2.3 Cationic surfactants

2.2.3.1 Quaternary ammonium compounds (QAC)

Tetradecyl trimethyl ammonium bromide (TDTMA), ethonium (see below), and benzethonium chloride (benzyl diisobutyl phenoxyetoxy-etoxydimethyl ammonium chloride) were used. QAC have a wide range of applications. Cationic surfactants belonging to the class of QAC are found in the sewage waters of such industries as oil (borehole drilling and well operation, oil mining, and storage), petrochemicals (production of latex items), gas (drilling of gas boreholes), chemicals (production of fertilizers, synthetic resins and plastics, chemical fibers and paints, photo and cinema materials and magnetic tapes), automobiles, aircraft, mechanical engineering (machining of metals), pulp and paper (production of cellulose, production of paper, and specialty papers manufacturing), construction engineering (production of asphalt and bitumen mixtures), textiles, foods, medical and biological technology, and in leather tanning; QAC are applied in agriculture and different types of transport (water, aviation, automobiles). An important example of this class of compounds is alkyl trimethyl ammonium chloride (ATM chloride), whose application includes protection of materials against biological damage (Ilyichev et al. 1985). The base of the compounds is nirtan, dezan. Nirtan was tested for disinfection in public health and cattle-breeding complexes, and for the protection of metal construction in the oil mining industry against microbial corrosion. The chemical at a concentration of 25 mg/l suppresses sulfate-reducing bacteria and decreases the rate of corrosion by 25%. ATM chloride as preparation Kationat-10 (up to 2%) is used in restoration works for antiseptic protection of the paint layers of ancient paintings. ATM chloride mixed with other QAC is recommended for microbial protection in oil distillate fuels. An effective concentration in fuels is up to 0.1%. $LC_{50} = 870$ mg/kg (rats), 900 mg/kg (mice) (Ilyichev et al. 1985).

Ethonium. 1,2-N,N-*bis*(dimethyl)-N,N'-*bis*(decylacetate) ethylendiammonium dichloride. CMC = 3.2 mmol/l, $LC_{50} = 55$ mg/kg, $LC_{100} = 70$ mg/kg (white rats, intramuscularly). This compound is thermally stable up to 170°C, is stable in acidic media and in the presence of hardness salts. Applications: emulsifier, hydrophobizer, fabric softener, stabilizer of dispersions, flotation agent, hardener of clay suspensions, antistatic; in pharmaceutical industry, for producing greases, emulsions, solutions; re-charger of polymeric coatings, clays, epoxy resins. Is also used in pharmaceutical industry for production of ointments, emulsions, and solutions (Abramson and Gaevoy 1979, p. 294).

The values of CMC for some synthetic surfactants are given in Table 2.1 (selectively adopted from Abramzon and Gaevoy 1979; the catalogues of some companies manufacturing and supplying synthetic surfactants were also used).

Table 2.1 Critical micelle concentration (CMC) of some synthetic surfactants (after Abramzon and Gaevoy 1979; catalogs of synthetic surfactants manufacturers and suppliers).

Type of surfactant	Name	CMC	Note
Anionic	Sodium dodecyl sulfate	8.47 mM	20°C
Anionic	Sodium dodecyl sulfate	1.49 mM	21°C, 100 mM NaCl
Anionic	Sodium dodecyl sulfate	8.1 mM	–
Nonionogenic	Triton X-100	0.24–0.9 mM	–
Nonionogenic	Decaethylene glycol nonylphenyl ether	0.0068% (w/v)	–
Nonionogenic	Decaethylene glycol nonylphenyl ether	0.0046% (w/v)	Na_2SO_4 0.5 n
Cationic	Hexadecyl trimethylammonium bromide	0.8 mM	25°C
Cationic	Hexadecyl trimethylammonium bromide	2 µM–0.2 mM	25°C, 1 M KNO_3

2.2.4 Mixed preparations, synthetic detergents, and foam detergents

Synthetic and foam detergents along with synthetic surfactants contain other components including phosphorus-containing salts, silicates, bicarbonates, optical bleachers, etc. The compositions of some synthetic and foam detergents studied in this work are given in Tables 2.2–2.4. Below, we present information on the preparations used, including the manufacturers and available information about the compositions (from incomplete data given in the product label).

Table 2.2 Composition of detergent Bio-S.

No.	Components	Fraction of total weight, %
1	Alkyl benzene sulfonate	8
2	Alkyl sulfates	about 4
3	Oxyethylated fatty alcohols	3
4	Synthetic fatty acids, 17–20°C	4
5	Sodium tripolyphosphate	40
6	Sodium silicate	5
7	Sodium bicarbonate	10
8	Crystalline microcellulose (prevents re-precipitation of dirt during the washing)	10
9	Optical bleacher	0.1–0.3
10	Fragrance	0.1–0.2
11	Proteolytic enzymes	2
12	Sodium sulfate	up to 100

Table 2.3 Composition of detergent Kristall.

No.	Components	Fraction of total weight, %
1	Alkyl benzene sulfonate (sulfonol) or alkyl sulfates or their mixture	20
2	Oxyethylated fatty alcohols	3
3	Synthetic fatty acids, 17–20°C	4
4	Sodium tripolyphosphate (eliminates water hardness)	40
5	Sodium silicate (improves detergent effect)	3
6	Alkylolamide (foam stabilizer)	2
7	Crystalline microcellulose	2
8	Optical bleacher	0.1
9	Fragrance	0.1–0.2
10	Sodium sulfate	about 27 or less

Table 2.4   Composition of detergent Kashtan (surfactant content, ~24%).

No.	Components	Fraction of total weight, %
1	Alkyl sulfonates (volgonate OST 6-01-35-79)	6–10.0
2	Oxyethylated fatty alcohols syntanol (TU 6-14-577-77)	6.0
3	Sodium salts of secondary alkyl sulfates (liquid washing substance Progress), TU 38-10719-77	8.0–12.0
4	Sodium tripolyphosphate, technical grade, GOST 13493-77	3.0
5	Potassium tripolyphosphate, TU 6-25-26-80	4.0
6	Optical bleacher (imported, Rilux VPA)	0.1
7	Fragrance for soap and detergents, TU 18-16-121-77	0.1
8	Potable water, GOST 2874-73	up to 100

Lotos-Extra, TU 6-39-1-89 (Vinnitsa Branch of Khimprom; 4 Frunze Street, Vinnitsa, 287100, Ukraine). Recommended concentration for use, 5–7 g/l.

Losk-Universal (Henkel), TU 2381-007-04831040-96 (Era JSC, 1 Moskovskoe Shosse, Tosno, Leningrad Region, 187020, Russia). Composition: surfactants, enzymes, optical bleacher, sodium carbonate, sulfate, phosphates, silicates, polymers, sodium citrate, fragrance. Recommended concentration in hard (6–9 mg-eq/l) water, 5–12 g/l.

OMO Intelligent Avtomat (Unilever Polska S.A. Oddizial Detergentow I Kosmetykow, ul. Kraszewskiego 20, 85-954, Bydgoszcz, Poland). Composition: anionic surfactants, 5–15%; nonionogenic surfactants, less than 5%; phosphate, 15–30%.

Deni-Avtomat, TU 2381-011-04831040-98 (Era JSC, 1 Moskovskoe Shosse, Tosno, Leningrad region, 187020, Russia; Henkel-Yug, 48 Pr. Stroitelei, Engels, Saratov Region, 413116, Russia). Recommended concentration, 6–7 g/l.

Vesna-Delikat, TU 2381-001-00336496-98 (cosmetic company Vesna, 33 Neverov Street, Samara, 443036, Russia). Composition: surfactants, sodium tripolyphosphate, sodium carbonate, sodium sulfate, sodium silicate, special softeners, fragrance, crystalline microcellulose, optical bleacher.

Lanza Avtomat (Benckiser) (52/11 Kosmodamianskaya nab., Moscow, 113054, Russia).

IXI Bio-Plus (Cussons Polska S.A.; ul. Krakowska 112/116. 50-427 Wroclaw, Poland). Composition: nonionogenic surfactants, < 5%; anionic surfactants, 5–15%; bleaching agent; phosphate, > 15%; sulfate, carbonate, silicate, crystalline microcellulose, enzyme, optical clarifier, fragrance. The product label indicates that the preparation was tested dermatologically.

Avon Herbal Care (normal hair shampoo [lavender and honey with milk proteins]). Composition: water, sodium laurate sulfate, ammonium lauryl sulfate, cocamidopropyl hydroxysultain, lauramide MEA, benzyl alcohol, cocamide MIPA, aromatizer, methyl paraben, sodium chloride, quaterny-80, disodium salt of ethylenediamine tetraacetic acid; sodium phosphate, sodium citrate, propylene glycol, benzoic acid, honey, quaterny-79, milk protein hydrolyzate, glycerine, phosphoric acid, lavender oil, methylated alcohol, phenoxyethanol, sodium benzoate, propyl paraben, CI 420990, CI 14700, CI60730 (Avon Cosmetics, NNI 5PA England; New York, Lisboa, London, etc.).

Foam detergent Verbena (Krasnogorsk experimental plant of domestic chemistry, 1987). Fennel oil is included in the formula.

Fairy (dish washing liquid; Procter & Gamble Ltd. Essex RM16 1AL, United Kingdom; Novomoskovskbytkhim Co., 64 Komsomolskoye Shosse, Novomoskovsk, 301670 Russia). Composition: surfactants, aromatic additives, color additives (green pigment), water, etc.

E Lemon (dish washing liquid; made for Cussons International Ltd.; England; Cussons Polska S.A., Poland). Composition: biodegradable surfactants, dyer (yellow-lemon), preservative, aromatizer.

Dish washing concentrate Mila, TU 2381-001-51102363-99 (Khimakom Co., 6/57 Second Peschanaya Street, Moscow, Russia). Composition: anionic surfactant, nonionogenic surfactants, pigment (yellow), aromatizer, preservative.

Additional information about the organisms and substances used in the study is given in our publications and references therein.

In the subsequent chapters, the concentrations of synthetic surfactants given in tables and text are initial concentrations at the onset of the experiment. The concentrations used in the experiments were chosen based on (1) the search for the minimum concentrations that have a noticeable biological impact; (2) existing information about real concentrations of synthetic surfactants found in natural, polluted and sewage waters (for some categories of synthetic surfactants, this information in relation to the reservoirs in Russia is absent); (3) the search for high values of concentrations with the aim to use the results for remediation of polluted aquatic ecosystems; (4) preliminary experiments, i.e., the concentrations used in the study are the result of scientific research; (5) experience of previous work with similar substances and test systems.

Methodological problems were analyzed in Ostroumov (1990, Vestn. MGU), Ostroumov and Khoroshilov (1992), Ostroumov et al. (1997), and Ostroumov (2000, Doklady RAS).

3 Biological Activity of Waters Containing Anionic Surfactants

The effects of aquatic media containing anionic surfactants on organisms is of great hydrobiological and ecological interest as these xenobiotics are released to the hydrosphere in largest amounts as compared with other surfactants. In most if not all countries including Russia, anionic surfactants predominate among the synthetic surfactants produced and used. In some countries, production and consumption of anionic surfactants reached 65% (on a weight basis) of the total consumption of surfactants (for the U.S.: Greek and Layman 1989).

Anionic surfactants are the lion's share of that 2.5 g of synthetic surfactants per citizen of the Russian Federation that daily enter the sewage and drainage systems (the other components: 7–8 g of ammonium salts, 1.5–3.3 g of phosphates, 8.5–9 g of chlorides, and 1.8–4.4 g of sulfates) (Akulova and Bushtuyeva 1986). Upon biological purification, the water at the exit from the water treatment plants can contain 1.6 mg/l of synthetic surfactants (on a chlorine sulfonol basis), as shown for one large water treatment plant in Moscow (Gordeyeva and Kozlova 1980). There were reports of anionic surfactants at about 200 mg/l in the sewage water of light industry enterprises, up to 15 mg/l in mixed municipal sewage waters, up to 0.25 mg/l in the water of the Volga River (an average at $n = 54$, i.e., some individual measurements yielded significantly larger values), 0.33 mg/l in the Klyazma River (Mozhayev 1989), and 2.4–2.8 mg/l in the Black and Azov Seas (Review of the State ... 1976).

There are data about the negative impact of anionic surfactants on aquatic invertebrates (Koskova and Kozlovskaya 1979), fish (Metelev et al. 1971; Mozhayev 1976; Huber 1985), algae (Leonova and Stavskaya 1980; Lenova et al. 1980; Lewis 1986; Nyberg 1988; Parshikova and Negrutsky 1988; Lipnitskaya et al. 1989; Sirenko 1991; Lewis 1991; Parshikova et al. 1994) and other organisms (Rebandel and Dryl 1980; Potapov and Galagan 1983; Passet 1985). A literature review up to the end of the 1980s was made in Stavskaya et al. (1988), Mozhayev (1989), Lewis (1991), Painter (1992), Fendinger et al. (1994).

Among the anionic surfactants, the most important are alkyl sulfates and alkyl benzene sulfonates (ABS) (Bock and Stache 1982; Industrial ... 1984; Surfactants 1984; Berth and Jeschke 1989; Kouloheris 1989; Ainsworth 1992; Facts ... 1992; Steinberg et al. 1995).

3.1   Biological Effects of Alkyl Sulfates. Sodium Dodecyl Sulfate (SDS)

Alkyl sulfates are widely used in many industries including textile, chemical, and nonferrous metallurgy. They are used in production of concrete, pesticides, and many cosmetics components. Alkyl sulfates are good wetting agents in the textile industry, in production of cellulose, and in agriculture; they constitute active bases of many washing compositions and textile auxiliaries. There are data on the effect of alkyl sulfates on algae and animals. For example, for some fish species alkyl sulfates were noted to be lethal at concentrations of 8.5–10 mg/l (18–23°C) (see Metelev et al. 1971). A representative of alkyl sulfates is sodium dodecyl sulfate (SDS, sodium lauryl sulfate). SDS is used in a wide range of applications: it is a good frother, emulsifier, wetting agent, dispersant, and the base for cleaning compositions in cosmetics (Abramzon and Gaevoy 1979). In some widely used compositions (for instance, shampoos), the content of SDS is up to 40%.

The action of SDS on hydrobionts was studied in Lenova et al. (1980), Nyberg and Koskimies-Soininen (1984), Röderer (1987), Stavskaya et al. (1988), Lenova and Stupina (1990), Belanger et al. (1995a,b), and other works. Lenova et al. (1980) studied the effect of SDS on several *Chlorella* strains. Under experimental conditions, concentrations up to 50 mg/l did not inhibit the algae significantly; at 50 mg/l some strains were inhibited but others grew without noticeable inhibition. At concentrations of 100 mg/l and higher, growth of all strains was observed to be inhibited.

Nyberg and Koskimies-Soininen (1984) showed SDS at concentrations of 5 mg/l to change the fatty-acid composition of glycolipids of red alga *Porphyridium purpureum*. At SDS concentrations greater than 5 mln^{-1} or 17 μM, the algal cells died.

Studies on green algae (Goryunova and Ostroumov 1986; Maksimov et al. 1988) and chrysophytes (Röderer 1987) also showed a negative impact of SDS-containing aquatic media; however, the sensitivity of algae to SDS was lower.

SDS at a concentration of 1 mg/l caused the death of *Scenedesmus quadricauda* cells and, if the concentration decreased to 0.01–0.1 mg/ml, it stimulated the growth of these algae (Goryunova and Ostroumov 1986).

The growth of chrysophyte algal culture *Poterioochromonas malhamensis* was suppressed by an SDS concentration in aqueous medium, exceeding 100 μM. Living cells disappeared completely at SDS concentrations greater than 250 μM (Röderer 1987).

Significant information on the biological effects of SDS is available in the databank of the Russian Register of Potentially Hazardous Chemical and Biological Substances.

Problems of biodegradation of synthetic surfactants are beyond the scope of this work. However, for the sake of completeness of the discussion, we should note some of the works on biodegradation of alkyl sulfates, including SDS (Stavskaya et al. 1988). In the degradation by *Pseudomonas* bacteria, SDS under the action of primary alkyl sulphatase was shown (Stavskaya et al. 1988; Thomas and White 1989) to be transformed to 1-dodecanol; the latter under the action of primary alcohol dehydrogenase is oxidized to dodecanal; dodecanal is oxidized by aldehyde dehydrogenase to dodecanoic acid. The latter can be transformed in two ways: (1) beta oxidation to acetate and carbon dioxide; (2) elongation to tetradecanoic acid followed by further elongation and desaturation to form saturated and unsaturated fatty acids (C_{14}, C_{16}, and C_{18}). These fatty acids can be included in the pathway of phospholipid synthesis (Thomas and White 1989).

Interestingly, the SDS biodegradation rates differ by orders of magnitude in the works by different authors, which does not prevent the authors of summaries from including this or that data in their tables without any comments and without indicating the experimental conditions that affect the observed rate of biodegradation so significantly. Our experience of work with bacteria isolated from active sludge and degrading the anionic surfactant dialkyl sulfosuccinate confirmed that the bacteria capable of biodegrading a synthetic surfactant are subject to inhibition by certain concentrations of the surfactant. Therefore, the literature data on the biodegradation rate of this or that synthetic surfactant including SDS (e.g., Kaplin 1979) should be considered with a reservation that the determined rate of biodegradation is related only to a particular situation with a particular concentration of surfactant and bacterial cells.

According to the data by Kaplin (1979), the coefficient k_1 (days^{-1}) of SDS bio-chemical oxidation in natural water at 20°C is 0.99 (for comparison, for chlorine sulfonol it is 0.15; and for m-cresol, 0.21). The coefficient k_1 characterizes the regularity of depletion of the SDS concentration according to the relation

$$S_t = S_0 \exp(-k_1 t),$$

where S_t is the concentration of SDS at time t.

This relation is widely used, but it does not take into account that (as was mentioned above) the rate of biodegradation can be inhibited by those concentrations of synthetic surfactants that suppress the bacteria responsible for the process.

It should be noted that Braginsky et al. (1980) revealed a much slower degradation of SDS. The existence of their degradation does not decrease the significance of information on the biological effects of surfactants, since the processes of synthetic surfactants' input into water systems with a flux of anthropogenic pollution get to dynamic equilibrium with the processes of their degradation. The chemical assays of natural water and measurements of surfactant concentrations register only the final balance concentration established as a result of that dynamic equilibrium.

Additional analysis of the data on biodegradation of alkyl sulfates and, in particular, SDS is given in the works by Cain (1987), Swisher (1987), Stavskaya et al. (1988), and others. The variants of the degrading strain *P. aeruginosa* 1C were noted to differ significantly by their resistance to the lytic action of SDS. The smooth S-strain was the most resistant, while the filamentous strain was less stable, and the least resistant was the rough R-strain (Stavskaya et al. 1988). Interestingly, the smooth strain is not capable of degrading SDS; in the presence of SDS it is subject to splitting and transition to the initial form or to the R-strain.

One of the fast methods to obtain information on the biological activity of substances or the degree of pollution of an aquatic medium is to register the state of bivalve mollusks and their activity as filter feeders. The filtration activity was observed to decrease under the action of heavy metals (for swan mussel *Anodonta cygnea* L., Shalanki 1985; zebra mussel *Dreissena polymorpha*, Stuijfzand et al. 1995) and organic xenobiotics including pesticides (for *Mytilus edulis*, Donkin et al. 1997). By the time this work began, nothing was known about whether alkyl sulfates (e.g., SDS) can inhibit water filtration by marine bivalve mollusks *M. edulis*. We carried out experiments in this direction and present the results in this chapter after the section on the action of SDS on plants.

The importance of choosing SDS to assess its effects on organisms not used earlier for such studies was recently confirmed again when we found that there are recommendations by the U.S. Environmental Protection Agency to use SDS as a reference toxicant (Fendinger et al. 1994). Before our investigations, practically nothing was known about the action of this substance on vascular land plants.

3.1.1   Effect of SDS on vascular plants

Growing seedlings of plants are of interest for studying the effect of SDS on organisms (Ivanov 1982; U.S. EPA 1982; Organization ... 1984; Wang 1985, 1987; U.S. Food ... 1987; Wang and Williams 1990). Elongation of the seedlings depends on two fundamental processes: the rate of cell division in the meristem and the rate of cell stretching in a corresponding zone (Ivanov 1974; Obroucheva 1992). "The experience of biology shows a surprising similarity of individual elementary processes of cell life, including the processes associated with cell growth and multiplication in the entire living world" (Ivanov 1974). Therefore, the data on the effect of earlier-not-studied substances on seedlings are of interest not only in the sense of interaction of a particular substance and a species of organisms but also in broader terms. Biotesting of the impacts of polluted waters on seedlings was recommended by the meeting of the leaders of the water management bodies of most Eastern European countries (Unified methods ... 1975); similar methods were recommended by the U.S. EPA.

Before our studies began, there was almost no data on the effect of SDS on plant seedlings. The need for such data is dictated not only by general theoretical interest but also by the fact that polluted and sewage waters in some cases enter ecosystems that include vascular plants. Surfactants are not in the list of indicators of the quality of water used in agriculture for irrigation (Unified Criteria of Water Quality 1982).

Our experiments showed that aquatic media containing 0.01 mg/ml SDS slightly stimulated the growth of *Sinapis alba* seedlings. After 72 h of incubation, their mean length was 110.1% of the control seedlings. Aquatic media with an SDS concentration of 0.1 mg/ml inhibited the growth of *Zea mays* seedlings (by 30–40%) but suppressed the seedlings of *Cucumis sativus* only insignificantly – their length after three days was 93% of that of the controls. At a concentration of 1 mg/ml, SDS had a lethal effect on all three species of the test objects.

We studied the effect of SDS on *Fagopyrum esculentum* and used two variants of the method: germination of seeds immediately in SDS solution (i.e., without transferring the seedlings from water) (Figures 3.1 and 3.2, Tables 3.1 and 3.2) and transfer of preliminarily prepared seedlings into SDS solution (Tables 3.3 and 3.4).

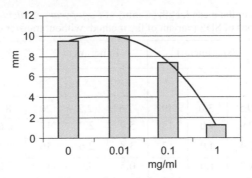

Figure 3.1   Effect of SDS (40 h) on the mean length of *F. exculentum*, in mm (see text for details). The fourth order polynomial trend line is shown.

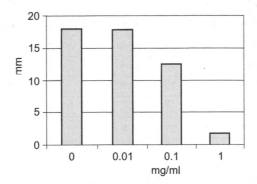

Figure 3.2   Effect of SDS (50 h) on the mean length of *F. exculentum*, in mm (see text for details).

In experiments with the transfer of seedlings into SDS solution (1 mg/ml) the length of the seedlings was approximately 1.5 times smaller than the length of controls. This effect was observed already after 10 h of the effect of SDS (Table 1 in Maksimov et al. (1986)). The effect of lower concentrations became noticeable after a longer exposure (at times t_3 and t_4 in the table mentioned, i.e. after 24 and 34 h of exposure).

The rate of elongation of seedlings as an effect of SDS was determined. It increased in all variants of the experiments (the control and various SDS concentrations), except for the variant with the highest concentration (1 mg/ml). As compared with the controls, at concentrations of 0.01 and 0.1 mg/l the average rate of elongation of the seedlings was found to decrease for all three time intervals. The effect was more pronounced at a concentration of 0.1 mg/ml. A comparison of these rates with the earlier obtained values for seedlings germinated directly in SDS solution showed that in this modification of the experiment (i.e., with the transfer of the seedlings from water to SDS solution) the rate is slightly higher.

According to our data, in an aquatic medium with an SDS concentration of 0.01 mg/ml, the growth rate of these seedlings for 50–64 h was 94.4% of the control; at an SDS concentration of 0.1 mg/ml it was 82.5% of the control.

Table 3.1 Estimate of the mean length (x, mm) of *Fagopyrum esculentum* seedlings at different concentrations of SDS (variant of the method: without transfer of seedlings).

Time, h	SDS, mg/ml	Experiment 1			Experiment 2			Experiment 3		
		x	s	n	x	s	n	x	s	n
40	0.0	9.5	3.0	19	7.6	2.2	20	8.3	4.9	15
	0.01	9.9	4.0	15	5.9	2.7	17	6.6	2.9	15
	0.1	7.4	4.0	17	3.8	2.6	16	8.3	3.1	13
	1.0	1.2	1.0	14	0.5	0.9	12	2.0	2.0	7
50	0.0	18.0	5.4	19	14.8	4.0	20	17.5	7.1	15
	0.01	17.8	6.5	15	11.9	4.0	17	14.4	5.0	15
	0.1	12.5	6.6	17	7.9	4.7	16	15.0	4.6	13
	1.0	1.6	1.1	14	0.8	1.2	12	2.6	2.2	7
64	0.0	28.3	7.9	19	28.6	8.5	20	38.1	12.2	15
	0.01	30.3	11.7	15	22.4	6.4	17	32.9	11.0	15
	0.1	19.9	9.9	17	16.8	9.8	16	28.3	6.2	13
	1.0	1.9	1.2	14	1.2	1.5	12	3.3	2.7	7
74	0.0	37.8	10.9	19	41.8	11.3	20	55.2	15.7	15
	0.01	44.1	16.4	15	32.2	8.9	17	49.8	17.0	15
	0.1	27.6	13.6	17	25.0	15.3	16	41.3	8.8	13
	1.0	2.1	1.3	14	1.4	1.6	12	3.4	2.8	7

Note: s, standard deviation; n, number of seedlings. For details, see *Biol. Nauki*, **12**: 81–84 (1987) (in Russian).

Table 3.2 Averaged results of three experiments to determine the mean length (x, mm) of *Fagopyrum esculentum* seedlings at different SDS concentrations (variant of the method: without transfer of seedlings).

Time, h	SDS, mg/ml	Combined data of three experiments		
		Mean length, x	Confidence interval	n
40	0.0	8.4	0.93	54
	0.01	7.4	1.06	47
	0.1	6.4	1.11	46
	1.0	1.1	0.47	33
50	0.0	16.7	1.50	54
	0.01	14.6	1.65	47
	0.1	11.6	1.80	46
	1.0	1.5	0.53	33

Table 3.2 (continued)

Time, h	SDS, mg/ml	Combined data of three experiments		
		Mean length, x	Confidence interval	n
64	0.0	31.1	2.77	54
	0.01	28.3	3.12	47
	0.1	21.2	2.94	46
	1.0	2.0	0.64	33
74	0.0	44.1	3.84	54
	0.01	41.6	4.66	47
	0.1	30.6	4.29	46
	1.0	2.1	0.67	33

Note: Confidence interval for 5% significance. For details, see *Biol. Nauki*, **12**: 81–84 (1987) (in Russian).

Table 3.3 Growth of *Fagopyrum esculentum* seedlings transferred to an SDS-containing medium.

SDS, mg/ml	40 h		50 h		64 h		74 h	
	x	CI	x	CI	x	CI	x	CI
0	7.1	0.61	15.1	1.04	32.8	2.22	49.2	3.29
0.01	7.0	0.67	14.7	1.06	31.3	2.11	47.1	3.14
0.1	8.0	0.71	15.1	0.87	29.6	1.59	44.2	2.69
1	7.4	0.66	10.5	0.63	13.5	0.77	15.4	1.00

Note: x, mean lengths of seedlings; CI, confidence interval for 5% significance.
Here and in the next table, 56 seedlings were measured at zero concentration of SDS; 55 seedlings each, at 0.01 and 0.1 mg/ml; 58 seedlings of *Fagopyrum esculentum* (cultivar Shatilovskaya 5), at 1 mg/ml. For details, see *Problems of Ecological Monitoring and modeling of ecosystems*, Vol. 9, pp. 87–97 (in Russian).

Table 3.4 Average rate of elongation of *Fagopyrum esculentum* transferred into an SDS-containing medium.

SDS, mg/ml	40 h	50 h	64 h
	40–50	50–64	64–74
0	0.80	1.26	1.64
0.01	0.77	1.19	1.58
0.1	0.71	1.04	1.46
1	0.31	0.21	0.19

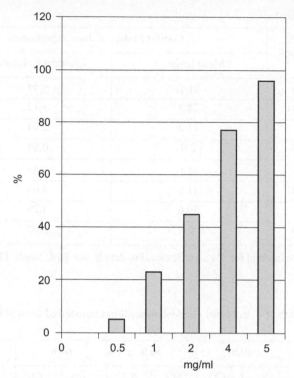

Figure 3.3 Inhibition by SDS of the water filtration rate by *Mytilus edulis* for the first 30-min period, % (see text for details). 100% inhibition means complete absence of water filtration.

It is useful to determine the value of the inhibition coefficient *I* calculated in the same way as in Ivanov (1974). This coefficient quantitatively characterizes the degree of inhibiting the growth of seedlings under the influence of a xenobiotic studied.

It was shown that the degree of inhibition at a concentration of 0.01 mg/ml was about 4–6%; at 0.1 mg/ml, about 11–18%; and at 1 mg/ml, about 61–88%. The degree of inhibition changed as a function of both the concentration of xenobiotic and the time of experiment, but the change in time was smaller than in transition from one concentration to another.

The dynamics of *I* was similar in both modifications of the method: within 64–74 h from the time of soaking the seeds (i.e., the concluding period of experiment) at concentrations of 0.01 and 0.1 mg/ml the degree of inhibition was observed to decrease as compared with the previous period of 50–64 h.

In general, comparison of the results obtained from different modifications of the method indicates the special importance of the first stage of development of the seedlings: if the xenobiotic (SDS) was added at this stage, the inhibitory effect of its sublethal concentrations was significantly higher than in the variant with the "transfer," when SDS acted on the seedlings initially grown on water without xenobiotic, and then the seedlings were transferred to the medium with SDS.

3.1.2 Effect of SDS on mollusks

As we noted, the effect of SDS on the rate of water filtration by mollusks, including *M. edulis*, was not investigated before. We studied the effect of SDS in the range of concentrations from 0 up to 5 mg/l (Figure 3.3). Experiments with higher concentrations appeared of smaller interest, as filtration stopped virtually completely at 5 ml/l.

If the initial concentration of SDS was 0.5 mg/l, we observed no significant effect on the mean concentration of algal cells (that remain unfiltered by *M. edulis*) (Table 3.5). As a preliminary comment, we should note that the variability of data between measurements in individual beakers (Nos. 1, 3, 5, 7) where the mussels were

Table 3.5 Number of *Isochrysis galbana* cells in beakers with *Mytilis edulis* mussels after a 30-min filtration of water by the mollusks during the action of anionogenic surfactant SDS (0.5 mg/l).

No. of beaker	Presence (+) or absence (−) of SDS, 0.5 mg/l	Number of algal cells per 0.5 ml	Average number of algal cells per 0.5 ml
1	+	5237, 5239, 5226	5234.0
3	+	3787, 3831, 3941	3853.0
5	+	3808, 3788, 3972	3856.0
7	+	8258, 8169, 8238	8221.7
1, 3, 5, 7	Average number of cells for four beakers (1, 3, 5, 7) with SDS	5291.0 (standard error, 537.9)	
2	−	4669, 4574, 4764	4669.0
4	−	4102, 4030, 4104	4078.7
6	−	5517, 5532, 5589	5546.0
8	−	5319, 5383, 5435	5379.0
2, 4, 6, 8	Average number of cells for four controls (2, 4, 6, 8)	4918.0 (standard error, 177.3)	

kept during the experiment increased as seen from the value of standard error. For instance, after a 60-min filtration in the variant with SDS, the standard error was 177.3 compared to 110.8 in the control (Table 3.6). After a 90-min filtration in the variant with SDS, the standard error was 68.8 as compared with 46.7 in the control (though the difference is not great and not significant alone, it fits the entire tendency of the observed increase of the standard error in the variants with SDS). The average number of cells per unit volume in the variant with SDS was slightly smaller than in the control (Table 3.7). The increased variability of data in different measurements in the beakers containing SDS (also noted in other experiments) reflects an increase in individual differences between particular species of mussels in the rate of water filtration by them. This means that in the variant with SDS some members of the group filter water much faster than the mean value, while others filter water much slower; this "slowliness" is larger than that in the subgroup of "slow" mussels in the control group. Slowing down of the filtration rate decreases the potential of these

Table 3.6 Number of *Isochrysis galbana* cells in beakers with *Mytilis edulis* mussels after a 60-min filtration of water by the mollusks during the action of anionogenic surfactant SDS (0.5 mg/l).

No. of beaker	Presence (+) or absence (−) of SDS, 0.5 mg/l	Number of algal cells per 0.5 ml	Average number of algal cells per 0.5 ml
1	+	2302; 2066; 2070	2146.0
3	+	1515; 1357; 1263	1378.3
5	+	913; 904; 1071	962.7
7	+	2447; 2416; 2380	2414.3
1, 3, 5, 7	Average number of cells for four beakers (1, 3, 5, 7) with SDS		1725.3 (standard error, 177.3)
2	−	2290; 2199; 2170	2219.7
4	−	1210; 1352; 1268	1276.7
6	−	2130; 1991; 2078	2066.3
8	−	1661; 1695; 1893	1749.7
2, 4, 6, 8	Average number of cells for four controls (2, 4, 6, 8)		1828.1 (standard error, 110.8)

Table 3.7 Number of *Isochrysis galbana* cells in beakers with *Mytilis edulis* mussels after a 90-min filtration of water by the mollusks during the action of anionogenic surfactant SDS (0.5 mg/l).

No. of beaker	Presence (+) or absence (−) of SDS, 0.5 mg/l	Number of algal cells per 0.5 ml	Average number of algal cells per 0.5 ml
1	+	755; 643; 587	661.7
3	+	712; 706; 724	714.0
5	+	385; 389; 336	370.0
7	+	1067; 973; 965	1001.7
1, 3, 5, 7	Average number of cells for four beakers (1, 3, 5, 7) with SDS		686.8 (standard error, 68.8)
2	−	945; 902; 862	903.0
4	−	562; 501; 480	514.3
6	−	823; 759; 830	804.0
8	−	587; 640; 792	673.0
2, 4, 6, 8	Average number of cells for four controls (2, 4, 6, 8)		723.6 (standard error, 46.7)

Note: Experimental conditions: Initial concentration of cells, 19465; 19390; 19491; 19593; 19700; 19486 per 0.5 ml (average, 19520.8); temperature, 16°C. The cells were counted using a Coulter counter. In the experiment, 16 mussels were used; 8 of them were subjected to surfactant action. Capacity of beakers, 2 l. The total wet weight of the mussels (with shells), in g: beaker 1, 18.22; beaker 2, 17.64; beaker 3, 17.82; beaker 4, 18.01; beaker 5, 17.09; beaker 6, 18.43; beaker 7, 17.61; beaker 8, 17.81. $p = 0.3315$.

species in obtaining energy for growth and formation of sexual products (in the variant with SDS). Thus, even if the mean values were the same in the control and in the variant with SDS, some of the animals in the presence of SDS prove to be in an unfavorable position. This should prevent us from making overoptimistic conclusions even when the mean values of experiment and control groups coincide.

Effects of higher SDS concentrations (1, 2, 4, and 5 mg/l) on mussels were studied. We should note that, to obtain more statistically reliable data, we tried to perform an experiment with a sufficient number of repeats. We had to compensate that by having to study only one concentration in one experiment. That is, in contrast with the experiments with algae or microorganisms (where the effect versus concentration dependence was studied in one experiment), we had to stage a series of experiments to reveal a regularity of how an effect depends on concentration.

At an initial SDS concentration of 1 mg/l, the rate of seawater filtration by mussels slightly decreased. This was evident from the fact that the number of phytoplankton cells remaining unfiltered was greater than in the control where filtration and removal of cells from water were at a normal rate (Tables 3.8–3.10). For instance, after 1 h of filtration the number of cells per unit volume of water in the control decreased almost 20-fold (down to a final concentration of 1090 cells/0.5 ml) and

Table 3.8 Number of *Isochrysis galbana* cells in beakers with *Mytilis edulis* mussels after a 30-min filtration of water by the mollusks during the action of anionogenic surfactant SDS (1 mg/l).

No. of beaker	Presence (+) or absence (−) of SDS, 1 mg/l	Number of algal cells per 0.5 ml	Average number of algal cells per 0.5 ml
1	+	9576; 9567; 9536	9560
3	+	6805; 6732; 6789	6775
5	+	5018; 5142; 5081	5080
7	+	4861; 5046; 4967	4958
1, 3, 5, 7	Average number of cells for four beakers (1, 3, 5, 7) with SDS		6593 (standard error, 1072)
2	−	7288; 7225; 7213	7242
4	−	3150; 3127; 3163	3147
6	−	4641; 4663; 4725	4676
8	−	4408; 4467; 4509	4461
2, 4, 6, 8	Average number of cells for four controls (2, 4, 6, 8)		4882 (standard error, 856)
9 (no mussels)	−	18860; 19149; 19101	19036.7

Note: Assessment of average significance of difference in control and experiment: $p = 0.013$; significance level, > 95%. Experimental conditions: Initial concentration of cells, 17024; 16651; 16747 per 0.5 ml (average, 16807.3); temperature, 16°C. The cells were counted using a Coulter counter. In the experiment, 16 mussels were used; 8 of them were subjected to surfactant action. Capacity of beakers, 2 l. The total wet weight of the mussels (with shells), in g: beaker 1, 16.97; beaker 2, 17.61; beaker 3, 18.23; beaker 4, 18.23; beaker 5, 18.29; beaker 6, 17.95; beaker 7, 18.72; beaker 8, 17.03.

almost 10-fold (down to a concentration of 2199 cells/0.5 ml) in the variant where filtration was inhibited by the presence of surfactant (Table 3.9). Hereafter, the concentrations of cells are given per water volume of 0.5 ml, as these are the features of the technique using a Coulter counter. Thus, the difference between the control and the experiment with SDS was approximately two-fold. After a 90-min filtration, the variability of the filtration rates in individual beakers containing SDS increased

Table 3.9   Number of *Isochrysis galbana* cells in beakers with *Mytilis edulis* mussels after a 60-min filtration of water by the mollusks during the action of anionogenic surfactant SDS (1 mg/l).

No. of beaker	Presence (+) or absence (–) of SDS, 1 mg/l	Number of algal cells per 0.5 ml	Average number of algal cells per 0.5 ml
1	+	4471; 4413; 4437	4440
3	+	2211; 2159; 2139	2170
5	+	1072; 1156; 1125	1118
7	+	1056; 1100; 1049	1068
1, 3, 5, 7	Average number of cells for four beakers (1, 3, 5, 7) with SDS		2199 (standard error, 789)
2	–	1756; 1771; 1776	1768
4	–	544; 568; 633	582
6	–	1011; 1042; 1010	1021
8	–	986; 1007; 974	989
2, 4, 6, 8	Average number of cells for four controls (2, 4, 6, 8)		1090 (standard error, 247)
9 (no mussels)	–	20055; 19716; 19437	19736

Note: Assessment of average significance of difference in control and experiment: $p = 0.012$; significance level, $> 95\%$. Experimental conditions: Initial concentration of cells, 17024; 16651; 16747 per 0.5 ml (average, 16807.3); temperature, 16°C. The cells were counted using a Coulter counter. In the experiment, 16 mussels were used; 8 of them were subjected to surfactant action. Capacity of beakers, 2 l. The total wet weight of the mussels (with shells), in g: beaker 1, 16.97; beaker 2, 17.61; beaker 3, 18.23; beaker 4, 18.23; beaker 5, 18.29; beaker 6, 17.95; beaker 7, 18.72; beaker 8, 17.03.

significantly. In the presence of SDS, the standard error was 448 cells as compared with 67 cells n the control (the difference between the mean values was less than 3 times) (Table 3.10). It should be noted that the maximum measured level of pollution with anionic surfactants in natural aquatic ecosystems reached and exceeded the concentration of anionic surfactants in this experiment. Thus, in the Black Sea the concentrations of anionic surfactants repeatedly exceeded 1–2 mg/l (Review of the State of Pollution of the Black and Azov Seas 1976).

At an increase of the initial SDS concentration up to 2 mg/l, a significant difference as compared with the control was observed already after the first 30-min

Table 3.10   Number of *Isochrysis galbana* cells in beakers with *Mytilis edulis* mussels after a 90-min filtration of water by the mollusks during the action of anionogenic surfactant SDS (1 mg/l).

No. of beaker	Presence (+) or absence (−) of SDS, 1 mg/l	Number of algal cells per 0.5 ml	Average number of algal cells per 0.5 ml
1	+	2221; 2271; 2186	2226
3	+	1083; 1084; 1200	1122
5	+	328; 296; 302	309
7	+	370; 332; 363	355
1, 3, 5, 7	Average number of cells for four beakers (1, 3, 5, 7) with SDS		1003 (standard error, 448)
2	−	533; 522; 509	521
4	−	178; 201; 209	196
6	−	415; 348; 350	371
8	−	385; 413; 380	393
2, 4, 6, 8	Average number of cells for four controls (2, 4, 6, 8)		370 (standard error, 67)
9 (no mussels)	−	19554; 19892; 19153	19533

Note: Assessment of average significance of difference in control and experiment: $p = 0.0102$; significance level, > 95%. Experimental conditions: Initial concentration of cells, 17024, 16651, 16747 per 0.5 ml (average, 16807.3); temperature, 16°C. The cells were counted using a Coulter counter. Readings of Coulter counter in filtered sea water: 164, 109, and 122 cells (average, 132). In the experiment, 16 mussels were used; 8 of them were subjected to surfactant action. Capacity of beakers, 2 l. The total wet weight of the mussels (with shells), in g: beaker 1, 16.97; beaker 2, 17.61; beaker 3, 18.23; beaker 4, 18.23; beaker 5, 18.29; beaker 6, 17.95; beaker 7, 18.72; beaker 8, 17.03.

incubation (Table 3.11). After 1-h filtration (Table 3.12), the concentration of cells in the experiment was 3491 cells/0.5 ml, i.e., almost three times higher than in the control (where the similar value was 1229 cells/0.5 ml). This indicates a significant disturbance of normal seawater filtration under the action of the surfactant. We also noted a difference in the variability of individual differences between the species. In the beakers with surfactant, the standard error was 340.8 cells, while a similar value in the control was 90.4 cells.

Additional control was set in the experiment with an SDS concentration of 2 mg/l. This experiment showed no significant change in the concentration of algae due to their natural growth. Table 3.11 shows that in beaker 9 (containing only algae without mussels) the concentration of cells during the period of incubation (30 min) virtually did not change (at the end of the period, 17817.7 cells/0.5 ml; in the beginning of the period, 17809.7 cells/0.5 ml). The same was observed in the subsequent period of time (Table 3.12). At the end of experiment, the concentration of cells in beaker 9 was 17,914 cells/0.5 ml, which differs only insignificantly from the initial value of 17,809.7 cells/0.5 ml (Table 3.13).

Table 3.11 Number of *Isochrysis galbana* cells in beakers with *Mytilis edulis* mussels after a 30-min filtration of water by the mollusks during the action of anionogenic surfactant SDS (2 mg/l).

No. of beaker	Presence (+) or absence (−) of SDS, 2 mg/l	Number of algal cells per 0.5 ml	Average number of algal cells per 0.5 ml
1	+	9307; 9537; 9311	9385.0
3	+	7475; 7455; 7384	7438.0
5	+	5880; 5779; 5678	5779.0
7	+	10608; 10508; 10467	10527.7
1, 3, 5, 7	Average number of cells for four beakers (1, 3, 5, 7) with SDS		8282.4 (standard error, 549.0)
2	−	5019; 4993; 4843	4951.7
4	−	4007; 3938; 4007	3984.0
6	−	4806; 4739; 4668	4737.7
8	−	3126; 2931; 3055	3037.3
2, 4, 6, 8	Average number of cells for four controls (2, 4, 6, 8)		4177.7 (standard error, 227.0)
9 (no mussels)	−	17875; 17881; 17697	17817.7

Note: Assessment of average significance of difference in control and experiment: $p = 2E - 06 = 2 \times 10^{-6}$; significance level, > 99.9%. Experimental conditions: Initial concentration of cells, 18119; 18101; 17209 per 0.5 ml (average, 17809.7); temperature, 16°C. The cells were counted using a Coulter counter. Average readings of Coulter counter in filtered sea water: 227 cells. In the experiment, 16 mussels were used; 8 of them were subjected to surfactant action. Capacity of beakers, 2 l. The total wet weight of the mussels (with shells), in g: beaker 1, 17.6; beaker 2, 17.7; beaker 3, 18.1; beaker 4, 17.6; beaker 5, 17.3; beaker 6, 16.9; beaker 7, 17.4; beaker 8, 18.0.

Table 3.12 Number of *Isochrysis galbana* cells in beakers with *Mytilis edulis* mussels after a 30-min filtration of water by the mollusks during the action of anionogenic surfactant SDS (2 mg/l).

No. of beaker	Presence (+) or absence (−) of SDS, 2 mg/l	Number of algal cells per 0.5 ml	Average number of algal cells per 0.5 ml
1	+	4813; 5084; 4897	4931.3
3	+	2905; 2890; 2897	2897.3
5	+	2002; 1923; 2061	1995.3
7	+	3991; 4252; 4173	4138.7
1, 3, 5, 7	Average number of cells for four beakers (1, 3, 5, 7) with SDS		3490.7 (standard error, 340.8)
2	−	1617; 1568; 1614	1599.7
4	−	1019; 1137; 1137	1097.7

Table 3.12 (continued)

No. of beaker	Presence (+) or absence (–) of SDS, 2 mg/l	Number of algal cells per 0.5 ml	Average number of algal cells per 0.5 ml
6	–	1413; 1376; 1420	1403.0
8	–	859; 784; 807	816.7
2, 4, 6, 8	Average number of cells for four controls (2, 4, 6, 8)		1229.3 (standard error, 90.4)
9 (no mussels)	–	17834; 18087; 18397	18106

Note: Assessment of average significance of difference in control and experiment: $p = 1.15E - 05 = 1.15 \times 10^{-5}$; significance level, > 99.9%. Experimental conditions: Initial concentration of cells, 18119; 18101; 17209 per 0.5 ml (average, 17809.7); temperature, 16°C. The cells were counted using a Coulter counter. Average readings of Coulter counter in filtered sea water: 227 cells. In the experiment, 16 mussels were used; 8 of them were subjected to surfactant action. Capacity of beakers, 2 l. The total wet weight of the mussels (with shells), in g: beaker 1, 17.6; beaker 2, 17.7; beaker 3, 18.1; beaker 4, 17.6; beaker 5, 17.3; beaker 6, 16.9; beaker 7, 17.4; beaker 8, 18.0.

Table 3.13 Number of *Isochrysis galbana* cells in beakers with *Mytilis edulis* mussels after a 90-min filtration of water by the mollusks during the action of anionogenic surfactant SDS (2 mg/l).

No. of beaker	Presence (+) or absence (–) of SDS, 2 mg/l	Number of algal cells per 0.5 ml	Average number of algal cells per 0.5 ml
1	+	2608; 2568; 2556	2577.3
3	+	1217; 1482; 1322	1340.3
5	+	981; 960; 931	957.3
7	+	1732; 1697; 1801	1743.3
1, 3, 5, 7	Average number of cells for four beakers (1, 3, 5, 7) with SDS		1654.6 (standard error, 182.1)
2	–	1023; 974; 980	992.3
4	–	530; 488; 410	476.0
6	–	955; 770; 674	799.7
8	–	379; 242; 227	282.7
2, 4, 6, 8	Average number of cells for four controls (2, 4, 6, 8)		637.7 (standard error, 86.0)
9 (no mussels)	–	17826; 18008; 17908	17914.0

Note: Assessment of average significance of difference in control and experiment: $p = 5.9E - 05 = 5.9 \times 10^{-5}$; significance level, > 99.9%. Experimental conditions: Initial concentration of cells, 18119; 18101; 17209 per 0.5 ml (average, 17809.7); temperature, 16°C. The cells were counted using a Coulter counter. In the experiment, 16 mussels were used; 8 of them were subjected to surfactant action. Capacity of beakers, 2 l. The total wet weight of the mussels (with shells), in g: beaker 1, 17.6; beaker 2, 17.7; beaker 3, 18.1; beaker 4, 17.6; beaker 5, 17.3; beaker 6, 16.9; beaker 7, 17.4; beaker 8, 18.0.

A further increase in the initial concentration of SDS up to 4 mg/l resulted in even greater differences in the concentration of algal cells in experiment and control samples. Thus, after the first period of filtration the concentration of cells in the experiment was 3 times greater than that in the control (Table 3.15); after the second period of filtration it was 6 times greater (Table 3.16); and after the third period it was almost 14 times greater (Table 3.17). A rapid manifestation of the differences in variability between the species is significant. For instance, the filtration rates in 5 min after the beginning of filtration were measured (Table 3.14). The difference of the mean values of the concentrations of algal cell after a 5-min filtration was barely noticeable. It was 17,273 cells/0.5 ml in the beakers with SDS as compared with a slightly smaller value of 14,877 cells/0.5 ml in the control. At the same time, the variability of data in the beakers with SDS was already much greater than in the control. The standard error in the variant with SDS was 240 cells as compared with 78 cells in the control. In the subsequent periods of time, the variability of the indices in the beakers with SDS was also greater than in the control.

Table 3.14 Number of *Isochrysis galbana* cells in beakers with *Mytilis edulis* mussels after a 5-min filtration of water by the mollusks during the action of anionogenic surfactant SDS (4 mg/l).

No. of beaker	Presence (+) or absence (−) of SDS, 4 mg/l	Number of algal cells per 0.5 ml	Average number of algal cells per 0.5 ml
1	+	16303; 16603; 16428	1644.7
3	+	16904; 16296; 16456	16552
5	+	18269; 18232; 18036	18179
7	+	17848; 18046; 17849	17914.3
1, 3, 5, 7	Average number of cells for four beakers (1, 3, 5, 7) with SDS		17273 (standard error, 240.2)
2	−	14713; 14653; 14577	14647.7
4	−	14820; 14692; 14499	14670.3
6	−	15421; 14949; 15055	15141.7
8	−	15217; 14974; 14959	15050
2, 4, 6, 8	Average number of cells for four controls (2, 4, 6, 8)		14877 (standard error, 78.4)
9 (no mussels)	−	18143; 17889; 17918; 18047	17999.3

Note: Assessment of average significance of difference in control and experiment: $p = 2E - 07 = 2 \times 10^{-7}$; significance level, > 99.9%. Experimental conditions: Initial concentration of cells, 18143; 17889; 17918; 18047 per 0.5 ml (average, 17999.25); temperature, 16°C. The cells were counted using a Coulter counter. Readings of Coulter counter in clean filtered water without algae were 315, 309, 279, and 302 cells (average, 301.25). In the experiment, 16 mussels were used; 8 of them were subjected to surfactant action. Capacity of beakers, 2 l. The total wet weight of the mussels (with shells), in g: beaker 1, 17.4; beaker 2, 18.4; beaker 3, 18.4; beaker 4, 17.6; beaker 5, 18.7; beaker 6, 17.7; beaker 7, 17.1; beaker 8, 16.3.

Table 3.15 Number of *Isochrysis galbana* cells in beakers with *Mytilis edulis* mussels after a 35-min filtration of water by the mollusks during the action of anionogenic surfactant SDS (4 mg/l).

No. of beaker	Presence (+) or absence (−) of SDS, 4 mg/l	Number of algal cells per 0.5 ml	Average number of algal cells per 0.5 ml
1	+	10312; 15280; 14520	13370.7
3	+	8312; 10392; 9531.5	9411.8
5	+	15497; 17170; 15682	16116.2
7	+	12935; 13072; 12696	12901.0
1, 3, 5, 7	Average number of cells for four beakers (1, 3, 5, 7) with SDS		12950 (standard error, 809.2)
2	−	4048; 5271; 5123.5	4814.2
4	−	3630; 4357; 3916	3967.7
6	−	2952; 3469; 3437.5	3286.2
8	−	4037; 4096; 3981	4038.0
2, 4, 6, 8	Average number of cells for four controls (2, 4, 6, 8)		4027 (standard error, 191.7)
9 (no mussels)	−	17167; 18525; 17554; 18828	18018.5

Note: Assessment of average significance of difference in control and experiment: $p = 8.31E -08 = 8.31 \times 10^{-8}$; significance level, > 99.99%. Experimental conditions: Initial concentration of cells, 18143; 17889; 17918; 18047 per 0.5 ml (average, 17999.25); temperature, 16°C. The cells were counted using a Coulter counter. Readings of Coulter counter in clean filtered water without algae were 315; 309; 279; and 302 cells (average, 301.25). In the experiment, 16 mussels were used; 8 of them were subjected to surfactant action. Capacity of beakers, 2 l. The total wet weight of the mussels (with shells), in g: beaker 1, 17.4; beaker 2, 18.4; beaker 3, 18.4; beaker 4, 17.6; beaker 5, 18.7; beaker 6, 17.7; beaker 7, 17.1; beaker 8, 16.3.

Table 3.16 Number of *Isochrysis galbana* cells in beakers with *Mytilis edulis* mussels after a 65-min filtration of water by the mollusks during the action of anionogenic surfactant SDS (4 mg/l).

No. of beaker	Presence (+) or absence (−) of SDS, 4 mg/l	Number of algal cells per 0.5 ml	Average number of algal cells per 0.5 ml
1	+	14115; 14195; 14032	14114
3	+	7018; 7295; 7350	7221
5	+	14965; 14996; 14706	4889
7	+	9628; 9464; 9461	9517.7
1, 3, 5, 7	Average number of cells for four beakers (1, 3, 5, 7) with SDS		11435 (standard error, 960.5)
2	−	2352; 2329; 2443	2374.7

Table 3.16   (continued)

No. of beaker	Presence (+) or absence (−) of SDS, 4 mg/l	Number of algal cells per 0.5 ml	Average number of algal cells per 0.5 ml
4	−	1637; 1535; 1638	1603.3
6	−	1402; 1459; 1480	1447.0
8	−	1708; 1654; 1732	1698.0
2, 4, 6, 8	Average number of cells for four controls (2, 4, 6, 8)		1781 (standard error, 107.6)
9 (no mussels)	−	19433; 19288; 19241; 19314	19319

Note: Assessment of average significance of difference in control and experiment: $p = 4E - 07 = 4 \times 10^{-7}$; significance level, > 99.9%. Initial concentration of cells, 18143; 17889; 17918; 18047 per 0.5 ml (average, 17999.25); temperature, 16°C. The cells were counted using a Coulter counter. Readings of Coulter counter in clean filtered water without algae were 315; 309; 279, and 302 cells (average, 301.25). In the experiment, 16 mussels were used; 8 of them were subjected to surfactant action. Capacity of beakers, 2 l. The total wet weight of the mussels (with shells), in g: beaker 1, 17.4; beaker 2, 18.4; beaker 3, 18.4; beaker 4, 17.6; beaker 5, 18.7; beaker 6, 17.7; beaker 7, 17.1; beaker 8, 16.3.

Additional control (beaker 9) was also set in the experiment with an SDS concentration of 2 mg/l. This beaker contained only algae without mussels. At an initial algal concentration of 17,999.3 cells/0.5 ml of the medium, a similar value measured after 35 min was 18,018.5 (Table 3.15); after 65 min, 19,319 (Table 3.16); after 95 min, 18,391.5 (Table 3.17). Thus, it was again confirmed that the concentration of algae alone does not significantly change during the experiment; therefore, the algal cell concentration dynamics observed in other variants of the experiment is due to the filtration activity of the mussels in the corresponding beakers.

Table 3.17   Number of *Isochrysis galbana* cells in beakers with *Mytilis edulis* mussels after a 95-min filtration of water by the mollusks during the action of anionogenic surfactant SDS (4 mg/l).

No. of beaker	Presence (+) or absence (−) of SDS, 4 mg/l	Number of algal cells per 0.5 ml	Average number of algal cells per 0.5 ml
1	+	13235; 13048; 13203	13162
3	+	4376; 4408; 4462	4415.3
5	+	11161; 10679; 11040	10960.0
7	+	6080; 5696; 5675	5817.0
1, 3, 5, 7	Average number of cells for four beakers (1, 3, 5, 7) with SDS		8588.6 (standard error, 1084.2)
2	−	807; 771; 738	772.0
4	−	514; 557; 538	536.3
6	−	529; 518; 477	508.0

Table 3.17 (continued)

No. of beaker	Presence (+) or absence (−) of SDS, 4 mg/l	Number of algal cells per 0.5 ml	Average number of algal cells per 0.5 ml
8	−	505; 526; 516	515.7
2, 4, 6, 8	Average number of cells for four controls (2, 4, 6, 8)		583 (standard error, 33.6)
9 (no mussels)	−	18797; 18215; 18422; 18132	18391.5

Note: Assessment of average significance of difference in control and experiment: $p = 7E − 06 = 7 \times 10^{-6}$; significance level, > 99.9%. Initial concentration of cells, 18143; 17889; 17918; 18047 per 0.5 ml (average, 17999.25); temperature, 16°C. The cells were counted using a Coulter counter. Readings of Coulter counter in clean filtered water without algae were 315; 309; 279; and 302 cells (average, 301.25). In the experiment, 16 mussels were used; 8 of them were subjected to surfactant action. Capacity of beakers, 2 l. The total wet weight of the mussels (with shells), in g: beaker 1, 17.4; beaker 2, 18.4; beaker 3, 18.4; beaker 4, 17.6; beaker 5, 18.7; beaker 6, 17.7; beaker 7, 17.1; beaker 8, 16.3.

Table 3.18 Number of *Isochrysis galbana* cells in beakers with *Mytilis edulis* mussels after a 30-min filtration of water by the mollusks during the action of anionogenic surfactant SDS (5 mg/l).

No. of beaker	Presence (+) or absence (−) of SDS, 5 mg/l	Number of algal cells per 0.5 ml	Average number of algal cells per 0.5 ml
1	+	8742; 8732; 8627	8700.3
2	+	11205; 11575; 11355	11378.3
5	+	12076; 11983; 12259	12106.0
1, 2, 5	Average number of cells for three beakers (1, 2, 5) with SDS		10728 (standard error, 519.3)
3	−	3937; 3584; 3830	3783.7
4	−	3948; 3955; 3988	3963.7
6	−	3255; 3002; 3289	3182.0
3, 4, 6	Average number of cells for four controls (3, 4, 6)		3643.1 (standard error, 124.8)

Note: Assessment of average significance of difference in control and experiment: $p = 2E − 07 = 2 \times 10^{-7}$; significance level, > 99.99%. Experimental conditions: Temperature, 16°C. The cells were counted using a Coulter counter. In the experiment, 12 mussels were used; 6 of them were subjected to surfactant action. Capacity of beakers, 2 l. Initial concentration of cells (number of cells per 0.5 ml): 12277; 12374; 12305 (average, 12318.7). The total wet weight of the mussels (with shells), in g: beaker 1, 16.1 (7.68; 8.37); beaker 2, 16.5 (9.94; 6.52); beaker 3, 16.3 (8.94; 7.32); beaker 4, 15.7 (9.08; 6.67); beaker 5, 19.23 (9.9; 9.33); beaker 6, 19.9 (10.31; 9.57).

The data of experiments with an SDS concentration of 5 mg/l (Tables 3.18 to 3.21) were also indicative of a strong inhibition of filtration and were in agreement with the results of the previous experiments. Already after 30 min of filtration, the difference in the mean concentration of algal cells between the experiment and control was more than two times, while that of the standard errors (which reflects the difference in the variability of individual responses of the organisms) was more than four times (Table 3.18). After 60 min of filtration, the difference in the mean concentration of cells was more than 6-fold, while that in the standard error was more than 17-fold (Table 3.19). After 90 min of filtration, the difference in the mean concentration of cells was more than 13-fold, while that in the standard error was more than 34-fold (Table 3.20).

Table 3.19 Number of *Isochrysis galbana* cells in beakers with *Mytilis edulis* mussels after a 60-min filtration of water by the mollusks during the action of anionogenic surfactant SDS (5 mg/l).

No. of beaker	Presence (+) or absence (−) of SDS, 5 mg/l	Number of algal cells per 0.5 ml	Average number of algal cells per 0.5 ml
1	+	7277; 7217; 7075	7189.7
2	+	10177; 10068; 10104	10,116.3
5	+	11794; 11936; 12078	11936.0
1, 2, 5	Average number of cells for three beakers (1, 2, 5) with SDS		97,47.3 (standard error, 691.9)
3	−	1484; 1386; 1402	1424.0
4	−	1395; 1686; 1515	1532.0
6	−	1521; 1305; 1582	1469.3
3, 4, 6	Average number of cells for three controls (3, 4, 6)		1475.1 (standard error, 38.8)

Note: Assessment of average significance of difference in control and experiment: $p = 1E - 06 = 1 \times 10^{-6}$; significance level, > 99.99%. Experimental conditions: Temperature, 16°C. The cells were counted using a Coulter counter. In the experiment, 12 mussels were used; 6 of them were subjected to surfactant action. Capacity of beakers, 2 l. Initial concentration of cells (number of cells per 0.5 ml): 12277; 12374; 12305 (average, 12318.7). The total wet weight of the mussels (with shells), in g: beaker 1, 16.1 (7.68; 8.37); beaker 2, 16.5 (9.94; 6.52); beaker 3, 16.3 (8.94; 7.32); beaker 4, 15.7 (9.08; 6.67); beaker 5, 19.23 (9.9; 9.33); beaker 6, 19.9 (10.31; 9.57).

Table 3.20 Number of *Isochrysis galbana* cells in beakers with *Mytilis edulis* mussels after a 90-min filtration of water by the mollusks during the action of anionogenic surfactant SDS (5 mg/l).

No. of beaker	Presence (+) or absence (−) of SDS, 5 mg/l	Number of algal cells per 0.5 ml	Average number of algal cells per 0.5 ml
1	+	6163; 6020; 6027	6070.0
2	+	9516; 9385; 9603	9501.3
5	+	11742; 11780; 11908	11810.0
1, 2, 5	Average number of cells for three beakers (1, 2, 5) with SDS		9127.1 (standard error, 834.2)

Table 3.20 (continued)

No. of beaker	Presence (+) or absence (−) of SDS, 5 mg/l	Number of algal cells per 0.5 ml	Average number of algal cells per 0.5 ml
3	−	764; 591; 655	670.0
4	−	601; 640; 634	625.0
6	−	783; 755; 688	742.0
3, 4, 6	Average number of cells for three controls (3, 4, 6)		679.0 (standard error, 24.1)

Note: Assessment of average significance of difference in control and experiment: $p = 3.87E - 06 = 3.87 \times 10^{-6}$; significance level, > 99.99%. Experimental conditions: Temperature, 16°C. The cells were counted using a Coulter counter. In the experiment, 12 mussels were used; 6 of them were subjected to surfactant action. Capacity of beakers, 2 l. Initial concentration of cells (number of cells per 0.5 ml): 12277; 12374; 12305 (average, 12318.7). The total wet weight of the mussels (with shells), in g: beaker 1, 16.1 (7.68; 8.37); beaker 2, 16.5 (9.94; 6.52); beaker 3, 16.3 (8.94; 7.32); beaker 4, 15.7 (9.08; 6.67); beaker 5, 19.23 (9.9; 9.33); beaker 6, 19.9 (10.31; 9.57).

Table 3.21 Number of *Isochrysis galbana* cells in beakers with *Mytilis edulis* mussels after a 120-min filtration of water by the mollusks during the action of anionogenic surfactant SDS (5 mg/l).

No. of beaker	Presence (+) or absence (−) of SDS, 5 mg/l	Number of algal cells per 0.5 ml	Average number of algal cells per 0.5 ml
1	+	4391; 4577; 4486	4484.7
2	+	9029; 8883; 8703	8871.7
5	+	11181; 11212; 11260	11217.7
1, 2, 5	Average number of cells for three beakers (1, 2, 5) with SDS		8191.3 (standard error, 987.1)
3	−	900; 1133; 842	958.3
4	−	308; 286; 286	293.3
6	−	223; 218; 188	209.7
3, 4, 6	Average number of cells for three controls (3, 4, 6)		487.1 (standard error, 121.2)

Note: Assessment of average significance of difference in control and experiment: $p = 3E - 05 = 3 \times 10^{-5}$; significance level, > 99.9%. Experimental conditions: Temperature, 16°C. The cells were counted using a Coulter counter. In the experiment, 12 mussels were used; 6 of them were subjected to surfactant action. Capacity of beakers, 2 l. Initial concentration of cells (number of cells per 0.5 ml): 12277; 12374; 12305 (average, 12318.7). The total wet weight of the mussels (with shells), in g: beaker 1, 16.1 (7.68; 8.37); beaker 2, 16.5 (9.94; 6.52); beaker 3, 16.3 (8.94; 7.32); beaker 4, 15.7 (9.08; 6.67); beaker 5, 19.23 (9.9; 9.33); beaker 6, 19.9 (10.31; 9.57).

We also conducted an experiment to study the effect of an SDS concentration equal to 25 mg/l. In this case, mollusks virtually stopped filtering water. We observed the closure of the mussels valves. This effect was registered only 20 min after the onset of the experiment.

The rate of water filtration in the beakers with SDS (from 0 to 5 mg/l) was calculated (Table 3.22).

Table 3.22 Water filtration rate (l/min) in beakers with *Mytilis edulis* mussels at various concentrations of anionic surfactant SDS (0.5, 1, 2, 4, and 5 mg/l).

Filtration periods	SDS concentration, mg/l									
	0.5		1		2		4		5	
	A	B	A	B	A	B	A	B	A	B
1	4.59	4.81	4.04	5.23	2.71	4.90	1.22	5.26	0.18	4.20
2	4.06	4.02	4.95	6.13	3.58	4.95	0.59	3.30	0.43	3.60
3	4.15	3.71	3.74	4.24	3.02	2.94	1.36	4.46	0.32	3.11

Note: A, variant with SDS (at a given concentration); B, control. Each filtration period was 30 min.

Table 3.23 Inhibition of the rate of water filtration by *Mytilis edulis* mussels by various concentrations of SDS (in %, measured by the depletion of *Isochrysis galbana*).

Time intervals, min	SDS concentration, mg/l					
	0	0.5	1	2	4	5
5–35	0	4.6	22.8	44.7	76.8	95.7
35–65	0	−1.0	19.2	27.7	82. 1	88.1
65–95	0	−11.9	11.8	−2.7	69.5	89.7

A comparison of the results of experiments carried out at different concentrations of SDS shows that they agree well and reveal a sufficiently clearcut regularity of the increasing inhibitory effect at increasing concentrations of SDS (Table 3.23). Statistical analysis of the data confirmed the significance of the differences between the mean values in experiment and control samples (Table 3.24). The statistical analysis also included the calculation of the regression coefficients (Table 3.25).

It should be noted that comparison of various methods of processing the data of filtration-rate experiments indicates that it is reasonable to use different techniques. It is useful to recalculate the data characterizing the concentration of cells in the experimental beakers before and after water filtration into the filtration-rate parameters expressed in liters per time unit. In this recalculation, the logarithms of the data on the concentration of cells are taken (see Chapter 2), which leads to a numerical decrease in the differences and a decrease of the apparent effect of the xenobiotic tested. Besides, in transition from the values characterizing the abundance of algal cells in the water column to the values characterizing the rate of water passing through the organism of a mollusk (measured in liters per minute), one's attention shifts from the ecological parameter (abundance of phytoplankton in an aquatic system) to a physiological parameter (which characterizes the physiological activity of a given species of bivalve mollusks). At the same time, it is the ecological aspect of suppressing the physiological activity of filter feeders by a xenobiotic that is of particular interest. By focusing attention on the ecologically important result of filtration activity – the effect on the concentration of plankton cells in water – we arrive at a new parameter, which can be called "the effect on the efficiency of removal" (EER) of suspended matter from water (Table 3.26).

The value of EER calculated via the method suggested shows how many times the concentration of suspension matter in water increases under the influence of the substance studied (as compared with the conditions of normal, uninhibited filtration

Table 3.24 Statistical significance of the effect of SDS on the efficiency of water filtration by mussels *Mytilus edulis* (differences between the variants with SDS and the controls).

Concentration, mg/l	Time, min	p	Value of effect	Level of significance
1	0 (before filtration)	0.487	No significant effect	95%
	30	0.013	Significant effect	> 95%
	60	0.012	Significant effect	> 95%
	90	0.0102	Significant effect	> 95%
2	30	2×10^{-6}	Significant effect	> 99.9%
	60	1.15×10^{-5}	Significant effect	> 99.9%
	90	5.9×10^{-5}	Significant effect	> 99.9%
4	5	2×10^{-7}	Significant effect	> 99.9%
	30	8.3×10^{-8}	Significant effect	> 99.9%
	60	4×10^{-7}	Significant effect	> 99.9%
	90	7×10^{-6}	Significant effect	> 99.9%
5	30	2×10^{-7}	Significant effect	> 99.9%
	60	2×10^{-6}	Significant effect	> 99.9%
	90	3.87×10^{-6}	Significant effect	> 99.9%
	120	3×10^{-5}	Significant effect	> 99.9%

Table 3.25. Regression coefficients $Y = a + bX$ (Y, filtration rate as percentage of control; X, concentration of SDS, mg/l) and calculated magnitudes of EC_{50}.

Periods	Coefficient a	Coefficient b	Number of observations, n	R^2	EC_{50}, mg/l
1	100	−19.526	23	0.844	2.56
2	100	−18.55	23	0.764	2.70
3	100	−15.357	23	0.684	3.26

Note: Each period was 30 min.

activity). The results of our experiments and of other authors testify that this parameter is useful in characterizing the biological activity and potential ecological hazard of a chemical substance. It is seen that at SDS concentrations of 1 and 2 mg/l, the values of EER reached more than 250% in 90 min, and at a concentration of 4 mg/l more than 1400%.

These results prompt a question as to the consequences of a longer effect of anionic surfactants on mollusks. We failed to find any literature data on the long-term effect of SDS, but work was done on another anionic surfactant, linear alkyl benzene sulfonate (LAS) in experiments that lasted up to 160 days. The effect of LAS on the growth of mussels was shown (Bressan et al. 1989). The growth increment along the main axis of the shell at concentrations of only 0.025 and 0.5 mg/l was found to decrease. However, a significant time of up to 70 days was needed to detect the effect. No significant effects were revealed for 30 days. If the experiment was continued for 160 days and more, the growth was observed to decrease two-fold at a surfactant concentration of 0.25 mg/l. The water filtration by mollusks was also observed to decrease under the influence of LAS at a concentration of 1 mg/l (1 ppm in the notation of the authors), but no less than 7 days was required to detect the effect as compared with 1.5 h in our work (at the same time, those data indicate the absence of the adaptation or adjustment of mollusks to surfactant at a long-term incubation). A significant difference between our work and the study by Bressan et al. (1989) was also the use of different anionic surfactants.

Table 3.26   Effect of SDS on the efficiency of removal of *Isochrysis galbana* algal cells from water by *Mytilus edulis* mussels as a result of inhibition of filtration.

SDS concentration, mg/l	Filtration period, min	EER*, %
1	30	135.0
	60	201.7
	90	271.1
2	30	198.3
	60	308.4
	90	259.5
4	5	116.1
	35	321.6
	65	642.1
	95	1473.2
5	30	294.5
	60	660.8
	120	1681.6

*EER = effect on the efficiency of removal = A/B, where A and B are concentrations of algal cells remaining in the water column (i.e., unfiltered) in the beakers upon filtration; A, concentration in the presence of filtration-inhibiting SDS; B, in the control.

Summing up the comparison of the results obtained in Bressan et al. (1989) and in our work, we can note that the data of both studies agree with each other. One can state the concordance of the experiments of different duration – short-term and those that lasted several weeks. Quite logical is also the fact that long exposures revealed the effect of low concentrations of SDS (0.025 mg/l) on mussels. This is also consistent with the observations that water filtration is the most sensitive component of the processes that form the scope for growth (SFG) (Smaal and Widdows 1994). SFG determines the potential of the organisms for further growth – the index that decreased under the influence of anionic surfactants.

Our data on the decrease of the physiological activity of bivalve mollusks under the influence of sublethal surfactant concentrations agrees well with the works of investigators who studied the action of other pollutants – heavy metals and pesticides. Sublethal concentrations of copper (0.1 mg/l) and zinc (1 mg/l) caused a decrease in the activity of the bivalve mussel *Scrobicularia plana* (Akberali and Trueman 1985).

Interestingly, the value of EC_{50} calculated based on our data for inhibition of water filtration by *M. edulis* with SDS (2.56 mg/l for the first 30-min period of filtration) was smaller than the estimate of the similar value for pesticide carbaryl (synonyms: alpha naphthyl-*N*-methyl carbamate; sevin). An estimate of EC_{50} for inhibition of the filtration rate by *M. edulis* under the influence of carbaryl is more than 4 mg/l according to Donkin et al. (1997). Hence, in this case SDS had a greater effect on the physiological activity of the hydrobiont than the pesticide does (at the same concentrations).

Registering the filtration rates makes it possible to reveal the responses of hydrobionts to much smaller concentrations of substances than when recording the lethal effects. This was confirmed by our experiments with SDS. We also incubated mussels in a medium with an increased SDS concentration of 25 mg/l. Even this concentration was not lethal for the mollusks. A much lower concentration (25 times smaller) evoked noticeable responses registered in observations of water filtration rates (see above).

We also studied the effect of SDS on another mollusk species, *M. galloprovincialis*. Filtration activity was shown to be inhibited. Thus, SDS concentrations of 1.7 mg/l already after 18 min were registered to give an almost 2-fold difference (192%) in the optical density of algal suspension in the experimental and control beakers. After 29 min of exposure, the difference was 2.8-fold (283%) (Table 3.27).

Similar results were obtained with another mollusk species, the oysters *Crassostrea gigas*. At the action of SDS (0.5 mg/l), only 4 min after the onset of the experiments the difference in the optical densities of the water medium between the experiment and control was shown to be 1.5-fold (155%). In 12 min, the difference became more than two-fold (211%) (Table 3.28).

Thus, for all mollusk species investigated, SDS concentrations of less than 3 mg/l induced a two-fold decrease in the efficiency of removing the suspension from water under certain registration conditions already after a testing time of less than 1 h. For two species, the effect was observed at SDS concentrations of 1.7 mg/l and less; for one species, a concentration of 0.5 mg/l was sufficient. We emphasize that such significant effects could be registered in a short period of time (a 30-min exposure).

Table 3.27 Effect of SDS (1.7 mg/l) on the efficiency of water filtration by juvenile bivalves *Mytilus galloprovincialis* and the removal of the algae *Dunaliella viridis* from water, leading to a change of OD_{658}.

Measuring period, No.	Time from the onset of incubation, min	OD_{658}				
		Experiment (mussels, algae, SDS)	Control 1 (mussels, algae, no SDS)	Control 2 (no mussels, algae, SDS)	Control 3 (no mussels, algae, no SDS)	Experiment/ Control 1 ratio (EER*), %
1	9	–	0.088	–	–	143.2
	10	0.126	–	–	–	
	12	–	–	0.171	–	
	13	–	–	–	0.189	
2	17	–	0.060	–	–	191.7
	18	0.115	–	–	–	
	21	–	–	0.170	–	
	23	–	–	–	0.175	
3	26	–	0.036	–	–	283.3
	29	0.102	–	–	–	
	31	–	–	0.172	–	
	32	–	–	–	0.172	
4	36	–	0.054	–	–	198.2
	37	0.107	–	–	–	
	38	–	–	0.170	–	
	40	–	–	–	0.181	
5	46	–	0.037	–	–	246.0
	49	0.091	–	–	–	
	50	–	–	0.168	–	
	52	–	–	–	0.174	

*EER = effect on the efficiency of removal of the suspension. Temperature, 20°C.

It is of interest to compare these numbers with the sensitivity of well-studied test objects, fish, to anionic surfactants. For instance, the toxicity of anionic surfactant alkyl benzene sulfonate (ABS) is well studied for a number of fish species. In some cases, it shows a greater hazardous effect than SDS. The values of LC_{50} (96 h) for ABS are over 3 mg/l at the impact on *Cyprinus carpio* (18.0 mg/l); *Mugil cephalus* (6.8 mg/l); *Pleuronectes flesus* (6.5 mg/l); *Gadus morhus* (3.5 mg/l) (Abel 1974; cited by Stroganov 1976a). We emphasize that in experiments with fish the exposure exceeded that in our experiments with mollusks 100 and more times. This indicates a greater dispatch of the test system with the filtration activity of mollusks and their high sensitivity.

Table 3.28 SDS (0.5 mg/l) inhibits the filtration capability of the *Crassostrea gigas* mollusks and the removal of suspended particles by them from water.

Measurement No.	Time of exposure, min	Optical density at 550 nm			B/A, % (EER*)
		Variant A (no SDS)	Variant B (with SDS)	Variant C (only *S. cerevisiae*, no mollusks, no SDS)	
1	4	0.117	0.181	0.176	154.70
2	12	0.074	0.156	0.179	210.81
3	20	0.048	0.111	0.174	231.25
4	29	0.035	0.074	0.164	211.43

*EER = effect on the efficiency of removal of the suspension. The beakers contained 16 mollusks each (total wet weight, including the shells: 23.5 g, beaker A; 23.6 g, beaker B). Age, 1 year. Incubation temperature, 23°C. Initial concentration of *S. cerevisiae*, 100 mg/l (dry weight). Volume of water in the beakers, 250 ml.

Note that the concentration of SDS that evoked a pronounced suppression of the filtration activity of *C. gigas* (0.5 mg/l) is remarkable also because these (and higher) concentrations of synthetic anionic surfactants in natural water systems were measured many times: in Lake Chernoye (Shatsk National Park), the Don River, the Poltava River (near Lvov), in the rivers of the U.S. and Germany, in the rivers and estuaries of Malaysia, in many regions of the Black Sea, in the Azov Sea and its firths, in the surface film of water bodies and in the water that went through water treatment plants (see above, Section 1.6; Steinberg et al. 1995). It is worth noting that a concentration of 0.5 mg/l is admissible for potable water according to the regulations and standards of a number of countries (e.g., U.S.) and World Health Organization (WHO) (WHO norms admit a greater concentration, 1 mg/l) (see above, Section 1.5).

It is appropriate to put forward a question as to what hydrobiological importance the decrease of filtration activity and removal of suspended matter (specifically, phytoplankton) from water can have. Owing to filtration of water, there occur processes in the ecosystem that are important for self-purification of water and regulation of the processes involved in it (Vinberg 1973, 1980; Alimov and Finogenova 1976; Kuzmenko 1976; Lowe and Pilesbury 1995; Ogilvie and Mitchell 1995; Palaski and Booth 1995; Stoeckman and Garton 1995; Savarese et al. 1997; Ostroumov 2004). Below, we give a list of nine processes (see also Ostroumov et al. 1997, 1998; Ostroumov and Donkin 1997; Ostroumov 2000b–d):

1. Adsorbed and assimilated pollutants are sedimented together with suspended matter (Kuzmenko 1996).
2. Water turbidity decreases; conditions for penetration of light and UV are improved, their effects on the organisms, photolysis or photooxidation of organic matter (Steinman and McIntire 1987; Wetzel et al. 1995) increases.
3. The content of high dispersion suspended matter in water decreases, which is favorable for increasing the fishery value of the reservoirs. In the opposite

case, when the content of high dispersion suspended matter increases above 25 mg/l, a decrease in fish catches, a decrease in the intensity of their nutrition, and a decrease in the rate of fish growth are observed (Alabaster and Lloyd 1984). In addition, when the content of suspended matter in water increases, the rate of filtration decreases by all filter feeders studied including mollusks and crustaceans (Sushchenya 1975, Alimov 1981; Mitin and Voskresensky 1982; Mitin 1984; Gorbunova 1988).

4. Water mixing increases, which results in aeration of water and produces effects on phytoplankton and zooplankton. If the water column is not sufficiently mixed, stratification of nutrients and chlorophyll is established. It was shown in the experiments that in a permanently mixed water body higher concentrations of phytoplankton are observed, and a decrease in the concentration of nutrients and zooplankton compared to poorly mixed water bodies (Oviatt 1981). A regular effect of the water column stability on phytoplankton is shown in Zhang and Prepas (1996).

5. Aeration of water and conditions for oxygen consumption are improved, which facilitates oxidation of organic matter by bacteria.

6. Species composition and abundance of specific species of algobacterial community are regulated (Officer et al. 1982; Buskey et al. 1997). In turn, the rate of generation and decomposition of hydrogen peroxide (Palenik et al. 1987) and the rate of free radical self-purification depends on the previously mentioned factors.

7. Components of dissolved organic matter (DOM) are excreted.

8. Sedimentation of organic matter is accelerated owing to the assimilation of phytoplankton and bacteria plankton by benthic and plankton filter feeders, excretions of pellets and pseudofaeces by the filter feeders. The rate of gravitation-driven precipitation of plankton is often tens of centimeters per day. For example, for diatoms *Melosira italica* and *Stephanodiscus astraea* it varied from 20 to 113 cm/day, the average value being 35 cm/day (Gibson 1984). Biological precipitation can be equal to gravitatrion-driven precipitation or surpass it (Parsons et al. 1982). The rate of particles' precipitation is proportional to the squared radius of the particle (Parsons et al. 1982), which leads to a sharp acceleration in the rate of sedimentation of pellets compared to the cells of phytoplankton and bacterial plankton as well as suspended organic matter, on which basis filter feeders form pellets of faeces and pseudofaeces. Bivalve mollusks can sediment approximately 0.267 g of carbon daily on an area of 1 m^2 in the form of biological deposits (calculated for *Crassostrea virginica* in Chesapeake Bay in the period previous to mass fishing of mollusks (see papers by Newell [Newell and Ott 1999]).

9. Intensive growth and functional activity of filter feeders facilitates the development and functioning of heterotrophic bacteria in the lower zone of the ecosystem. These bacteria are active in CO_2 fixation in the dark and assimilate organic compounds (Kuzhinovsky and Mitskevich 1992).

It is seen from this list that normal filtration activity of mollusks is very important for maintaining the stable state of the aquatic ecosystem and links between pelagic and benthic zones as well as for maintaining water quality and preserving the habitats of hydrobionts. Disturbance of the filtration activity cannot but affect the state of the aquatic ecosystem.

A working hypothesis from this consideration is that other classes of synthetic surfactants can also have similar effects (a verification of this hypothesis is described in Chapters 4 and 5) as well as the hypothesis that the mixtures containing anionic surfactants can also inhibit the filtration activity (a verification of this hypothesis is described in Chapter 6).

At the end of the section on the effect of SDS we mention some other data on its effect, which are useful for a more complete understanding of the interaction of this substance with organisms and cell structures. As shown by the example of many systems, the molecular mechanism of the effect of SDS on biological objects is related to the membrane-affecting property of surfactants and their effect on biomembranes (although this mechanism can be complicated by many other mechanisms; the multiplicity of the mechanisms of xenobiotics was established for many substances subjected to detailed investigation). Some data indicate that SDS can act as an uncoupler of coupling membranes (Skulachev V.P., personal communication).

Elucidation of the details of interaction of SDS with phospholipid bilayers of liposomal membranes has shown (Fukuda et al. 1987) that molecules of SDS (2 mM) introduced into the phospholipid layers (13.6 mM) probably form clusters and evoke phase separation in the bilayers (the phenomenon of intrasegregation). Besides, it was found that the movement of the spin label near the polar regions of a bilayer was subject to the action of SDS molecules, but in the inner layers of the bilayer it proved not to depend on the SDS molecules incorporated into the bilayer. There are also data on the effect of SDS on microtubules, namely on the inhibition of microtubules' assembly and the ability of SDS to cause disintegration of previously formed microtubules (Röderer 1987).

As noted above, another large group of anionic surfactants widely used in practice and entering the environment in large amounts are alkyl benzene sulfonates (ABS). The biological effects of ABS are discussed in the next section. Before moving on, the author would like to thank P. Donkin, who took part in some of the above experiments with *M. edulis*.

3.2   Biological Effects of Alkyl Benzene Sulfonates (ABS)

The importance of ABS for the state of aquatic media (McEvoy and Giger 1985; Holysh and Paaterson 1986; Marcomini et al. 1988; Gledhill et al. 1991; Conzalez-Mazo and Gomez-Parra 1996) is related to the fact that they are the most important and most widespread classes of surfactants. Their discharge to water is ubiquitous as they are used in almost all branches of economy as frothers, emulsifiers, and dispersants in preparation, processing, and coloring fibers; in production of pesticides and washing compositions; in emulsion polymerization of vinyl chloride, vinylidene chloride, and raw rubber; in production of photographic films; and also as de-emulsifiers of crude oil, floating agents, corrosion inhibitors, and in other applications (Abramzon et al. 1979). Total production of ABS in the U.S. only is no less than 300,000 tons, and production of alkyl sulfates is no less than 150,000 tons (Dean 1985). According to the data by other authors, 290,000 tons (640 million pounds) of LAS (linear alkyl benzene sulfonates) was sold in the U.S. in 1978, and 259,000 tons (570 million pounds) was sold in 1982 (Holish et al. 1986).

Data on the effect of ABS on some hydrobionts was already discussed above. Information on the effect of ABS on algae is given in Chawla et al. (1986), Lenova and Stupin (1990). Inhibition (limiting) of algal growth is recorded depending on the species of the organism at ABS concentrations of 0.08 mM (*Chlamydomonas reinhardii* Dang 11-32a), 0.1 mM (*Scenedesmus armatus*), 0.15 mM (*Chlorella sorokiniana, Ch. vulgaris* 211-1e, *Haematococcus vulgaris*), 0.25 mM (*Ch. saccharophila*), 1 mM (*Scenedesmus obliquus* 276-3a) or 1.5 mM (*Ch. fusca*) (Biedlingmaier et al. 1987; see Lenova and Stupina 1990). Possibly, it is not by chance that at these concentrations some of synthetic surfactants (SDS) exert an uncoupling effect on coupling membranes (V.P. Skulachev, personal communication). It was shown that at similar concentrations the growth of *Chlamydomonas reinhardii* Dang (20 mg/l) and *Plectonema boryanum* (30 mg/l) was inhibited (Azov et al. 1982, cited from Lenova and Stupin 1990).

The effect of linear ABS (LAS) on fish was studied in sufficient detail (e.g., Mozhayev 1976; Huber 1985). LC$_{50}$ (95 h) for *Pimephales notatus* was 0.86–1.23 mg/l. Other authors (Abel 1974, cited from Stroganov 1975a) give greater values of LC$_{50}$ (96 h) for alkyl benzene sulfonate: *Cyprinus carpio* (18.0); *Mugil cephalus* (6.8); *Pleuronectes flesus* (6.5); *Gadus morhus* (3.5); *Pseudopleuronectes americanus* (2.5); *Anguilla rostrata* (2.3); *Menidia menidia* (2.1 mg/l). These results agree with earlier data reviewed by Metelev et al., (1971). A lethal concentration of sodium dodecyl sulfate was 3.7 mg/l (48 h) for *Lepomis* sp. (a weight up to 5 g, at 20°C). A damaging concentration in 10-day experiments with alkylaryl sulfonate was 10 mg/l (at 15–18°C) (Mann 1955, cited from Metelev et al. 1971).

For comparison, we should note that in some cases a lower toxicity with respect to fish is given for heavy metals than the toxicity of LAS. For instance, LC$_{50}$ (96 h) for cadmium chloride was, according to Stroganov (1976a), 21.0 for *Fundulus majalis*, 55.5 for *Fundulus heteroclitus*, and 50.0 for *Cyprinodon variegatus*.

The Indian scientists Lal et al. (1983) obtained data about a much higher toxicity of ABS for fish than that indicated in Huber (1985). They showed that ABS were toxic for important Indian commercial fish species, mrigala *Cirrhina mrigala* (Family Cyprinidae, carps), even at concentrations of 0.014 mg/l. They consider a safe concentration to be 0.0015 mg/l, which is a few orders of magnitude lower than that indicated in Huber (1985). In the same paper the authors demonstrate toxicity of concentrations 0.04–0.06 mg/l for *Culex pipiens, Lymnaea vulgaris, Rana cynaphlyctis, Daphnia magna.* LAS at concentrations of 0.3–0.5 mg/l inhibited the response of β-adrenoreceptors of gills of *Parasalmo mykiss* (=*Salmo gairdneri*) to the effect of β-adrenergetic antagonist of isoprenalin, which Stagg and Shuttleworth consider an indication of the effect of LAS on cell membranes (Stagg and Shuttleworth 1987).

A comparison of the toxicity of LAS and branched ABS to daphnia *Daphnia magna* and danio-rerio fish *Brachydanio rerio* (Cyprinidae) showed a three times greater toxicity of LAS to both species of test organisms. The French investigators Gard-Terech and Palla compared the toxicity data with the kinetics of ABS biodegradation (Gard-Terech and Palla 1986).

LC$_{50}$ of LAS for aquatic invertebrates varies within broader limits, from 1 to 100 mg/l (Faba et al. 1979; Sivak et al. 1982, cited from Ostroumov 1981).

LAS was shown to have a negative effect on the larvae of the mussels *Mytilus edulis* (Hansen et al. 1997). The specific growth rate of the larvae decreased twice at LAS concentrations of 0.82 mg/l (9 days). In the experiment with mesocosms, the abundance of larval populations decreased significantly even at smaller concentrations of LAS (0.08 mg/l, 2 days) (Hansen et al. 1997). The recent data agree with earlier studies with larvae of *Ostrea edulis* and *Crassostrea gigas* mollusks (Renzoni 1971, cited from Stroganov 1976a). ABS at a concentration of 0.05 mg/l (6 h) inhibited the growth of one-day old mollusks of these species.

Pittinger et al. (1989) studied the effect of ABS dodecyl benzene sulfonate on *Chironomus riparitus*. At a concentration of 18.9 mg/l, no significant decrease of spawning was observed (the maximum concentration at which this effect was not observed). However, larvae were more sensitive than eggs. LC$_{50}$ (72 h) for just-spawned larvae was 2.2 mg/l. Interestingly, the sensitivity of larvae to anionic surfactant, ABS, was paradoxically greater than that to cationic surfactants, dodecyl trimethyl ammonium bromide (DDTMA) and distearyl dimethyl ammonium bromide. LC$_{50}$ for DDTMA was 14.6 mg/l (48 h) and for distearyl dimethyl ammonium bromide it was 11.3 mg/l (72 h). It is interesting that a greater ecological hazard of ABS compared to cationic surfactants manifested itself more strongly in those experiments where the larvae were incubated with surfactants in the presence of autoclaved natural stream bottom sediments from the clean Rapid Creek (South Dakota) (Pittinger et al. 1989).

Sakunthala et al. (1990) studied the effect of linear and branched ABS, sodium sulfonate and soap on shrimp *Metapenaeus dobsoni*, which is of commercial importance. It is interesting that linear ABS appeared most toxic (Sakunthala et al. 1990), which contradicts a widespread opinion about a greater toxicity of branched ABS.

Binding of ABS to dissolved humic substances can change the biological availability of these synthetic surfactants (Traina et al. 1996).

For self-purification of water, which is discussed in detail in Chapter 7, it is important that ABS inhibits by 20–25% the biochemical consumption of oxygen (BCO) dissolved in water at concentrations of 5 mg/l compared to control; at 10 mg/l the inhibition is by 36–48% (Garshenin 1965, cited from Mozhayev 1989).

3.2.1 Effect of alkyl benzene sulfonates on algae

Biological activity (BA) of ABS on algae was investigated, in particular the effect of sulfonol on *Scenedesmus quadricauda* and *Dunaliella asymmetrica* (with participation of T.N. Kovaleva and E.V. Borisova). Sulfonol was shown to inhibit the growth of the algae (Ostroumov et al. 1990).

Using microalgae to test the BA of aquatic media and aqueous solutions of xenobiotics, a scientist can encounter serious difficulties when comparing the results of various experiments. In order to make the results of different experiments more compatible quantitatively, we used the approach developed in Maksimov et al. (1988), which is based on the calculation of inhibition coefficients at different stages of culture growth. The calculation demonstrated that at an initial concentration of sulfonol equal to 0.05 ml/ml, the inhibition coefficient of the culture *Sc. quadricauda* was 61.5–65.0% in the first 40 h of growth; during the period from 40 to 50 h, it was 81.3–87.5%, and then sharply decreased down to 19.2–43.2% in the period from 50 to 64 h. At the further stages of the experiment (from 64 to 74 h) it remained low: 19.1–36.9%.

As a matter of discussion of the data, we should mention a work that studied the effect of a slightly different structurally but still similar anionic surfactant on green algae (Parshikova et al. 1994). That work showed that anionic surfactant, a sodium salt of dodecyl sulfonated acid, at a concentration of 5 mg/l had no noticeable effect on the growth of *Chlorella vulgaris* Beijer, strain HPDP-19, in the first 24 h, but evoked a slight inhibition (approximately 15%) in 48 h. A slight inhibition of the photosynthetic evolution of oxygen was observed under the influence of 10 mg/l of anionic surfactant on the algal cells at their biomass concentration of 79 mg/l. Interestingly, at an algal biomass of 56 mg/l the same concentration of anionic surfactant caused a certain stimulation of the photosynthetic evolution of oxygen. Recording of delayed fluorescence made it possible to reveal the effect of a sharp increase of delayed fluorescence upon addition of 1–5 mg/l of anionic surfactant (Parshikova et al. 1994). A comparison of the action of anionic surfactants, nonionogenic surfactants, and cationic surfactants carried out in that study demonstrated a comparatively higher biological activity (e.g., a negative effect on the photosynthetic activity) of cationic surfactants.

On the whole, information on the biological activity of sulfonol obtained in our studies is consistent with the above-mentioned literature data on the effect of ABS on green algae.

3.2.2   Effect of alkyl benzene sulfonates on vascular plants

Attempts have been made to use new, alternative methods for estimating the BA of waters containing anionic surfactants (Devi and Devi 1986). Ya. Devi and S. Devi from the National Botanical Research Institute (India) detected a high sensitivity of fern sporulation to this xenobiotic. At a concentration of anionic surfactant equal approximately to 0.01 mg/l the spores of *Diplazium esculentum* did not germinate. A decrease in the proportion of germinated spores compared to controls was observed at an anionic surfactant concentration of 0.0004 mg/l (Devi and Devi 1986).

V.B. Ivanov and his laboratory investigated the effects of more than 100 substances on plant seedlings (e.g., Ivanov 1982, 1992). The effects of heavy metals on seedlings were also studied in this laboratory, and also by other investigators (e.g., Chernenkova 1987; and others). The action of synthetic surfactants on seedlings of angiosperm plants (especially from the point of view of using this approach to characterize polluted waters in view of their remediation) has not been studied by the time our research was started.

We studied the effect of sulfonol-containing waters on the seedlings of various, aquatic and terrestrial, plants. It is noteworthy that among the consequences of pollution of the hydrosphere there are some aspects that make necessary the study of the effects of xenobiotics not only on the aquatic plants but also on the terrestrial plants. One of the reasons is the fact that sewage sludge is formed during biological purification of waters. Removal of this sewage sludge is a very difficult problem. The scale of the problem is illustrated by the following figure: only in the U.S., 16 million tons of sewage sludge is transported to landfills, which is approximately 5.5% of the total amount of municipal solid waste (Alexander 1993). The most drastic approach to the removal of sewage sludge is incineration, which is expensive (drying and incineration of one ton of dry sewage sludge, which is a waste in the production of cellulose, requires US $400 to 800 in Russia (*Financial Izvestiya*, #102 (336) October 31, 1996, page 8). In addition, incineration does not solve the problem completely, as dioxins are formed. Therefore, introduction of sewage sludge into the soil is used in practice, and persistent pollutants, including some of synthetic surfactants are transferred into soil and later can come into contact with plants. Another way of synthetic surfactants entry from polluted waters to plants is the use of polluted and sewage waters for irrigating lands and watering agricultural plants.

Oryza sativa is an example of a plant that spends a significant part of its onto-genesis in aquatic medium as a hydrobiont during this time. Our investigation using a rice culture showed that its growth is significantly inhibited in aquatic medium containing sulfonol (A.E. Golovko participated in this study).

A peculiarity of the method for investigating seedlings was the use of a new integral morphogenetic index, which was introduced by the author to characterize the degree of disturbances in the ontogenesis of plants. This is the so-called apparent average length (AAL) of the roots of seedlings (Ostroumov 1990). This integral index summarizes two types of data about the degree of germination of seeds and the length of the roots of seedlings.

We studied the effect of sulfonol at concentrations of 0.031, 0.063, 0.125, and 0.25 mg/l in the tested aquatic solution (Tables 3.29–3. 31). Inhibition of the growth of seedlings roots was observed at tested concentrations of surfactants. No decrease in the degree of germination of seeds was observed. Inhibition of the growth of the roots of seedlings was more significant at the concentration of sulfonol equal to 0.25 mg/l (at 26°C). Additional experiments showed that even at decreased concentrations of sulfonol (0.03 mg/l) the AAL of the roots was approximately 65–70% of the control level, i.e., the degree of inhibition was 30% or even greater.

Part of the experiment on testing the seedlings of plants to estimate the biological activity of surfactants was carried out using *Fagopyrum esculentum* (Ostroumov et al. 1990; Ostroumov and Tretyakova 1990; and others). We tested various variants of biological testing methods including the method where 40-hour-old seedlings were transferred from non-polluted water to water with surfactant (Table 3.32). During the incubation in tested medium during the interval from 50 to 64 h from the onset of germination of the seeds it was found that the mean rate of the seedlings growth at a concentration of sulfonol equal to 0.25 ml/l compared to the control was 82.8%, and at a concentration of sulfonol equal to 0.5 ml/l the rate was 60% of the control (Table 3.33). During the next time interval (from 64 to 74 h) the rate of growth at the same concentration of sulfonol was 83.5 and 54.1% of the control, respectively. This experiment is also interesting in that it demonstrated the possibility of revealing the effect of inhibition of this xenobiotic in only 10 h of incubation of the seedlings in the tested aquatic medium.

Table 3.29   Effect of anionic surfactant sulfonol on AAL* of *Oryza sativa* roots ($n = 20$).

Statistical parameters	Sulfonol content, ml/l		
	0 (control)	0.0625	0.0312
Incubation time, 73 h			
AAL, mm	7.05	2.80	4.60
Standard deviation	3.44	2.14	3.73
Standard error	0.77	0.48	0.83
CV, %	48.80	76.50	81.15
Incubation time, 95 h			
AAL, mm	20.05	9.00	14.05
Standard deviation	9.83	4.24	4.51
Standard error	2.20	0.95	1.01
CV, %	49.00	47.10	32.10

*AAL = apparent average length

Table 3.30   The rate of AAL increase for *Oryza sativa* roots during the effect of sulfonol.

Sulfonol content, ml/l	Mean increase rate, mm/h		Inhibition coefficient, %	
	$\Delta t = 73 - 0$ h	$\Delta t = 95 - 73$ h	$\Delta t = 73 - 0$ h	$\Delta t = 95 - 73$ h
0 (control)	0.097	0.590	0.0	0.0
0.0312	0.063	0.430	34.8	27.3
0.0620	0.038	0.280	60.3	52.3

Table 3.31   Effect of sulfonol on AAL of *Oryza sativa* roots ($n = 20$; incubation time, 71 h).

Statistical parameters	Sulfonol content, ml/l		
	0 (control)	0.25	0.125
AAL, mm	16.35	1.35	2.90
Standard deviation	8.00	0.67	2.29
Standard error	1.79	0.15	0.51
CV, %	48.90	49.70	79.00

Table 3.32 Change of length x of buckwheat seedlings (*Fagopyrum esculentum*, cultivar Shatilovskaya 5) in various periods of time. The transfer of the seedlings from water without sulfonol to water with sulfonol was made 40 h after the start of the experiment (addition of water to dry seeds).

Time, h	Sulfonol concentration, µl/ml	x		σ	Confidence interval	Number of seedlings
		mm	%			
40	0	6.9	100	2.4	0.68	49
	0.05	6.9	100	2.2	0.64	50
	0.25	6.9	100	2.3	0.65	50
	0.5	6.9	100	2.9	0.82	50
	0.8	6.6	100	2.4	0.68	50
	1.0	6.4	100	1.8	0.52	50
	5.0	5.8	100	2.0	0.58	50
50	0	11.9	172.5	2.6	1.00	29
	0.05	13.0	188.4	3.3	1.23	30
	0.25	12.6	182.6	2.8	1.06	30
	0.5	11.6	168.1	3.2	1.19	30
	0.8	10.5	159.1	2.9	1.07	30
	1.0	10.7	167.2	4.0	1.51	30
	5.0	6.3	108.6	2.0	0.74	30
64	0	29.5	427.5	6.9	1.98	49
	0.05	29.4	426.1	7.8	2.21	50
	0.25	26.7	387.0	7.2	2.04	50
	0.5	21.9	317.4	5.5	1.54	50
	0.8	18.7	283.3	5.1	1.44	50
	1.0	17.4	271.9	5.6	1.59	50
	5.0	9.2	158.6	3.1	0.87	50
74	0	42.8	620.3	9.8	2.81	49
	0.05	42.0	608.7	9.7	2.74	50
	0.25	37.8	547.8	10.5	3.00	50
	0.5	29.1	421.7	7.4	2.09	50
	0.8	24.2	366.7	6.5	1.83	50
	1.0	21.2	331.3	7.3	2.07	50
	5.0	10.1	174.1	3.6	1.01	50

Table 3.33   Change of the mean rate of elongation (mm/h) of *Fagopyrum esculentum* (cultivar Shatilovskaya 5) seedlings during the action of sulfonol in various intervals of time.

Sulfonol concentration, μl/ml	Time interval, h		
	40–50	50–64	64–70
0	0.5	1.22	1.33
0.05	0.61	1.17	1.26
0.25	0.57	1.01	1.11
0.5	0.47	0.74	0.72
0.8	0.39	0.59	0.55
1.0	0.33	0.48	0.38
5.0	0.05	0.21	0.09

For details, see *Problems of Ecological Monitoring and Modeling of Ecosystems*, 1986, Vol. 9, pp. 87–97 (in Russian).

The concentrations which did not evoke a pronounced inhibition of elongation of seedlings of *F. esculentum* were established. These were the concentrations within 0.05–0.5 ml/ml (for 10 h after the transfer into the medium containing sulfonol), 0.05–0.25 ml/ml (for 24 h), and 0.05 ml/ml (for 34 h) (Table 3.32). The latter concentration is equivalent to 50 mg/l. This information can be applied for practical usage in phytoremediation.

We also studied the effect of waters containing sulfonol on the seedlings of another test object *Sinapis alba* VNIIMK-162. At a sulfonol concentration of 0.05 ml/l and greater, the mean length of the seedlings was observed to decrease as compared to the control, and at 0.25 ml/l their mean length was approximately 59.9% as compared with the control after 74 h of incubation. The inhibitory effect at a concentration of 0.05 ml/l was weaker but still noticeable. The estimate of EC_{50} in the measurement of the mean length of the seedlings indicated that this value depended on the exposure and varied from 0.33 to 0.90 ml/ml. When exposure increased, the value of EC_{50} regularly decreased (Table 3.34). We note that significant concentrations of sulfonol were required to inhibit the growth of the seedlings, which agrees with a comparatively low sensitivity of the seedlings to

Table 3.34   Effect of sulfonol on *Sinapis alba* seedlings.

No.	Measurement time, h	EC_{50}, μl/ml
1	40	0.90
2	50	0.62
3	64	0.40
4	74	0.33

Note: Times are after seeds were soaked. For details, see *Gidrobiol. Zhurn.* **9 (4)**: 54 (in Russian).

another anionic surfactant, SDS (see above) and allows us to make a conclusion about the prospects of using these organisms for phytoremediation.

It is interesting to study the effect of sulfonol on animals. Taking into account a high chemoreceptor sensitivity of aquatic animals (studied in detail for fish (Pavlov and Kasumyan 1990; Kasumyuan 1995, 1997) we carried out experiments using an experimental setup (developed by A.Ya. Kaplan), which allows us to record the electrographic reaction of *C. carpio* (Kaplan 1987, 1988).

3.2.3 Effect of ABS on the receptors of C. carpio

Our joint work with A.Ya. Kaplan (Moscow State University) found that at the concentration of sulfonol in water equal to 1 mg/l the olfactory function of *Cyprinus carpio* is totally inhibited (Ostroumov et al. 1988). The destruction (inhibition) of electrographic reaction of the olfactory bulb by the standard stimulus at a sulfonol concentration in water equal to 0.5, 0.25, and 0.05 mg/l was revealed. *L*-alanine was the standard stimulus.

For comparison we note the value of LC_{50} (96 h) during the action of ABS on *C. carpio*: it is equal to 18.0 mg/l (Abel 1974, cited from Stroganov 1976a). It is seen that concentrations smaller by three orders of magnitude than LC_{50} (96 h) exert a clearly unfavorable effect, which is important for understanding the effects of the chemical pollutant on aquatic organisms.

This study is also important because the negative effect of sulfonol at its concentration in water below the recommended allowable concentration was revealed. In addition, the postulate was confirmed that "anthropogenic substances polluting the biosphere can distort the interaction of organisms in various ways" (Ostroumov 1986) (in this case they can distort the information interactions by means of the olfactory channel of communication). The works by A.O. Kasumyan on the investigation of the effect of other pollutants, heavy metals, at low sublethal concentrations on the behavior of fish also confirm this thesis (Kasumyan 1995, 1997).

Recently, high-polymer synthetic surfactants (HPSS) have begun to be increasingly used. In the next section we describe the results of our experiments to study the effect of HPSS on organisms.

3.3 Biological Effects of High-Polymer Synthetic Surfactants

High-polymer synthetic surfactants (HPSS) enter reservoirs in significant volumes with sewage waters and also because they are used in textile, radiotechnical, construction engineering, cosmetics, pharmaceutical, perfumery, and food industry as dispersants of dyes, as emulsifiers and dispersants of herbicides, cement setting retarders, agents for improving the quality of drilling fluids for boreholes, and in other industries (Abramzon and Gaevoy 1979).

Table 3.35 AAL of *Fagopyrum esculentum* (cultivar Shatilovskaya 5) seedlings in an aqueous medium containing the surfactant CHMA.

Time, h	CHMA, ml/l	Active substance concentration, g/l	AAL	σ	Standard error	CV, %	M
43	0	0	11.0	12.2	2.03	110.91	11
	2	0.44	4.5	8.8	1.47	195.56	20
	4	0.88	6.2	10.5	1.75	169.35	18
50	0	0	13.3	14.7	2.45	110.53	11
	2	0.44	5.7	11.2	1.87	196.49	20
	4	0.88	8.2	13.7	2.28	167.07	18
67	0	0	26.0	27.8	4.63	106.92	11
	2	0.44	10.0	19.7	3.28	197.00	20
	4	0.88	14.9	22.9	3.82	153.69	18

Note: In Petri dishes 12 seeds each, 10-ml test solution or STW (control), incubation temperature $28 \pm 1°C$, M is the number of non-germinated seeds at a sample size $n = 36$. In more detail, see *Ekologiya* **6**: 50–55 (1993) (in Russian).

Table 3.36 Mean lengths (x) of the roots of *Fagopyrum esculentum* (cultivar Shatilovskaya 5) seedlings after the transfer of the seedlings from STW into an aqueous medium containing the surfactant CHMA (0.44 and 0.88 g/l).

Time, h	CHMA, ml/l	Active substance concentration, g/l	x		σ	CV, %	Standard error
			mm	%			
25	0	0	7.0	100	2.2	31.43	0.64
	2	0.44	7.0	100	1.3	18.57	0.38
	4	0.88	7.2	100	2.0	27.78	0.58
44	0	0	22.7	324.3	12.5	55.07	3.61
	2	0.44	22.0	314.3	8.5	38.64	2.45
	4	0.88	20.9	290.3	8.0	38.28	2.31
49	0	0	29.8	425.7	16.0	53.69	4.62
	2	0.44	30.7	438.6	10.1	32.90	2.92
	4	0.88	24.4	338.9	10.5	43.03	3.03
90	0	0	49.3	704.3	21.9	44.42	6.32
	2	0.44	51.2	731.4	18.4	35.94	5.31
	4	0.88	40.4	561.1	17.6	43.56	5.08

Note: Seeds were germinated on STW. After 25 h, 12 seedlings each and 10-ml test solution or STW (control) were put into Petri dishes. After 44 h, 10 ml of the same solution was added into each dish. Incubation at $28 \pm 1°C$. In all tests, sample size $n = 12$.

As suggested by V.V. Bocharov, we studied a new substance from the HPSS group, which was developed and recommended for application at the Surfactant Research Institute (VNIIPAV) in Shebekino. The problems of potential ecological hazard of this representative of HPSS were not studied at all by the time of this study. We analyzed the effect of water solutions of copolymer of hexene and maleic aldehyde (CHMA, mol. mass ~20,000) on seedlings of plants (Tables 3.35–3.38). A slight inhibition of the growth of buckwheat *F. esculentum* seedlings in aquatic medium containing 0.88 g/l of the substance was shown. When 25-hour old seedlings were transferred to such a medium 90 h after soaking the seeds, the apparent average length of the roots was 40.4 mm and 49.3 mm in the control (Table 3.36) (postgraduate student N.A. Semykina participated in this work). At smaller concentrations of 0.44 g/l, inhibition was not manifested. It was interesting to compare these data with the experiments on the effect of the same substance on other types of organisms.

The apparent average length of watercress *Lepidium sativum* in aquatic medium with CHMA at a concentration of 2.2 g/l (with respect to the active substance) after 50 h of incubation was 10.5 mm compared with 15.2 in the control.

We studied the effect of CHMA on the seedlings of rice *Oryza sativa* during time intervals up to 69, 92, 117, and 124 h. Inhibition of seedlings growth was found after the transfer of the 69-hour old seedlings of rice *O. sativa* (cultivar Ognevsky) from non-polluted water to the water containing 2.2 g/l of active CHMA. For instance, in 124 hours after germinating the seeds, the mean length of the roots was 40.6 mm compared to 52.2 mm in the control, which was 77.8%.

A relatively low biological activity of CHMA compared to low-polymer surfactants is clearly seen in the experiments; still, the conclusion about low ecological hydrobiological hazard of CHMA would be untimely due to its low biological degradation. CHMA is stable in water medium and is not practically subject to bio-

Table 3.37 AAL of the roots of *Fagopyrum esculentum* (cultivar Shatilovskaya 5) seedlings in an aqueous solution containing the surfactant CHMA (0.88 and 2.2 g/l).

Time, h	CHMA, ml/l	Active substance concentration, g/l	AAL, mm	σ	Standard error	M
48	0	0	10.1	14.9	2.48	20
	4	0.88	11.8	10.3	1.72	11
	10	2.20	5.8	9.7	1.62	20
54	0	0	13.3	18.6	3.10	20
	4	0.88	16.0	14.0	2.33	11
	10	2.20	7.0	10.9	1.82	20
70	0	0	23.9	31.6	5.27	20
	4	0.88	21.9	22.6	3.77	11
	10	2.20	10.9	16.7	2.78	20

Note: In Petri dishes 12 seeds each and 10-ml test solution or STW, incubation at $28 \pm 1°C$, M is the number of non-germinated seeds; sample size $n = 36$.

logical destruction in the system with active sludge (Bocharov et al. 1988, cited from Ostroumov and Semykina 1991).

It is not to be ruled out that the cause of low biological activity of the high-polymer surfactant we investigated is the fact that the penetration of the molecules of this substance through biological membranes of test organism is worse than the penetration of low polymeric molecules.

Low biological decomposition of CHMA makes it a hazardous pollutant for water reservoirs. Thus, revealing the organisms which are relatively tolerant to this substance and can be used in the systems of phytoremediation, seems a useful result.

Table 3.38 Mean lengths (x) of the roots of *Fagopyrum esculentum* (cultivar Shatilovskaya 5) seedlings after their transfer from STW into an aqueous medium containing the surfactant CHMA (up to 5.5 g/l) (generalized data, n = 36).

Time, h	CHMA, ml/l	Experiment 1				Experiment 2			
		x, mm	%	s	n	x, mm	%	s	n
28	0	5.4	100	1.3	12	5.6	100	1.0	12
	15	5.8	100	1.9	12	5.5	100	1.0	12
	25	5.6	100	0.9	12	5.5	100	0.9	12
48	0	25.5	472.2	12.7	12	25.0	446.4	11.5	12
	15	14.7	253.2	4.7	12	14.8	269.1	4.9	12
	25	7.4	132.1	2.3	12	8.0	145.4	3.0	12
53	0	32.9	609.3	16.7	12	32.7	583.9	12.2	12
	15	17.8	306.9	6.6	12	18.8	341.8	9.1	12
	25	8.2	146.4	2.8	12	8.4	168.0	3.5	12
90	0	67.6	1251.9	24.0	12	64.3	1148.2	24.2	12
	15	24.1	415.5	8.8	12	26.3	478.2	10.0	12
	25	8.25	147.3	3.5	12	8.6	156.4	2.8	12

Time, h	CHMA, ml/l	Experiment 3				Generalized data			
		x, mm	%	s	n	x, mm	%	s	Standard error
28	0	5.8	100	1.1	12	5.7	100	1.3	0.22
	15	5.8	100	1.1	12	5.5	100	1.1	0.18
	25	5.6	100	0.8	12	5.6	100	0.9	0.15
48	0	24.3	419.0	14.6	12	24.9	436.8	12.6	2.10
	15	14.0	241.4	6.4	12	14.5	263.8	5.3	0.88
	25	7.8	139.3	2.3	12	7.8	139.3	2.5	0.42
53	0	28.6	493.1	8.3	12	31.4	550.4	12.4	2.07
	15	17.5	301.7	7.8	12	22.6	410.9	7.8	1.30
	25	9.0	160.7	3.3	12	8.5	151.8	3.2	0.53
90	0	53.8	927.6	20.3	12	61.9	1086.0	21.9	3.70
	15	23.4	403.4	11.0	12	24.6	447.3	9.9	1.65
	25	9.2	164.3	3.7	12	8.7	155.4	33.0	5.50

3.4 Concluding Remarks

The study of the effects of various anionic synthetic surfactants on a wide range of biological objects demonstrated the presence of a wide spectrum of disturbances. The negative effect of anionic synthetic surfactants on marine and freshwater species was demonstrated. Some of the studied species are actively involved in self-purification of the ecosystems. The diversity of negative consequences of the effect of anionic synthetic surfactants on organisms found in our experiments prompts us to make a conclusion that surfactants can be a more hazardous group of pollutants than considered earlier. This conclusion agrees with some data from scientific literature. For example, a comparison of the data on LC_{50} of anionic surfactant ABS (mg/l, for 96 h) for seven fish species demonstrated that these values fall within the interval from 2.1 to 18.0, while for six species of fish the interval is 2.1–6.8. These values are smaller than similarly determined (96 h) values of LC_{50} for cadmium chloride for fish (*Fundulus majalis, F. heteroclitus, Cyprinodon variegates*), which were in the range from 21.0 to 55.0 mg/l (Stroganov 1976b). Taking into account that one of the economically reasonable (low cost) methods of processing polluted waters is irrigation of agricultural lands with these waters, new quantitative information about the effect of water containing anionic synthetic surfactants on plants becomes significant. The results of these experiments allow us to develop in more detail the methods of biological testing alternative to biological testing on animals, which is now attracting attention taking into account humane considerations.

We note in conclusion the link between our new data about the effect of anionic surfactants on water filtration and the important problem of self-purification of water under conditions of anthropogenic impact (Ostroumov 1998). Many other filter feeders are also very sensitive to pollution of aquatic medium. For example, such filter feeders like sponges and Bryozoa (Skalskaya and Flerov 1999) as well as Brachiopoda (Zezina 1997) show a sensitivity to pollution. Another ecological process, mineralization of xenobiotics by microorganisms, is very important for self-purification. This aspect of self-purification can be also disturbed under the effect of surfactants. It was shown in experiments with microcosms that seven different substances from the surfactant group inhibited mineralization of phenanthrene by microorganisms destructing polyaromatic hydrocarbons (Tsomides et al. 1995).

The relative sensitivities of organisms to anionic surfactants are compared in Table 3.39.

The filtration activity of mollusks is part of the process in aquatic ecosystems which leads to self-purification of water (e.g., Konstantinov 1979). Thus, a comparatively high sensitivity of this activity of mollusks to synthetic surfactants puts forward the conclusion that these effects pose a potential hazard if they occur due to pollution of water.

Taking into account the data on surfactants and the literature data (Chapter 1) we cannot rule out that in the case of respective concentrations, summation of biological effects of pollutants of different types can cause a hazard of inhibition of the natural ability of aquatic ecosystems for self-purification (additional argument is given in Chapter 1 and subsequent Chapters 4 and 7), which confirms the previously formulated concept (e.g., Ostroumov 1986, 1991; Yablokov and Ostroumov 1983, 1985; Yablokov and Ostroumov 1991, Ostroumov 1998) that a disturbance in the self-purifying function of ecosystems under the effect of pollutants is a more hazardous factor than it was considered before.

A significant part of our data characterizes the effects of anionic surfactants on vascular plants. The latter are the main biological components of phytoremediation systems (McCutcheon et al. 1995). Their application requires the knowledge of the limiting concentrations to which the plants are tolerant. Our data on the effect of anionic surfactants on *F. esculentum, S. alba, L. sativum* and other plants (see above) contain this information.

Table 3.39 Comparison of the general sensitivity of organisms to anionic surfactants: general assessment of sensitivity (at inhibitory effects) to anionic surfactants: L, low; M, moderate; H, high, S, stimulating effect.

Organisms	Sensi-tivity	References/comments
Various Gram-negative heterotrophic bacteria	L	General assessment of anionic surfactants as comparatively low toxic substances for Gram-negative bacteria (Stavskaya et al., 1988)
Various Gram-positive heterotrophic bacteria	L or M	Survived at concentrations up to 20 mg/l of sulfonol (ABS); were able to grow at SDS concentrations of up to 80 mg/l (Serov, 1981, cited by Stavskaya et al., 1988)
Various species of green algae, including *Chlorella* sp. and *Scenedesmus* sp.	L, S	Different authors including Lenova et al. (1980) and Parshikova et al. (1994); some strains grow at SDS concentrations > 50 mg/l without inhibition (Lenova et al., 1980); SDS at 10–100 mg/l stimulated the growth (Goryunova and Ostroumov, 1986)
Seedlings of *Lepidium sativum, Fagopyrum esculentum, Sinapis alba*, and others	L	Sections 3.1.1 and 3.2.2; comparatively low sensitivity enables the use of these organisms for phytoremediation (Ostroumov, 2000)
Mollusks of *Mytilus edulis, M. galloprovincialis, Crassostrea gigas*	H	Section 3.1.2; comparatively high sensitivity to sublethal concentrations can cause danger of decreasing the removal of suspended particles and and cells of plankton from water (Ostroumov et al. (1997, 1998), Ostroumov (2000b, c, d))
Larvae of *M. edulis*	H	Low concentrations of LAS sharply decreased the specific rate of growth in 2–9 days (Hansen et al. (1997)), which can cause danger for the reproduction of the population
Cyprinus carpio (olfactory reception; electrographic reaction)	H	Section 3.2.3; high sensitivity causes danger of disrupting the chemical channel of communication in aquatic ecosystems, and disturbances of normal behavior of fish (Ostroumov, Kaplan et al. (1988))

On the whole, our data make it necessary to comment on the publications (Fendinger et al. 1994 and others) where the problem of the ecological hazard of anionic surfactants is reduced to revealing and analysis of LC_{50} for sharp and chronic toxicity, or NOEC (No Observed Effect Concentration) estimated on the basis of lethality as well as their effect on reproduction in the experiments on a limited number of test species, which do not include bivalve mollusks (Fendinger et al. 1994). The basis of facts and concepts for the conclusions about ecological safety of anionic surfactants including alkyl sulfates and LAS made in the investigations similar to Fendinger et al. (1994) seems to be incomplete because the data on the effect of surfactants on filtration activity of hydrobionts (the example is bivalve mollusks) indicate that negative effects can be observed at concentrations much lower than LC_{50} or NOEC estimated on the basis of lethality.

The results of the study allow us to formulate new working hypotheses that non-ionogenic surfactants and cationic surfactants can cause similar effects during their effect on mollusks and that angiosperm plants are significantly more tolerant than mollusks with respect to nonionogenic and cationic surfactants. New experiments described in Chapters 4 and 5 are intended to test these hypotheses.

4 Biological Activity of Waters Containing Nonionogenic Surfactants

Nonionogenic surfactants are second to anionic surfactants by the amount of their production and discharge into aquatic ecosystems (Surfactants 1984; Steinberg et al. 1995; Bailey 1996; Thiele et al. 1997). The main classes of nonionogenic surfactants are alcohol ethoxylates and oxides of fatty amines. In 1988, the worldwide annual consumption of nonionogenic surfactants belonging to the alkyl phenol ethoxylates was approximately equal to 360,000 tons (Ahel et al. 1993). Along with the worldwide use of nonionogenic surfactants in industry and other branches of the economy, chemicals of this class have other applications: nonionogenic surfactant nonoxynol-9 (NP-9) is frequently used as intravaginal spermicide (Meyer et al. 1988). This chemical is a usual component used for lubricating individual AIDS protection devices. By its structure, NP-9 is nonylphenoxyl poly(ethylenoxyl)$_9$ ethanol (Meyer ct al. 1988). Nonionogenic surfactants belonging to the group of alkyl phenol derivatives are also used as hair dyes (Meyer et al. 1988).

The problems of environmental pollution with nonionogenic surfactants were discussed in Stavskaya et al. (1988), Lewis (1991) and Holt et al. (1992). The amount of nonionogenic surfactants in sewage waters reaches significant concentrations up to 30 g/l (Stavskaya et al. 1988). Nonionogenic surfactants were found in natural aquatic systems at concentrations of up to 1 or even 2.6 mg/l (Holt et al. 1992). We emphasize, however, that owing to the property of nonionogenic surfactants to form complexes with many compounds a great proportion of nonionogenic surfactants can exist in "masked" states and cannot be revealed using analytical methods. Hence, the probability of obtaining underestimated results is high. The real amount of nonionogenic surfactants in aquatic ecosystems can be even higher than that given by the water analysis.

The efficiency of water purification as far as nonionogenic surfactants are concerned is low at waste water treatment plants with mechanical and biological purification systems. Approximately 60% of nonylpolyethoxylates, which enter water purification installations with polluted water, are released to the environment with the so-called purified waters; 85% of these chemicals can be somewhat transformed (Ahel et al. 1993), which hampers quantitative analysis of environment pollution with nonionogenic surfactants.

Although some negative effects of nonionogenic surfactants were demonstrated earlier (see below), by the beginning of our studies nonionogenic surfactants were considered chemicals of comparatively low hazard (Meyer et al. 1988). For example, it was stated that "nonionogenic surfactants are not toxic or low toxic" (Stavskaya et al. 1988, p. 20). Comparatively low toxicity of nonionogenic surfactants was demonstrated in Sirenko (1991).

Monitoring of nonionogenic surfactant content in natural waters of Russia is not carried out, and the data on pollution of water bodies and streams in Russia with nonionogenic surfactants are practically absent. The disadvantages of the existing methods of determining nonionogenic surfactants are discussed in Stavskaya et al. (1988). The major disadvantages are poor reproducibility of the results and low selectivity. The presence of anionic surfactants, cationic surfactants, sulfates, proteins, and the presence of various organic and inorganic ions affect the results.

Nonionogenic surfactants are one of the components of many dispersants. For example, Corexits 9527, 7664, 8667, 9660, and 9550 contain nonionogenic surfactants. Corexit 8667 had a low LC_{50} value (i.e., manifested high toxicity) for daphnia *D. magna*: 3 mg/l (48 h, 5°C) and 0.03 mg/l (48 h, 20°C) (Bobra et al. 1989). Relative toxicity of this dispersant was approximately 200 times greater than water-soluble oil fraction. It was shown for all mixtures of Corexits that their toxicity was higher than the toxicities of physical oil dispersions without Corexits. In experiments, toxicity of oil pollution increased when dispersants containing nonionogenic surfactants were added to the system (Bobra et al. 1989).

Alkyl phenol ethoxylates or oxyethylated alkyl phenols are among the important classes of nonionogenic surfactants. The chemicals of this class enter aquatic systems with polluted waters because they are widely used as emulsifiers, solubilizers, penetrating agents, dispersants, and components of cleaning and degreasing compositions. They are used in petroleum refining, petrochemical, natural gas, and many other industries (Abramzon and Gaevoy 1979). Nonionogenic surfactants of this class are used, in particular, for emulsification for cellulose production. Production of 1 m^3 of cellulose requires approximately 1.5 kg of nonylphenolethoxylates (NPE). Washing off paint in paper recycling process by U.S. technologies requires 2–3 kg of alkyl phenol ethoxylates for 1 t of paper (Kouloheris 1989).

Nonionogenic surfactant Triton X-100 (TX100, oxyethylated alkyl phenol, polyoxyethylenoxyphenyl ether, a standard preparation with molecular mass 624.9) is widely used in studies of nonionogenic surfactants. It is one of the most widely used monoalkylphenyl ethers of polyethylene glycol (alkylaryl polyether).

The biological activity of nonionogenic surfactant TX100 was studied in many organisms. TX100 caused a certain increase in the degree of saturation of fatty acids of mono- and digalactosyldiglycerides of red algae *Porphyridium purpureum*. This effect was manifested at concentrations of TX100 in the range 5–10 million^{-1} (ppm), i.e., 5–15 mg/l. At concentrations of 5–20 mg/l (8–32) μM suppression of the growth of algal cells was observed (Nyberg and Koskimes-Soininen 1984). Nyberg and Koskimes-Soininen (1984) also demonstrated that TX100 increased the degree of saturation of fatty acids of phosphatidylcholine. This nonionogenic surfactant also changed the fatty acid composition of phosphatidyl ethanolamine and decreased the ratio of phosphatidylcholine / phosphatidyl ethanolamine.

According to the data by Röderer (1987), TX100 (preparation with molecular mass 654, 10 ethoxyl monomers) and Triton X-405 (preparation with molecular mass 1976, 40 ethoxyl monomers) at different concentrations exceeding 100 mg/l (72 h) caused the death of chrysophyte cells of *Poterioochromonas malhamensis*. The action of three types of nonionogenic surfactants on plankton algae of three types was studied (alcohol ethoxylates (AE), Yamane et al. 1984). The values of EC_{50} were equal to 2–50 mg/l, which meant a higher toxicity than in the case of five types of anionic surfactants. The actions of nonionogenic surfactant Hydropol and of cationic and anionic surfactants on *Chlorella vulgaris* Beijer strain HPDP-19 were compared (Parshikova et al. 1994). Hydropol did not have a significant effect on photosynthetic evolution of oxygen and on oxygen consumption by algal cells in the dark (Parshikova et al. 1994).

Mortality of fish under the influence of different nonionogenic surfactants including nonylphenolethoxylates was studied. LC_{50} for technical NPE with 8–10 ethoxyl monomers was equal to 4–12 mg/l (48 h), while the products of their biological decomposition with a smaller number of ethoxyl monomers had LC_{50} equal to 1–4 mg/l (Huber 1985). Thus, the products of biological decomposition of these nonionogenic surfactants can be a greater biological hazard than the initial chemicals.

The value of LC_{50} caused by the effect of Nconol nonionogenic surfactant on embryo and larva of the mud loach *Misgurnus fossilis* (Lesyuk et al. 1983) was determined. These values for Neonols 2B 1317-12, 2B 1315-12, and AF-14 were equal to 35.0, 34.7, and 21.7 mg/l, respectively.

Water medium with NP8 nonionogenic surfactant at a concentration equal to 1 µM inhibited β-adrenergic reactions of gills of rainbow trout. Alcohol ethoxylate A7 has a similar effect (Stagg and Shuttleworth 1987).

TX100 was used for the so-called demembranation of spermatozoids of rainbow trout (Okuno and Morisawa 1989). The spermatozoids lost mobility after 30 s of incubation in aquatic medium containing 0.04% (w/v) of TX100. However, under certain conditions these demembranated spermatozoids retained their mobile ability.

Demembranation of spermatozoids of starfish *Asterina pectinifera* was also studied. They were incubated for 10 min in aquatic medium containing 0.02% of TX100. Even after such processing the spermatozoids retained the ability for mitosis of starfish eggs (Yamada and Hirai 1986).

The effect of waters containing nonionogenic surfactants on fungi is not well studied. Emulgen 120 (polyoxyethylene lauryl ether; CMC 0.007%) inhibited the growth of *Puricularia oryzae* fungus by 50% at a concentration of 0.01%. Emulgen 909 (polyoxyethylene nonylphenol ether; CMC, 0.005%) inhibited the growth of the fungus approximately by 90% at a concentration of 0.005%. Emulgen 108 (CMC, 0.004%) affected the growth in a similar manner. A combined effect of fungicides polioxyne B and kitezin P together with nonionogenic surfactant Emulgen 120 led to significant synergism (Watanabe et al. 1988). Hence, nonionogenic surfactants can enhance a negative effect of other chemicals that enter the aquatic environment.

4.1   Biological Effects of Nonionogenic Surfactants in a System with Bacteria

Biofouling of solid surfaces in seawater environment including hydrotechnical constructions and vessels is a serious and unsolved problem. The initial stage of biological fouling formation is colonization of the surface by marine bacteria. Marine prosthecobacteria of *Hyphomonas* genus occupy one of the leading places in this process (Weiner et al. 1985). After attaching to a surface, the cells of these bacteria begin to germinate daughter cells, which also attach to the surface, and the process takes the avalanche form. Attempts to find the chemicals that specifically inhibit the film forming bacteria including *Hyphomonas* did not yield any results. Therefore, the search for chemicals that are capable of affecting the growth of *Hyphomonas* is continuing. At the same time, *Hyphomonas* are interesting because they perform a function important for ecosystems participating in the mineralization processes of organic matter, thus contributing to self-purification of the aquatic medium.

The degree of sensitivity and stability of *Hyphomonas* to synthetic surfactants including nonionogenic surfactants has not been studied well enough. Detailed data on whether synthetic surfactant TX100 can negatively affect *Hyphomonas* were previously absent. We studied for the first time how TX100 affects the growth of *Hyphomonas* bacterial cultures of strains MHS-3 and VP-6.

The inhibitory effect was shown to increase within the range of TX100 concentrations from 1 to 50 mg/l, which is also characteristic for strain MHS-3 (Tables 4.1 and 4.2) and for VP-6 (Tables 4.3 and 4.4).

The obtained values of EC_{50} depended on the time period during which the incubation was performed. Inhibition of growth was 10–20% (Tables 4.1 and 4.3) at concentrations of TX100 from 1 to 10 mg/l. An increase in the concentration up to 50 mg/l significantly increased the degree of inhibition. For strain MHS-3 the value of EC_{50} (inhibition of the culture growth; incubation for 24 h and longer) was approximately 50 mg/l (Table 4.2).

The effect of concentration of 50 mg/l on strain VP-6 was even more notable and the inhibition was greater than 50% (Table 4.4).

These data indicate that by their sensitivity to TX100 the *Hyphomonas* bacteria (both strains) occupy an intermediate position between two strains of marine cyanobacteria that we studied (Waterbury and Ostroumov 1994). TX100 at concentrations of 5 mg/l inhibited both strains of *Hyphomonas* to a lesser extent than *Synechococcus* 7805. Both strains of prosthecobacteria were more sensitive to TX100 than the other strain of cyanobacteria *Synechococcus* 8103, because the latter was not inhibited at all by the concentration mentioned. The comparison of the effect of TX100 on prokaryotes and on the filtration activity of mollusks (see below) demonstrates that the latter organisms are much more vulnerable to comparatively low concentrations of nonionogenic surfactants.

The data obtained testify to the existence of additional aspects of ecological hazards of polluting the environment with TX100 and probably by other alkyl phenols in the situations of mass pollution, because the role of *Hyphomonas* in the formation of biological film in marine ecosystems is significant. Chemicals of

TX100 type can enter marine environment during oil mining on shelves, washing of tankers, introduction of chemical means for treatment of oil spills, fire extinguishing, and other types of extraordinary or emergency situations.

Table 4.1 Growth of *Hyphomonas* (strain MHS-3) at Triton X-100 (TX100) concentrations from 0 up to 10 mg/l (OD_{600}, the optical path 10 mm).

Time, days	TX100, mg/l	Beaker 1	Beaker 2	Beaker 3	Mean OD_{600}	Mean OD_{600}, %
1	0	0.145	0.128	0.129	0.134	100
	1	0.137	0.120	0.125	0.127	94.8
	5	0.113	0.114	0.120	0.116	86.6
	10	0.097	0.118	0.126	0.114	85.1
2	0	0.230	0.184	0.163	0.192	100
	1	0.204	0.172	0.164	0.180	93.8
	5	0.175	0.163	0.175	0.171	89.1
	10	0.156	0.160	0.164	0.160	83.3
3	0	0.268	0.200	0.180	0.216	100
	1	0.234	0.190	0.182	0.202	93.5
	5	0.164	0.186	0.197	0.172	79.6
	10	0.196	0.187	0.188	0.190	88.0
4	0	0.305	0.231	0.207	0.248	100
	1	0.280	0.209	0.210	0.233	94.0
	5	0.245	0.215	0.212	0.224	90.3
	10	0.233	0.208	0.168	0.203	81.9

Note: After inoculation, $OD_{600} = 0.098$. Incubation: 25°C, no mixing, in polystyrene tissue-culture tubes 17×100 mm, with snap caps (Fisher Scientific, Pittsburgh). Initial volume of medium, 10 ml per tube. Inoculate: 5% v/v, one-day culture, $OD_{600} = 0.193$. After inoculation, $OD_{600} = 0.098$. Medium: S-1. After 4-day incubation and measurements, sterile TX100 solution was added to beakers with initial concentration of 1 mg/l to make a final surfactant concentration of 51 mg/l.

Table 4.2 Growth of *Hyphomonas* (strain MHS-3) at Triton X-100 (TX100) concentrations from 0 up to 50 mg/l (OD_{600}, the optical path 10 mm).

Time, days	TX100, mg/l	Beaker 1	Beaker 2	Beaker 3	Mean OD_{600}	Mean OD_{600}, %
5	0	0.339	0.286	0.246	0.290	100
5	5	0.262	0.227	0.214	0.234	80.7
5	10	0.273	0.237	0.209	0.240	82.8
5 = 4 (1 mg/l) + 1 (51mg/l)	51*	0.192	0.133	0.140	0.155	53.4
6	0	0.375	0.330	0.266	0.324	100
6	5	0.301	0.279	0.268	0.283	87.3
6	10	0.314	0.292	0.204	0.270	83.3

Table 4.2 (continued)

Time, days	TX100, mg/l	Beaker 1	Beaker 2	Beaker 3	Mean OD_{600}	Mean OD_{600}, %
6 = 4 (mg/l) + 2 (51 mg/l)	51*	0.170	0.130	0.130	0.143	44.1
7	0	0.460	0.395	0.328	0.394	100
7	5	0.324	0.352	0.282	0.319	81.0
7	10	0.407	0.399	0.272	0.359	91.1
7 = 4 (1 mg/l) + 3 (51 mg/l)	51*	0.224	0.164	0.136	0.175	44.3
9	0	0.615	0.573	0.456	0.548	100
9	5	0.480	0.545	0.460	0.495	90.3
9	10	0.580	0.618	0.462	0.553	100.9
9 = 4 (1 mg/l) + 5 (51 mg/l)	51*	0.381	0.319	0.259	0.286	52.2

Note: For conditions of the experiment, see Note to Table 4.1.
*The variants designated as "TX100 51 mg/l" contained 1 mg/l surfactant for the first 4 days of incubation; then sterile solution of TX100 was added (to a final concentration of 51 mg/l), and incubation continued.

Table 4.3 Growth of *Hyphomonas* VP-6 in the presence of Triton X-100 (TX100), 0–10 mg/l (OD_{600}, the optical path 10 mm).

Time, days	TX100, mg/l	Beaker 1	Beaker 2	Beaker 3	Mean OD_{600}	Mean OD_{600}, %
1	0	0.047	0.061	0.056	0.055	100
	1	0.050	0.049	0.050	0.050	90.9
	5	0.051	0.049	0.054	0.051	92.7
	10	0.055	0.042	0.046	0.048	87.3
2	0	0.077	0.080	0.075	0.077	100
	1	0.084	0.077	0.073	0.078	101.3
	5	0.075	0.069	0.075	0.073	94.8
	10	0.079	0.067	0.069	0.072	93.5
3	0	0.112	0.082	0.102	0.099	100
	1	0.121	0.110	0.100	0.110	111.1
	5	0.114	0.098	0.107	0.106	107.1
	10	0.111	0.096	0.102	0.103	104.0
4	0	0.159	0.155	0.139	0.151	100
	1	0.150	0.110	0.100	0.120	79.5
	5	0.148	0.100	0.134	0.127	84.1
	10	0.141	0.120	0.126	0.129	85.4

Note: Incubation: 25°C, no mixing, in polystyrene tissue-culture tubes 17 × 100 mm, with snap caps (Fisher Scientific, Pittsburgh). Initial volume of medium, 10 ml per tube. Inoculate: 5% v/v, one-day culture, OD_{600} = 0.193. After inoculation, OD_{600} = 0.098. Medium: S-1. After 4-day incubation and measurements, sterile TX100 solution was added to beakers with initial concentration of 1 mg/l to make a final surfactant concentration of 51 mg/l.

Table 4.4 Growth of *Hyphomonas* VP-6 in the presence of Triton X-100 (TX100), 0–50 mg/l (OD_{600}, the optical path 10 mm).

Time, days	TX100, mg/l	Beaker 1	Beaker 2	Beaker 3	Mean OD_{600}	Mean OD_{600}, %
5	0	0.195	0.221	0.199	0.205	100
5	5	0.178	0.145	0.144	0.157	76.6
5	10	0.154	0.134	0.150	0.146	71.2
5	51*	0.087	0.083	0.093	0.088	42.2
6	0	0.252	0.381	0.222	0.285	100
6	5	0.229	0.185	0.191	0.202	70.9
6	10	0.201	0.175	0.191	0.189	66.3
6	51*	0.113	0.106	0.107	0.108	37.9
7	0	0.323	0.513	0.327	0.388	100
7	5	0.314	0.244	0.226	0.261	67.3
7	10	0.261	0.210	0.243	0.238	61.3
7	51*	0.203	0.189	0.159	0.181	46.6
9	0	0.601	0.690	0.538	0.610	100
9	5	0.419	0.296	0.307	0.341	55.9
9	10	0.361	0.290	0.321	0.324	53.1
9	51*	0.445	0.407	0.363	0.405	66.4

Note: For conditions of the experiment, see Note to Table 4.3.
*The variants designated as "TX100 51 mg/l" contained 1 mg/l surfactant for the first 4 days of incubation; then sterile solution of TX100 was added (to a final concentration of 51 mg/l), and incubation continued.

The mechanism of the interaction of TX100 with bacterial cells requires further investigation. There are indications that a significant part of molecular mechanisms of TX100 interaction (similarly to many other surfactants) is related to the effect on biological membranes (Stavskaya et al. 1988). This suggestion agrees with the data of some later studies. TX100 and other substances of this class (whose molecules have 7–13 polymerized ethylene oxide monomers as structural components) increased the sensitivity of 22 strains of *Staphylococcus aureus*, *S. epidermis*, and *S. sciuri* to oxacyllin (Suzuki et al. 1997).

There is an opinion in literature that nonionogenic surfactants are comparatively low toxic for bacteria, which is manifested in the ability of many species to endure significant concentrations of nonionogenic surfactants (Stavskaya et al. 1988). However, the hazard of xenobiotics for organisms can manifest itself in the other form. The indications that nonionogenic surfactants can cause mutations of *Salmonella typhimurium* and *Bacillus subtilis* reading frames are interesting (Naumova et al. 1981). This work carried out at the Kazan State University also revealed the induction of prophage from lysogenic bacteria under the influence of surfactants both under conditions of laboratory cultivation and in the process of biological purification of sewage waters. The authors observed high titers of virulent particles of phage already after 4 h of contact of lysogenic bacteria with the tested chemical. The exact names of synthetic surfactants were not given in this paper. Direct correlation was found between mutagenic activity of the studied synthetic surfactants and their ability to induce prophages.

4.2   Biological Effects of Nonionogenic Surfactants on Phytoplankton Organisms

The effect of nonionogenic surfactants on marine cyanobacteria and diatom algae was studied.

4.2.1   Biological effects of nonionogenic surfactants in a system with cyanobacteria

Marine coccoid cyanobacteria contribute significantly to the total biomass and productivity of marine phytoplankton. Unicellular coccoid cyanobacteria of the genus *Synechococcus* frequently make up 20–80% of the total biomass of picoplankton (Sherr and Sherr 1991; cited from Waterbury and Ostroumov 1994), while their number can be as high as 10–100 thousand cells/ml. During the blooming of cyanobacteria in the eutrophic parts of such large estuaries as Chesapeake Bay in the U.S., the density of the cells exceeded 5 million in one ml of aquatic medium (Falkenhayn and Hass 1990, cited from Waterbury and Ostroumov 1994). Cyanobacteria of this genus accumulate such elements as Sn, Hg, and Pu with the concentration factor (v/v) of the order of 1 million (Fisher 1985, cited from Waterbury and Ostroumov 1994), which is important for self-purification of seawater. The concentration of hydrogen peroxide can be important for self-purification of marine ecosystems, and it is relevant in this relation that cyanobacteria *Synechococcus* accelerate the decomposition of hydrogen peroxide more effectively than the majority of other species of phytoplankton studied (Kim et al. 1992, cited from Waterbury and Ostroumov 1994).

        To date, the impact of nonionogenic surfactants on marine cyanobacteria has been studied insufficiently. We first studied the effect of nonionogenic surfactant TX100 on the strains of marine *Synechococcus*, which were different from each other in terms of their pigments and absorption spectra (Waterbury and Ostroumov 1994). Our experiments demonstrated that (Tables 4.5 and 4.6) the growth of cyanobacteria changed in the presence of nonionogenic surfactant TX100. The character of the changes depended on the cyanobacterial strain studied and on the concentration of nonionogenic surfactant. One of the strains (*Synechococcus* WH 7805) was more sensitive than the other (*Synechococcus* WH 8103). The strains were significantly less sensitive to TX100 than the filtration activity of mussels (see below).

        The demonstrated stimulation of the growth of phytoplankton cyanobacteria agrees with the data of another author who studied the effect of TX100 on phytoplankton organisms. In the experiments by Wong (1985) natural water from nine Canadian lakes with addition of 10% of Bristol's medium was used as the medium

Table 4.5 Changes in the optical density of the culture *Synechococcus* sp. 7805 under the action of surfactant Triton X-100 (TX100) (measurements were made before and after sucrose addition).

Cultivation time, days (measurement conditions)	Concentration of TX100, mg/l	569–572 nm		679–681 nm	
		optical density units	%	optical density units	%
4 (before sucrose was added)	0	0.224	100	0.143	100
	0.5	0.249	111.2	0.175	122.4
	5.0	0.059	26.3	0.041	28.7
4 (after sucrose was added)	0	0.135	100	0.084	100
	0.5	0.162	120.0	0.110	131.0
	5.0	0.057	42.2	0.059	70.2
6 (before sucrose was added)	0	0.465	100	0.310	100
	0.5	0.379	81.5	0.269	86.8
	5.0	0.105	22.6	0.074	23.9
6 (after sucrose was added)	0	0.230	100	0.145	100
	0.5	0.217	94.3	0.144	99.3
	5.0	0.069	30.0	0.044	30.3
13 (before sucrose was added)	0	0.428	100	0.311	100
	0.5	0.963	225	0.675	217
	5.0	0.092	21.5	0.066	21.2
13 (after sucrose was added)	0	0.178	100	0.120	100
	0.5	0.441	247.8	0.294	245.0
	5.0	0.054	30.3	0.039	32.5

Table 4.6 Changes in the optical density of the culture *Synechococcus* sp. 8103 under the action of surfactant Triton X-100 (TX100) (measurements were made before and after sucrose addition).

Cultivation time, days (measurement conditions)	Concentration of TX100, mg/l	438–440 nm		679–681 nm	
		optical density units	%	optical density units	%
13 (before sucrose was added)	0	0.871	100	0.398	100
	0.5	1.173	134.7	0.557	139.9
	5.0	1.284	147.4	0.598	150.3
13 (after sucrose was added)	0	0.433	100	0.186	100
	0.5	0.595	137.4	0.252	135.5
	5.0	0.650	150.1	0.280	150.5

for growing *Chlorella fusca* Shihers et Krauses. An addition of TX (0.4–1.0 mM, i.e., approximately 240–600 mg/l) caused stimulation of the growth of *C. fusca*. A 10- to 20-fold increase was observed in the growth under the influence of TX100 compared to the medium without TX100 (i.e., the difference was 1000–2000%) (Wong 1985).

4.2.2 Biological effects of nonionogenic surfactants in a system with diatomic algae

A great role of diatoms in marine ecosystems makes interesting the study of how non-ionogenic surfactants affect them. We have chosen *Thalassiosira pseudonana* Hasle & Heimdal 1970 [=*Cyclotella nana* Guillard clone 3H (in Guillard and Ryther 1962)] as the test species. This species (Order Biddulphiales, Suborder Coscinodiscinae, Family Thalassiosiraceae) is a characteristic representative of diatoms. The family Thalassiosiraceae includes both marine and freshwater species of plankton diatoms. The genus includes more than 100 species.

The calculation of cells in a unit volume after certain periods of cultivation in the presence of TX100 showed that the concentration of this nonionogenic surfactant within 0.1–10 mg/l had a negative effect on algal growth (Table 4.7). A decrease in the specific rate of growth was demonstrated (Table 4.8).

Table 4.7   Effect of surfactant Triton X-100 on the density of *Thalassiosira pseudonana* Hasle & Heimdal 1970 culture (10^5 cells/ml; standard error is given in brackets).

Triton X-100, mg/l	Time, days			
	7	8	11	14
0	3.20 (0.26)	9.26 (2.27)	5.00 (0.46)	4.25 (0.29)
0.1	2.53 (0.14)	4.65 (0.35)	5.76 (0.72)	3.37 (0.42)
1	1.22 (0.31)	3.74 (0.35)	5.43 (0.22)	2.93 (0.33)
10	1.29 (0.22)	1.04 (0.23	nd	nd

Note: nd, not determined (as only 0–2 cells in a volume of 10^{-4} ml were determined in the counting chamber).

Comparison of the results of investigating the effect of TX100 on diatoms and cyanobacteria (see above) indicated that equal concentrations of nonionogenic surfactants had absolutely different effects. TX100 concentrations of 1 mg/l induced a pronounced inhibition of diatoms. On the other hand, a much greater concentration of 5 mg/l not only failed to inhibit but stimulated the growth of cyanobacteria *Synechococcus* sp. 8103.

As real algobacterial planktonic communities include representatives of both cyanobacteria and diatoms, one could not help devising an idea that under conditions of varidirectional effects at certain levels of aquatic ecosystem pollution there are prerequisites for changes in the relations between different groups of phytoplankton.

The changes in the composition of the communities and relations between the species are characteristic types of anthropogenic disturbances in the ecosystems.

Table 4.8 Change of the specific growth rate of *Thalassiosira pseudonana* (μ) at various concentrations of surfactant Triton X-100.

Concentration of Triton X-100, mg/l	μ for a given period of time, days	
	7	8
0 (control)	0.488 (100)	0.619 (100)
0.1	0.440 (90.16)	0.494 (79.81)
1	0.290 (59.43)	0.455 (73.51)
10	0.301 (61.68)	0.224 (36.19)

Note: In brackets, the growth rate μ as percentage of the control.

In the discussion of the results of investigating the effect of nonionogenic surfactants on phytoplankton, it is worth considering the data obtained from other organisms. In the experiments carried out by Röderer (1987), TX100 (preparation with molecular mass 654, 10 ethoxyl monomers) and TX405 (preparation with molecular mass 1976, 40 ethoxyl monomers) at different concentrations caused the death of the cells of *Poterioochromonas malhamensis* chrysophytes. No living cells were found in the aquatic culture of algae after 72 h of incubation when 124.3 mg/l of TX100 or 177 mg/l TX405 was present in the medium. We found a decrease in the biological activity of nonionogenic surfactants when the molecular mass of this chemical significantly increases. This agrees with the decrease in the biological activity of surfactants if we replace low polymeric surfactants with high polymeric ones (e.g., CHMA, see Section 3.3).

The effect of three types of nonionogenic surfactants on plankton algae was studied (AE, Yamane et al. 1984). The values of EC_{50} were equal to 2–50 mg/l, which means a greater toxicity than in the case of five types of anionic surfactants (AS, LAS) studied in the same work. The values of EC_{50} for anionic surfactants were 10–100 mg/l (Yamane et al. 1984).

Parshikova et al. (1994) compared the effects of cationic surfactant catamine AB (alkyl methyl benzyl ammonium chloride), anionic surfactant sodium salt of dodecyl sulfonated acid, and nonionogenic surfactant Hydropol on *Chlorella vulgaris* Beijer (strain HPDP-19). Under conditions of the experiments, Hydropol at concentrations of 10 mg/l had no significant effect on the photosynthetic evolution of oxygen and on the consumption of oxygen in the dark by the algal cells at their biomass of 79 mg/l and exerted a weak inhibition effect on both processes at a biomass of 56 mg/l. No significant changes in the content of chlorophyll *a* (μg/l) were observed in the first 24 h of culture growth under the influence of 5 mg/l of Hydropol. After 48 h, a slight inhibition was observed (approximately 2300 μg/l in the control and approximately 1800 μg/l in the experiment with Hydropol (Parshikova et al. 1994).

4.3 Biological Effects of Nonionogenic Surfactants on Higher Eukaryotes

4.3.1 Biological effects of nonionogenic surfactants in the systems with angiosperm plants

As we mentioned above, the search for and approbation of new systems of biological testing are required in order to decrease the cost of biotesting of new substances and conform to the humane considerations that require the application of systems alternative to traditional biological tests with animals. Therefore, one is interested in systems that use plant objects. We found no information in the literature concerning the impacts of aquatic media containing nonionogenic surfactants on angiosperm plants belonging to Angiospermae (Magnoliophyta).

We obtained some data from plant seedlings. It has been shown that germination of buckwheat seeds *Fagopyrum esculentum* in an aquatic medium containing non-ionogenic surfactant TX100 decreases (Table 4.9). The coefficient of inhibition of germination was used in the processing of the results of experiments. This coefficient (GIC = germination inhibition coefficient) was calculated by the following relation

$$GIC = \frac{M_x - M_0}{n - M_0} \cdot 100,$$

where M_x is the number of seeds that did not germinate at a given concentration x of the tested xenobiotic; M_0 is the number of seeds that did not germinate in the control; n is the number of seeds taken for testing at each concentration.

It was also shown in this series of research projects that TX100 at concentrations of 0.06 mg/l and greater inhibits elongation of the seedlings of *F. esculentum*. In the experiment with the transfer of 21-hour seedlings into the tested aquatic medium their mean increment 22 h after the transfer (the total time after the germination was 43 h = 21 + 22 h) was 666.8% at a concentration of nonionogenic surfactant equal to 0.0625 mg/l, while the increment in the control was 936.4%. Thus, inhibition of the rate of elongation of the seedlings was observed, which regularly increased with the increase in the concentration of TX100 (Table 4.10).

Studies of the effect of TX100 concentrations on the inhibition of the seedlings allowed us to estimate the EC_{50}. The value of EC_{50} for slowing down the growth compared to the control during a period of 21–26 h (i.e., during the first 5 h after transferring the seedlings into the solution of nonionogenic surfactant) was approximately 0.36 µl/ml (i.e., 0.36 ml/l or 360 mg/l). A comparatively large value of EC_{50} suggests that this type of plant can be used for phytoremediation. The value of EC_{50} slightly decreases with time: in 26–43 h (5–17 h after transferring the seedlings into the solution of nonionogenic surfactant) EC_{50} was approximately equal to 0.14 µl/ml.

We also showed inhibition of elongation of the seedlings of other species of plants under the influence of nonionogenic surfactant TX100, e.g., *Lepidium sativum* (Ostroumov 1999).

Table 4.9   Change of the share of germinating seeds under the action of Triton X-100 on seeds of *Fagopyrum esculentum* (cultivar Shatilovskaya 5).

Concentration of Triton X-100, µl/ml	Number of seeds			GIC, %
	total	germinated	ungerminated, %	
0 (control)	68	44	35.3	–
0.0625	67	37	44.8	13.95
0.125	68	18	73.5	59.09
0.25	68	7	89.7	84.09
0.5	68	5	92.7	88.64

Note: Seeds were germinated on Petri dishes (17 seeds per dish) containing 10 ml solution of nonionogenic surfactant in distilled water or 10 ml of distilled water (in the control). GIC = $[M_0 - M_c)/(N - M_c)] \times 100$, where N is the number of seeds taken for testing each concentration; M_c is the number of seeds ungerminated in the control; M_0 is the number of seeds ungerminated at a given concentration of surfactant. For details, see *Problems of Ecological Monitoring and Modeling of Ecosystems*, 1986, Vol. 9 (in Russian).

Table 4.10   Mean length x (mm) of *Fagopyrum esculentum* (cultivar Shatilovskaya 5) seedlings under the action of Triton X-100 after various periods of time.

Concentration of Triton X-100, µl/ml	$t_1 = 21$ h		$t_1 = 26$ h		$t_1 = 43$ h	
	x	confidence limit	x	confidence limit	x	confidence limit
0 (control)	5.0	0.46	10.9	0.74	46.8	3.05
	100%		218.4%		936.4%	
0.0625	5.4	0.62	10.6	0.98	35.7	4.03
	100%		197.2%		666.8%	
0.125	5.8	0.52	10.7	0.93	29.9	4.42
	100%		184.1%		515.5%	
0.25	5.45	0.46	8.9	1.04	18.33	3.72
	100%		163.8%		336.2%	
0.5	5.6	0.65	7.85	0.88	15.5	3.30
	100%		141.4%		279.7%	

Note: Seeds were germinated on distilled water. After 21 h, seedlings were transferred on TX100 solution. Means of two independent experiments are given. Total of 40 seedlings was tested at each concentration. For details, see *Problems of Ecological Monitoring and Modeling of Ecosystems*, 1986, Vol. 9 (in Russian).

In further investigations we have found that, along with the general slowing down of the growth of seedlings affected by TX100 and other surfactants, a destruction of morphogenesis of rhizoderm cells, which normally form root hairs, occurs. The morphogenesis of cells leads to a decrease in the ability to attach to the substrate under conditions of the experiment. The inhibition of the formation of root hairs and related disturbance of the ability to attach at the substrate was observed in several species including *F. esculentum* (Table 4.11) and *S. alba* (Table 4.12). It is characteristic that this effect was observed at concentrations that did not exert similar pronounced inhibitory effects on the elongation of seedlings (*S. alba*, Table 4.13). This indicated a greater sensitivity of the test system based on the recording of the effect of the formation of rhizoderm root hairs.

A distortion in the formation of root hairs was demonstrated during the incubation of the seedlings of buckwheat, wheat *Triticum aestivum* and other species in aquatic media containing surfactants. Among the effects revealed in the course of the analysis of the effect of nonionogenic surfactants on the plants is the disturbance of the ability of the seedlings to attach to a substrate (filter paper in the experiment) and the disturbance of the ability to keep hypocotyls in vertical position. The latter effect was observed, in particular, for the seedlings of gold-of-pleasure *Camelina sativa* (Ostroumov and Maksimov 1988).

Table 4.11 Number and percentage of *F. esculentum* (cultivar Shatilovskaya 5) seedlings unattached to substrate at various concentrations of Triton X-100.

Concentration of TX100, µl/ml	Variant of experiment					
	I			II		
	Unattached seedlings		Total seedlings, number	Unattached seedlings		Total seedlings, number
	number	%	number	number	%	number
0	12	18.75	64	0	0	39
0.0625	89	94.68	94	18	45	40
0.125	28	100	28	40	100	40
0.25	7	100	7	40	100	40
0.5	4	100	4	40	100	40

Note: For details, see *Izv. Akad. Nauk SSSR, Ser. Biol.*, **4**: 571–575 (in Russian).

Table 4.12 Number and percentage of *S. alba* VNIIMK seedlings unattached to substrate at various concentrations of Triton X-100.

Concentration of Triton X-100, µl/ml	Unattached seedlings (72 h)		Total seedlings, number
	number	%	
0	2	6.67	30
0.0625	30	100.00	30
0.125	29	100.00	29
0.25	30	100.00	30
0.5	30	100.00	30

Table 4.13 Effect of Triton X-100 on the growth of *S. alba* VNIIMK.

TX100, μl/ml	Measurement time								n
	45 h				51 h				
	x, mm	CL	SD	CV, %	*x*, mm	CL	SD	CV, %	
0	12.70	1.76	4.73	37.23	16.80	1.98	5.31	31.63	30
0.0625	11.00	1.47	3.85	35.04	14.51	1.73	4.53	31.24	29
0.125	9.34	1.04	2.74	29.34	11.86	1.25	3.29	27.75	29
0.25	6.90	1.02	2.75	39.80	8.57	1.22	3.27	38.13	30
0.5	4.53	0.60	1.61	35.58	5.47	0.77	2.08	38.05	30

Note: x, mean length; CL, confidence limit; SD, standard deviation; CV, coefficient of variation, %; *n*, number of seedlings.

Along with the study of the effect of nonionogenic surfactants on plants it was interesting to study the influence of nonionogenic surfactants on representatives of the animal world. Taking into account the important role of mollusks and their filtration activity, we studied the effect of nonionogenic surfactants on the ability of bivalve mollusks to filter water and remove plankton cells from water.

4.3.2 Biological effects of nonionogenic surfactants in a system with mollusks

4.3.2.1 Unio tumidus

The effect of nonionogenic surfactant TX100 on the ability of freshwater mollusks *Unio tumidus* to filter water and remove phytoplankton cells from water was studied. Three controls were set, in which the experimental system included: (1) mollusks and phytoplankton without additions of nonionogenic surfactants; (2) plankton and nonionogenic surfactant without mollusks; and (3) plankton without mollusks and without nonionogenic surfactants. The experiments demonstrated that normal filtration of water was observed in the first of the controls. As a result, phytoplankton cells were removed from water, and its optical density decreased.

No pronounced decrease in the optical density was observed in the second system, which indicated that the studied concentration of nonionogenic surfactants did not cause any notable destruction of the pigment apparatus of the cells during the experiment. A decrease in the optical density in the experiment can be related specifically to filtration activity. In the third control, no significant changes in the optical density were observed, which indicated that during the experiment the concentration of phytoplankton cells in water (in the absence of the effect of filter feeders) remained at a stable level. Thus, the results of the observations in all three variants of the experiment confirm the conclusion that the effect of nonionogenic surfactants on filtration activity of the mollusks was the sole cause for the decrease in the optical density of water observed in the experiment.

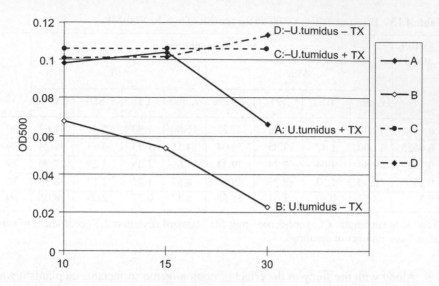

Figure 4.1   Effect of TX100 (5 mg/l) on water filtration by *Unio tumidus* (+ *Scenedesmus quadricauda*).

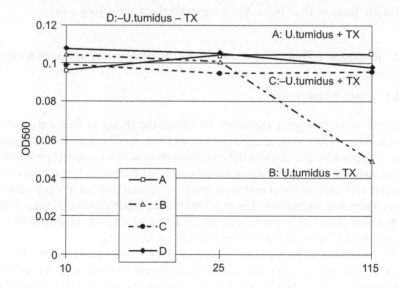

Figure 4.2   Effect of TX100 (5 mg/l) on water filtration by *Unio tumidus* (+ *Synechocystis* 6803).

In the experiment presented in Fig. 4.1, green algae (*Scenedesmus quadricauda*) were introduced into the system. The next experiment (Fig. 4.2) used cyanobacteria.

It is indicative that the main regularities were kept regardless of which phytoplankton organisms were present: green algae (Fig. 4.1) or cyanobacteria (Fig. 4.2). The author thanks Dr N.N. Kolotilova for assistance in the experiments.

4.3.2.2 *Mytilus edulis*

The effect of nonionogenic surfactants on the rate of filtration by mussels was not studied before. Our experiments demonstrated that the presence of nonionogenic surfactant TX100 in the medium for mussels' incubation caused significant decrease in the rate of water filtration. Inhibition of the rate of filtration depended on the concentration of nonionogenic surfactant. At a concentration of 0.5 mg/l the difference between the control and the experiment was seen only slightly after 30 min of exposure (Table 4.15). Thereafter (after 60 and 90 min of exposure) the difference between the control and experiment increased (Tables 4.16 and 4.17).

At a concentration of 1 mg/l the difference between the control and experiment samples became more notable. One can calculate that the effect of TX100 on the efficiency of the removal (EER, see Chapter 3) was equal to 163.1% after 30 min, (calculated based on the data from Table 4.18), equal to 236.2% after 60 min (calculated based on the data from Table 4.19), and equal to 309.4% after 90 min (calculated based on the data from Table 4.20).

Table 4.14 Effect of nonionogenic surfactant Triton X-100 on the development of *Camelina sativa* VNIIMK-P-17 seedlings (in brackets, percentage of seedlings with normal hypocotyls).

Time, h	Number of seedlings with vertical hypocotyls (numerator) with respect to the total number of seedlings (denominator)	
	control (distilled water)	Triton X-100 solution (0.25 µl/ml)
48	17 34 (50%)	0 29 (0.0%)
65	30 34 (88.2%)	1 29 (3.4%)

Note: Seeds (29–34) were put into Petri dishes (10 cm), 7-ml test solution was added to each dish; incubation in the dark, 27 °C; germination efficiency, 100%. For details, see Ostroumov 1990, *Vest. Mosk. Universiteta* (in Russian).

Table 4.15 Number of *Isochrysis galbana* algal cells in beakers with mussels *Mytilus edulis* after a 30-min water filtration by the mollusks under the action of nonionogenic surfactant Triton X-100 (0.5 mg/l).

Beaker No.	Presence or absence of TX100, 0.5 mg/l	Number of algal cells in 0.5 ml	Average number of algal cells in 0.5 ml
1	+	4266; 4165; 4139	4190.0
3	+	2712; 2508; 2516	2578.7
5	+	4850; 4708; 4663	4740.3
7	+	7890; 8070; 8117	8025.7
1, 3, 5, 7	Average number of cells for four beakers with surfactant (1, 3, 5, 7)		4883.7 (standard error, 597.6)

Table 4.15   (continued)

Beaker No.	Presence or absence of TX100, 0.5 mg/l	Number of algal cells in 0.5 ml	Average number of algal cells in 0.5 ml
2	–	5415; 5274; 5273	5320.7
4	–	3330; 3442; 3451	3407.7
6	–	3616; 3570; 3669	3618.3
8	–	4558; 4435; 4588	4527.0
2, 4, 6, 8	Average number of cells for four control beakers (2. 4. 6. 8)		4218.4 (standard error, 230.7)

Note: $p = 0.158$.
Experimental conditions: Initial concentration of cells per 0.5 ml, 13416, 13472, 13228 (average 13372). Temperature, 16°C. Number of cells was counted by a Coulter counter. Number of animals, 16. Eight 2-liter experimental beakers (Nos 1 to 8) contained 2 mussels each, total 16 mollusks. Average readings of the Coulter counter in filtered seawater, 673.67 (758; 643; 620). Total wet weight of mussels (with shells), in g: beaker 1, 17.30; beaker 2, 17.26; beaker 3, 18.02; beaker 4, 18.33; beaker 5, 14.81; beaker 6, 14.66; beaker 7, 14.24; beaker 8, 14.96.

Table 4.16   Number of *Isochrysis galbana* algal cells in beakers with mussels *Mytilus edulis* after a 60-min water filtration by the mollusks under the action of nonionogenic surfactant Triton X-100 (0.5 mg/l).

Beaker No.	Presence or absence of TX100, 0.5 mg/l	Number of algal cells in 0.5 ml	Average number of algal cells in 0.5 ml
1	+	1336; 1406; 1238	1326.7
3	+	709; 680; 621	670.0
5	+	1853; 1776; 1728	1785.7
7	+	4558; 4696; 4659	4637.7
1, 3, 5, 7	Average number of cells for four beakers with surfactant (1, 3, 5, 7)		2105 (standard error, 457.1)
2	–	1857; 1763; 1765	1795.0
4	–	808; 903; 879	863.3
6	–	1007; 977; 998	994.0
8	–	1656; 1598; 1680	1644.7
2, 4, 6, 8	Average number of cells for four control beakers (2, 4, 6, 8)		1324.3 (standard error, 121.6)

Note: $p = 0.0614$.
Experimental conditions: Initial concentration of cells per 0.5 ml, 13416, 13472, 13228 (average 13372). Average readings of the Coulter counter in filtered seawater, 673.67 (758; 643; 620). Total wet weight of mussels (with shells), in g: beaker 1, 17.30; beaker 2, 17.26; beaker 3, 18.02; beaker 4, 18.33; beaker 5, 14.81; beaker 6, 14.66; beaker 7, 14.24; beaker 8, 14.96. (The other conditions as in experiment at 0.5 mg/l and exposure 30 min, see above.)

Table 4.17 Number of *Isochrysis galbana* algal cells in beakers with mussels *Mytilus edulis* after a 90-min water filtration by the mollusks under the action of nonionogenic surfactant Triton X-100 (0.5 mg/l).

Beaker No.	Presence or absence of TX100, 0.5 mg/l	Number of algal cells in 0.5 ml	Average number of algal cells in 0.5 ml
1	+	427; 451; 468	448.7
3	+	335; 338; 362	345
5	+	795; 766; 819	793.3
7	+	2806; 2743; 2793	2780.7
1, 3, 5, 7	Average number of cells for four beakers with surfactant (1, 3, 5, 7)		1091.9 (standard error, 298.3)
2	–	727; 684; 716	709
4	–	347; 337; 348	344
6	–	359; 398; 456	404.3
8	–	638; 659; 716	671
2, 4, 6, 8	Average number of cells for four control beakers (2, 4, 6, 8)		532.1 (standard error, 48.9)

Note: Significance of difference of the means in the control and experiment: $p = 0.044$; the *difference is significant* (significance level, over 95%).
Experimental conditions: Initial concentration of cells per 0.5 ml, 13416, 13472, 13228 (average 13372). Average readings of the Coulter counter in filtered seawater, 673.67 (758; 643; 620). Total wet weight of mussels (with shells), in g: beaker 1, 17.30; beaker 2, 17.26; beaker 3, 18.02; beaker 4, 18.33; beaker 5, 14.81; beaker 6, 14.66; beaker 7, 14.24; beaker 8, 14.96. (The other conditions as in experiment at 0.5 mg/l and exposure 30 min, see above.)

Table 4.18 Number of *Isochrysis galbana* algal cells in beakers with mussels *Mytilus edulis* after a 30-min water filtration by the mollusks under the action of nonionogenic surfactant Triton X-100 (1 mg/l).

Beaker No.	Presence or absence of TX100, 1 mg/l	Number of algal cells in 0.5 ml	Average number of algal cells in 0.5 ml
1	+	12533; 12369; 12555	12485.7
3	+	9630; 9600; 9577	9602.3
5	+	6426; 6474; 6509	6469.7
7	+	6210; 6338; 6283	6277.0
1, 3, 5, 7	Average number of cells for four beakers with surfactant (1, 3, 5, 7)		8708.7 (standard error, 768.7)
2	–	–	–
4	–	5153; 5134; 5261	5182.7
6	–	5734; 5831; 5743	5769.3
8	–	5044; 5032; 5136	5070.7

Table 4.18   (continued)

Beaker No.	Presence or absence of TX100, 1 mg/l	Number of algal cells in 0.5 ml	Average number of algal cells in 0.5 ml
4, 6, 8	Average number of cells for three control beakers (4, 6, 8)		5340.9 (standard error, 109.7)

Note: Significance of difference of the means in the control and experiment: $p = 0.0006$; the *difference is significant* (significance level, >99.9%).
Experimental conditions: Initial concentration of cells per 0.5 ml, 18260.7 (standard error, 43.6). Total wet weight of mussels (with shells), in g: beaker 1, 17.7; beaker 3, 18.4; beaker 4, 17.9; beaker 5, 17.3; beaker 6, 18.2; beaker 7, 17.9; beaker 8, 18.35. (The other conditions as in experiment at 0.5 mg/l and exposure 30 min, see above.)

Table 4.19   Number of *Isochrysis galbana* algal cells in beakers with mussels *Mytilus edulis* after a 60-min water filtration by the mollusks under the action of nonionogenic surfactant Triton X-100 (1 mg/l).

Beaker No.	Presence or absence of TX100, 1 mg/l	Number of algal cells in 0.5 ml	Average number of algal cells in 0.5 ml
1	+	5530; 5483; 5200	5404.3
3	+	3158; 3149; 3238	3181.7
5	+	2300; 2347; 2199	2282.0
7	+	1682; 1698; 1733	1704.3
1, 3, 5, 7	Average number of cells for four beakers with surfactant (1, 3, 5, 7)		3143.1 (standard error, 425.2)
2	–	–	–
4	–	1379; 1304; 1227	1303.3
6	–	1480; 1560; 1569	1536.3
8	–	1124; 1177; 1157	1152.7
4, 6, 8	Average number of cells for three control beakers (4, 6, 8)		1330.8 (standard error, 58.0)

Note: Significance of difference of the means in the control and experiment: $p = 0.0007$; the *difference is significant* (significance level, >99.9%).
Experimental conditions: Initial concentration of cells per 0.5 ml, 18260.7 (standard error, 43.6). Total wet weight of mussels (with shells), in g: beaker 1, 17.7; beaker 3, 18.4; beaker 4, 17.9; beaker 5, 17.3; beaker 6, 18.2; beaker 7, 17.9; beaker 8, 18.35. (The other conditions as in experiment at 0.5 mg/l and exposure 30 min, see above.)

Table 4.20   Number of *Isochrysis galbana* algal cells in beakers with mussels *Mytilus edulis* after a 90-min water filtration by the mollusks under the action of nonionogenic surfactant Triton X-100(1 mg/l).

Beaker No.	Presence or absence of TX100, 1 mg/l	Number of algal cells in 0.5 ml	Average number of algal cells in 0.5 ml
1	+	2713; 2550; 2569	2610.7
3	+	1273; 1288; 1349	1303.3

Table 4.20 (continued)

Beaker No.	Presence or absence of TX100, 1 mg/l	Number of algal cells in 0.5 ml	Average number of algal cells in 0.5 ml
5	+	744; 722; 706	724.0
7	+	543; 543; 607	564.3
1, 3, 5, 7	Average number of cells for four beakers with surfactant (1, 3, 5, 7)		1300.6 (standard error, 464.7)
2	–	–	–
4	–	403; 484; 451	446.0
6	–	442; 417; 454	437.7
8	–	350; 419; 364	377.7
4, 6, 8	Average number of cells for three control beakers (4, 6, 8)		420.4 (standard error, 21.5)

Note: Significance of difference of the means in the control and experiment: $p = 0.002$; the *difference is significant* (significance level, >99%).
Experimental conditions: Initial concentration of cells per 0.5 ml, 18260.7 (standard deviation, 43.6). Total wet weight of mussels (with shells), in g: beaker 1, 17.7; beaker 3, 18.4; beaker 4, 17.9; beaker 5, 17.3; beaker 6, 18.2; beaker 7, 17.9; beaker 8, 18.35. (The other conditions as in experiment at 0.5 mg/l and exposure 30 min, see above.)

The increasing values of EER, when the concentration increased to 2 mg/l, can be calculated in a similar manner. After 30 min of EER exposure, the removal was equal to 193.9% (calculated based on the data from Table 4.21), and equal to 302.7% after 60 min, and to 474.0% after 90 min (calculated based on the data from Tables 4.22 and 4.23).

The effect of nonionogenic surfactant became even stronger at a concentration of up to 4 mg/l. The calculated values of EER were equal to 818.9% after 60 min of exposure, equal to 1505.6% after 90 min of exposure, and 2191% after 120 min of exposure (calculated based on the data from Tables 4.24–4.26).

Table 4.21 Number of *Isochrysis galbana* algal cells in beakers with mussels *Mytilus edulis* after a 30-min water filtration by the mollusks under the action of nonionogenic surfactant Triton X-100 (2 mg/l).

Beaker No.	Presence or absence of TX100, 2 mg/l	Number of algal cells in 0.5 ml	Average number of algal cells in 0.5 ml
1	+	8679; 8543; 8622	8614.7
3	+	11923; 11943; 11912	11926.0
5	+	11898; 12020; 11833	11917.0
7	+	10006; 10045; 10116	10055.7
1, 3, 5, 7	Average number of cells for four beakers with surfactant (1, 3, 5, 7)		10628 (standard error, 419.4)
2	–	7476; 7486; 7587	7516.3
4	–	4320; 4230; 4250	4266.7

Table 4.21   (continued)

Beaker No.	Presence or absence of TX100, 2 mg/l	Number of algal cells in 0.5 ml	Average number of algal cells in 0.5 ml
6	–	5568; 5534; 5619	5573.7
8	–	4576; 4632; 4498	4568.7
2, 4, 6, 8	Average number of cells for four control beakers (2, 4, 6, 8)		5481.3 (standard error, 383.4)

Note: Significance of difference of the means in the control and experiment: $p = 3 \times 10^{-9}$; the *difference is significant* (significance level, >99.99%).
Experimental conditions: Initial concentration of cells per 0.5 ml, 16373.3 (16449, 16229, 16442). Total wet weight of mussels (with shells), in g: beaker 1, 13.4; beaker 2, 13.7; beaker 3, 17.2; beaker 4, 18.4; beaker 5, 18.3; beaker 6, 17.2; beaker 7, 18.1; beaker 8, 17.9. (The other conditions as in experiment at 0.5 mg/l and exposure 30 min, see above.)

Table 4.22   Number of *Isochrysis galbana* algal cells in beakers with mussels *Mytilus edulis* after a 60-min water filtration by the mollusks under the action of nonionogenic surfactant Triton X-100 (2 mg/l).

Beaker No.	Presence or absence of TX100, 2 mg/l	Number of algal cells in 0.5 ml	Average number of algal cells in 0.5 ml
1	+	3908; 3892; 3791	3863.7
3	+	5505; 5477; 5590	5524.0
5	+	6682; 6640; 6637	6653.0
7	+	5357; 5437; 5478	5424.0
1, 3, 5, 7	Average number of cells for four beakers with surfactant (1, 3, 5, 7)		5366.2 (standard error, 299.6)
2	–	2915; 2799; 2918	2877.3
4	–	1264; 1152; 1173	1196.3
6	–	1635; 1567; 1656	1619.3
8	–	1413; 1458; 1326	1399.0
2, 4, 6, 8	Average number of cells for four control beakers (2, 4, 6, 8)		1773.0 (standard error, 198.0)

Note: Significance of difference of the means in the control and experiment: $p = 3 \times 10^{-9}$; the *difference is significant* (significance level, >99.99%).
Experimental conditions: Initial concentration of cells per 0.5 ml, 16373.3. Total wet weight of mussels (with shells), in g: beaker 1, 13.4; beaker 2, 13.7; beaker 3, 17.2; beaker 4, 18.4; beaker 5, 18.3; beaker 6, 17.2; beaker 7, 18.1; beaker 8, 17.9. (The other conditions as in experiment at 0.5 mg/l and exposure 30 min, see above.)

Table 4.23   Number of *Isochrysis galbana* algal cells in beakers with mussels *Mytilus edulis* after a 90-min water filtration by the mollusks under the action of nonionogenic surfactant Triton X-100 (2 mg/l).

Beaker No.	Presence or absence of TX100, 2 mg/l	Number of algal cells in 0.5 ml	Average number of algal cells in 0.5 ml
1	+	1947; 1974; 1936	1952.3

Table 4.23 (continued)

Beaker No.	Presence or absence of TX100, 2 mg/l	Number of algal cells in 0.5 ml	Average number of algal cells in 0.5 ml
3	+	2424; 2413; 2401	2412.7
5	+	3536; 3565; 3528	3543.0
7	+	2569; 2642; 2688	2633.0
1, 3, 5, 7	Average number of cells for four beakers with surfactant (1, 3, 5, 7)		2635.3 (standard error, 174.7)
2	–	951; 973; 935	953.0
4	–	393; 345; 378	372.0
6	–	475; 497; 514	495.3
8	–	403; 423; 385	403.7
2, 4, 6, 8	Average number of cells for four control beakers (2, 4, 6, 8)		556.0 (standard error, 70.6)

Note: Significance of difference of the means in the control and experiment: $p = 7 \times 10^{-9}$; the *difference is significant* (significance level, >99.99%).
Experimental conditions: Initial concentration of cells per 0.5 ml, 16373.3. Total wet weight of mussels (with shells), in g: beaker 1, 13.4; beaker 2, 13.7; beaker 3, 17.2; beaker 4, 18.4; beaker 5, 18.3; beaker 6, 17.2; beaker 7, 18.1; beaker 8, 17.9. (The other conditions as in experiment at 0.5 mg/l and exposure 30 min, see above.)

Table 4.24 Number of *Isochrysis galbana* algal cells in beakers with mussels *Mytilus edulis* after a 60-min water filtration by the mollusks under the action of nonionogenic surfactant Triton X-100 (4 mg/l).

Beaker No.	Presence or absence of TX100, 4 mg/l	Number of algal cells in 0.5 ml	Average number of algal cells in 0.5 ml
1	+	10270; 9961; 9860	10030.3
3	+	12186; 12067; 12013;	12088.7
5	+	11493; 11412; 11317	11407.3
7	+	7898; 7791; 7680	7789.7
1, 3, 5, 7	Average number of cells for four beakers with surfactant (1, 3, 5, 7)		10329 (standard error, 496.5)
2	–	927; 865; 891	894.3
4	–	1289; 1365; 1251	1301.7
6	–	1033; 1117; 1059	1069.7
8	–	1783; 1819; 1737	1779.7
2, 4, 6, 8	Average number of cells for four control beakers (2, 4, 6, 8)		1261.3 (standard error, 100.8

Note: Significance of difference of the means in the control and experiment: $p = 3 \times 10^{-10}$; the *difference is significant* (significance level, >99.99%).
Experimental conditions: Initial concentration of cells per 0.5 ml, 13207.3. Average readings of the Coulter counter in filtered seawater, 66.5. Total wet weight of mussels (with shells), in g: beaker 1, 14.22; beaker 2, 14.43; beaker 3, 14.14; beaker 4, 15.08; beaker 5, 15.01; beaker 6, 14.79; beaker 7, 15.36; beaker 8, 14.47. (The other conditions as in experiment at 0.5 mg/l and exposure 30 min, see above.)

Table 4.25 Number of *Isochrysis galbana* algal cells in beakers with mussels *Mytilus edulis* after a 90-min water filtration by the mollusks under the action of nonionogenic surfactant Triton X-100 (4 mg/l).

Beaker No.	Presence or absence of TX100, 4 mg/l	Number of algal cells in 0.5 ml	Average number of algal cells in 0.5 ml
1	+	9067; 9084; 8970	9040.3
3	+	12055; 12112; 12155	12107.3
5	+	11084; 11271; 11375	11243.3
7	+	5753; 5614; 5732	5699.7
1, 3, 5, 7	Average number of cells for four beakers with surfactant (1, 3, 5, 7)		9522.7 (standard error, 746.4)
2	–	287; 343; 313	314.3
4	–	559; 485; 561	535
6	– –	472; 665; 649	595.3
8	–	1159; 1055; 1042	1085.3
2, 4, 6, 8	Average number of cells for four control beakers (2, 4, 6, 8)		632.5 (standard error, 86.5)

Note: Significance of difference of the means in the control and experiment: $p = 7 \times 10^{-8}$; the *difference is significant* (significance level, >99.99%).
Experimental conditions: Initial concentration of cells per 0.5 ml, 13207.3. Average readings of the Coulter counter in filtered seawater, 66.5. Total wet weight of mussels (with shells), in g: beaker 1, 14.22; beaker 2, 14.43; beaker 3, 14.14; beaker 4, 15.08; beaker 5, 15.01; beaker 6, 14.79; beaker 7, 15.36; beaker 8, 14.47. (The other conditions as in experiment at 0.5 mg/l and exposure 30 min, see above.)

An estimate of the statistical significance of the differences between the control and experiment was made. At a concentration of 0.5 mg/l, the difference became statistically significant (at a level of significance 95%) after 90 min of exposure ($p < 0.05$); at a smaller time of exposure the difference did not reach the level of statistical significance. At higher concentrations the difference was statistically significant (the level of significance was 99% and greater) at all exposure times (Table 4.27).

The fact that the value of the standard deviation was greater in the beakers with additions of TX100 is worth our attention. For example, at concentrations of 0.5 ml/l the value of the standard error was equal to: 597.6 (230.7 in the control) after 30 min and was 2.5 times greater; 457.1 (121.6 in the control) after 60 min and was 3.8 times greater; 298.3 (48.9 in the control) after 90 min and was 6.1 times greater.

This trend remained at greater concentrations from 1 to 4 mg/l. For example, at a concentration of 1 mg/l, the standard error was equal to 464.7 (21.5 in the control) after 90 min and was 21.6 times greater. At a concentration of 2 mg/l, the standard error was equal to 174.7 (70.6 in the control) also after 90 min, i.e., was 2.5 times greater. At a concentration of 4 mg/l, the standard error was equal to 746.4 (86.5 in the control) and after 90 min it was 8.6 times greater (Table 4.25). An increase in the standard error in the beakers with nonionogenic surfactant can reflect an increase in the variability in the rate of filtration by the samples of *M. edulis* (compared to the control). This means that some individual mollusks, when exposed to the synthetic

Table 4.26 Number of *Isochrysis galbana* algal cells in beakers with mussels *Mytilus edulis* after a 120-min water filtration by the mollusks under the action of nonionogenic surfactant Triton X-100 (4 mg/l).

Beaker No.	Presence or absence of TX100, 4 mg/l	Number of algal cells in 0.5 ml	Average number of algal cells in 0.5 ml
1	+	7291; 6982; 6955	7076
3	+	11806; 11709; 11685	11733.3
5	+	10002; 10253; 10269	10174.7
7	+	4811; 4538; 4481	4610
1, 3, 5, 7	Average number of cells for four beakers with surfactant (1, 3, 5, 7)		8398.5 (standard error, 831.75)
2	–	202; 254; 320	258.7
4	–	377; 278; 269	308
6	–	281; 300; 491	357.3
8	–	625; 563; 640	609.3
2, 4, 6, 8	Average number of cells for four control beakers (2, 4, 6, 8)		383.3 (standard error, 44.7)

Note: Significance of difference of the means in the control and experiment: $p = 5.4 \times 10^{-7}$; the *difference is significant* (significance level, >99.99%).
Experimental conditions: Initial concentration of cells per 0.5 ml, 13207.3. Average readings of the Coulter counter in filtered seawater, 66.5. Total wet weight of mussels (with shells), in g: beaker 1, 14.22; beaker 2, 14.43; beaker 3, 14.14; beaker 4, 15.08; beaker 5, 15.01; beaker 6, 14.79; beaker 7, 15.36; beaker 8, 14.47. $p = 5.42 \times 10^{-7}$. (The other conditions as in experiment at 0.5 mg/l and exposure 30 min, see above.)

Table 4.27 Statistical significance of the effect of synthetic surfactant Triton X-100 on the efficiency of water filtration by the mussels *Mytilus edulis* (the differences between the variants with surfactant and the control).

Concentration, mg/l	Time, min	p	Significance of effect	Level of significance
0.5	30	0.158	–	95%
	60	0.061	–	95%
	90	0.044	+	95%
1	30	0.0006	+	>99.9%
	60	0.0007	+	>99.9%
	90	0.002	+	>99%
2	30	4×10^{-9}	+	>99.9%
	60	3×10^{-9}	+	>99.9%
	90	7×10^{-9}	+	>99.9%
4	60	3×10^{-10}	+	>99.9%
	90	7×10^{-8}	+	>99.9%
	120	5×10^{-7}	+	>99.9%

surfactant, filtered water significantly slower than the mean value obtained in the experiment (which was lower than the average value in the control).

The results of calculation of the rate of filtration in l/min are given in Table 4.28. In the control beakers, the rate of filtration was equal to 1.8–5.2 l/min on average. It was always greater than the rate of filtration in the beakers containing nonionogenic

Table 4.28 Filtration rate of *Mytilus edulis* and inhibition coefficient under the action of nonionogenic surfactant Triton X-100.

TX100, mg/l	Time, min	Filtration rate, l/h	Filtration rate in the control	% of the control	Inhibition coefficient, %
1	0–30	2.65; 3.67; 4.85; 4.97; average, 4.04	4.30; 6.16; 5.15; 5.31; average, 5.23	77.25	22.75
	30–60	3.07; 4.55; 6.06; 6.14; average, 4.95	5.64; 6.75; 6.09; 6.03; average, 6.13	80.75	19.25
	60–90	2.76; 2.64; 5.15; 4.41; average, 3.74	4.88; 4.35; 4.05; 3.69; average, 4.24	88.21	11.79
2	0–30	2.57; 1.27; 1.27; 1.95; average, 1.765	3.11; 5.38; 4.31; 5.11; average, 4.48	39.42	60.58
	30–60	3.21; 3.08; 2.33; 2.47; average, 2.77	3.84; 5.09; 4.94; 4.73; average, 4.65	59.62	40.38
	60–90	2.73; 3.31; 2.52; 2.89 average, 2.86	4.42; 4.67; 4.74; 4.97 average, 4.72	60.85	39.15
4	60–90	0.42; −0.01; 0.06; 1.25; average, 0.43	4.18; 3.56; 2.34; 1.98 average, 3.02	14.24	85.76
	90–120	0.98; 0.13; 0.40; 0.85; average, 0.59	0.78; 2.21; 2.04; 2.31; average, 1.84	32.06	67.94

surfactants. The coefficient of inhibition regularly increased with the increase in the concentration of the nonionogenic surfactant in the range from 1 to 4 mg/l. For example, at a concentration of 1 mg/l (the first 30-min period) the coefficient of inhibition was equal to 22.75%, which led to a significant effect of the removal of algae from water. The values of the effect on the efficiency of removal of the suspended matter are given in Table 4.29.

Table 4.29   Effect of TX100 on the efficiency of removal (EER) of *Isochrysis galbana* algal cells from water by *Mytilus edulis* mussels (the coefficients of filtration-rate inhibition are given for comparison).

Concentration of surfactant, mg/l	Period of filtration, min	EER, %, TX100	Coefficient of inhibition of filtration rate, l/h (time period is given in parenthesis)
1	30	163.1	22.75 (0–30)
	60	236.2	19.25 (30–60)
	90	309.4	11.79 (60–90)
2	30	193.9	60.58 (0–30)
	60	302.7	40.38 (30–60)
	90	474.0	39.15 (60–90)
4	60	818.9	–
	90	1505.6	85.76 (60–90)
	120	2191.1	67.94 (90–120)

Note: EER, effect on the efficiency of removal of suspended particles during water filtration. EER = A/B, where A, B are concentrations of algal cells in the beakers after the removal of cells during water filtration: A, in the presence of a filtration-inhibiting substance (SDS); B, in the control.

Note that the values of EER demonstrate in a more illustrative manner the biological effects of nonionogenic surfactants in a given system rather than the traditional index, the coefficient of inhibition of the filtration rate (in liters per hour, Table 4.29). This gives us grounds to use the value of EER in the analysis of experimental results in our further work.

Available information on the concentration of nonionogenic surfactants in natural reservoirs is obviously not sufficient. Attention is focused on the hazardous products of their biological degradation; see Section 4.4 in the end of the chapter. It is useful to focus attention on the fact that in many regions the entry of nonionogenic surfactants to the environment is only slightly less than the entry of anionic surfactants (on average, 6.71 g of anionic and 4.07 g of nonionogenic surfactants per German citizen is transferred to the environment; Steinberg et al. 1995). Nonionogenic surfactants are badly decomposed in purification installations and are relatively more stable in aquatic media than anionic surfactants. Thus, the predicted or expected concentrations of nonionogenic surfactants would be hardly lower than the concentrations of anionic surfactants, which are known to be very high (see Chapter 3).

It is not surprising that nonionogenic surfactants were found in natural reservoirs at concentrations up to 1 mg/l and even 2.6 mg/l (Holt et al. 1992), which exceeds the minimum concentrations at which the inhibition of filtration activity of mollusks and a decrease in the removal of suspension by them from water were demonstrated in our experiments.

It was already noted that the role of water filtration by hydrobionts including bivalve mollusks is very high in the ecosystems (see Chapter 3, Section 3.1.2;

Voskresensky 1948; Kondratyev 1977; Zaika et al. 1990; Alekseyenko and Alek-sandrova 1995; and others). Filtration of water significantly affects many aspects of the structure and functionality of the ecosystems including the concentration of suspended particles in water, penetration of visual light and UV radiation into water, formation of pellets (faeces and pseudofaeces) by mollusks, which sediment on the bottom and participate in the accumulation of organic matter by benthic deposits, etc. (e.g., Ostroumov et al. 1997; Ostroumov 2000). The significance of water filtration by mollusks in the formation of important parameters of ecosystems in the natural conditions was recently identified as a result of the analysis of the consequences of intrusion of a mollusk with high filtration activity into the water systems (Strayer et al. 1999). Hence, the inhibition of the processes of filtration leads to disturbances or changes in many aspects of the structure and functionality of an aquatic ecosystem.

4.3.2.1 *Mercenaria mercenaria* (Linne)

We observed the behavior of larvae of mollusks in a medium containing different concentrations of TX100. At a concentration of 1 mg/l we observed the changes in the character of the motion of some larvae after 4 h of incubation. Instead of the usual motion (frequently along an arch) we observed a chaotic motion with overturning. At a concentration of 5 mg/l, within 30 min after the onset of incubation, most larvae descended to the bottom and remained there until the end of the experiment. In the latter case (5 mg/l), their behavior was characterized by slowing down of the motion compared to the control. After 4 days in the variants with TX100 at a concentration of 1 mg/l some larvae at the bottom were completely immobilized (2.87%); in the variant with 0.5 mg/l 1.06% was completely immobilized compared to 0% in the control. It was found during the observation of the larvae that their orientation (position in the water column) was not the same, and some of them, especially in the presence of TX100, were not normally oriented and seemed to tumble down and tend to be in inclined positions.

On the sixth day of the experiment, only part of the larvae in all variants of the experiment (0, 0.5, and 1 mg/l) maintained normal orientation while swimming in the water column. A significant percentage of the larvae obtained an (abnormal) lateral position. The proportion of larvae with normal orientation decreased in the variants with surfactants compared to the controls. For example, the proportion of such larvae compared to the control was 49.3% in the variant with a concentration of 0.5 mg/l and 19.8% in the variant with a concentration of 1 mg/l. At TX100 concentrations of 5 mg/l, 10, and 25 mg/l no larvae achieved a normal position in water.

These observations provide additional information about the unfavorable effects of sublethal concentrations of synthetic surfactants.

Taking into account that sublethal concentrations of synthetic surfactants (and of some other pollutants) distort the normal rate of water filtration by mollusks (see above, and also the chapter about anionic surfactants) we can assume that the total effect of sublethal concentrations on the population is a combination of several components. Along with the other factors, they can include a decrease in the capabilities for obtaining nutrients for growth and formation of juvenile specimens (as a result of the decrease in the water filtration) and a decrease in the rate of the larvae development (as a result of distortions in the orientation and behavior and a decrease in the locomotive activity). Even small negative effects, when they sum up, can lead to a hazard of the extinction of a population.

4.4 Biological Effects of Nonionogenic Surfactants and their Hazards to Aquatic Ecosystems

There is a significant amount of evidence that molecular mechanisms of the effect of nonionogenic surfactants on organisms are related to their influence on the membranes (Isomaa and Hagerstrand 1988) and on protein–protein interactions (Weiner and Rudy 1988).

The information, which is important for the understanding of the molecular aspects of the interaction of nonionogenic surfactants with biological membranes in an aquatic medium, was obtained using the pigment–protein complexes of the thylakoids of pea *Pisum sativum* (Murphy and Woodrow 1984). The incubation of thylakoids in aquatic medium at a concentration of TX100 of 1 mg/l led to a sharp decrease in the peak of fluorescence emission at 735 nm (at 77 K). The incubation of thylakoids in aquatic medium with even greater concentration of TX100 led to the complete disappearance of this peak, although the emission peak at 680 nm remained. It is likely that nonionogenic surfactants cause dissociation of pigment–protein complexes LHCP-I and LHCP-II (light-harvesting chlorophyll–protein complexes) and photosystems PSI and PSII. It is likely that energy transfer between complexes is disturbed as a result of dissociation. Similar disturbances can occur in the cells of photosynthesizing hydrobionts.

There are additional data about the ambiguous character of the effect of nonionogenic surfactants on biological objects (Eilenberg et al. 1989; Tragner and Csordas 1987; King et al. 1988). For instance, Eilenberg et al. (1989) describe the favorable effect of 0.2% solution of TX100 on the activity of microsomal enzyme squalene epoxidase in microsomes that were previously treated with digitonin, which inhibited this enzyme.

Austrian scientists D. Tragner and A. Csordas found an unusual effect while investigating the behavior of erythrocytes in aquatic medium containing nonionogenic surfactant Triton X-45. Unlike other octyl phenoxyl ethylene ethers, this nonionogenic surfactant did not exert any hemolytic effect even at high concentrations, e.g., 1%. Even more, it protected cells from osmotic destruction (Tragner and Csordas 1987). The absence of a toxic effect of nonionogenic surfactant pluronic F-68 on the cells of aquatic culture of yeast *Saccharomyces cerevisae* was shown in King et al. (1988). The number of viable cells in a unit of aquatic medium with 10% of this nonionogenic surfactant, block copolymer polyoxyethylene polyoxypropylamine with 75 oxyethylene monomers, did not decrease. These data confirm once more the necessity for additional analysis of diverse and sufficiently complicated biological effect of nonionogenic surfactant.

Biological activity of amphoteric surfactants with respect to a number of bacteria and fungi was studied by the Japanese authors S. Osanai et al. (1989). In a number of cases, the minimum inhibitory concentration (mg/l) during the effect of N-dodecyl amino sulfonated acids on *E. coli*, *Salmonella typhi*, *Pseudomonas aeruginosa*, and *Aspergillus niger* was over 200. In the experiments with *Staphylococcus aureus*, *Bacillus subtilis*, *Micrococcus lutea*, *Candida albicans*, *S. cerevisiae*, *Trichophyton interdigital*, *Microsporium gypseum*, and *Penicillium chrysogenum*, the minimum inhibitory concentration varied and was sometimes as low as 25. The minimum inhibitory concentration of N-alkyl amino acids for *T. interdigital* and *M. gypseum* varied from 5 to 100, and for *S. typhi* and *P. aeruginosa* exceeded 200; for the other test organisms it varied from 10 to 200 and more.

In order to discuss the hydrobiological significance of our experiments on the effect of nonionogenic surfactants on mollusks, it is interesting to consider the works containing information on the effect of nonionogenic surfactants on other hydrobionts: biological filter feeders. The filter feeding hydrobionts include the larvae of insects Simuliidae. One of the research projects studied the effects of linear alcohol ethoxylate C_{14-15} AE-7 on aquatic invertebrates under conditions of mesocosms in flowing waters (Gillespie et al. 1996). In the experiments, which lasted 28 and 30 days, at a concentration of this nonionogenic surfactant equal to 0.16 mg/l and greater, the density of the population of Simuliidae midge larvae decreased. Their number in the drift increased. No significant effect of this surfactant at concentrations below 0.55 mg/l was found on the density of copepods, cladocers, chironomids, nematods, and annelids. The authors concluded that under conditions of their experiments the concentration of surfactant in water solutions that did not affect aquatic invertebrates was equal to 0.08 mg/l, while the minimum concentration that affected them was 0.16 mg/l (Gillespie et al. 1996). A high sensitivity of the larvae of this group of filter feeder hydrobionts to nonionogenic surfactants agrees with our experiments on another group of filter feeders. Note that the larvae of Simuliidae are very important in the removal of the suspension of organic matter from water, which was confirmed in Wotton et al. (1998). The authors of that paper determined the density of larvae in water, which reached $6 \cdot 10^5$ larvae/m^2. The larvae extract a significant amount of suspended matter, part of which is deposited on the bottom as faeces pellets. In this connection, a decrease in the density of the population due to the interaction with nonionogenic surfactants detected by Gillespie et al. (1996) seems to be a very important effect on the aquatic ecosystem.

Thus, the literature and experimental data confirm the existence of a wide range of effects characterizing the biological activity of aquatic media containing nonionogenic surfactants.

In the analysis of the problems related to the protection of the environment in the application to marine and aquatic ecosystems, it is important that nonionogenic surfactants are included in the composition of emulsifiers and dispersing agents used to clean the bottoms, decks, and engines of the ships from oil as well as for treatment of oil spills and removal of oil spill from the sandy, stone, and pebble coasts. A number of emulsifiers applied for these purposes contain 8–50% (on average 20%) of mainly nonionogenic surfactants (Nelson-Smith 1977). Nonionogenic surfactants are included into such preparations as ML-72 (Nesterova 1989) designed for chemical, mechanical and jet cleaning of oil tankers, reservoirs, cisterns, cargo and fuel tanks of the vessels from the remains of oil, oil products, and animal and vegetable fats.

The other important aspect related to the pollution of the environment with nonionogenic surfactants is related to the application of nonionogenic surfactants in the composition of technical pesticides. They can contain 1−20% nonionogenic surfactants produced by the industry, e.g., the so-called OP-7 and OP-10 (Chemical Means for Protecting Plants 1979). Preparations such as butifos (defoliant of cotton plants) can contain approximately 30% of OP-7 or OP-10.

Thus, synthetic nonionogenic surfactants can indeed enter the terrestrial and aquatic ecosystems in large amounts. The role of nonionogenic surfactants as pollutants of the environment is increasing due to the fact that their decay in the environment is slower than that of many other xenobiotics. The coefficients of the rate of biochemical oxidation of preparations OP-7 and OP-10 in the natural environment at a temperature of 20°C were equal to 0.007 and 0.006, respectively (Kaplin 1979). This means that biological degradation of these substances occurs slower than the degradation of alkyl sulfates, glucose, phenol, furaldehyde, formaldehyde, chlorine sulfonol, kerosene alkyl sulfonate, m-cresol, p-cresol, lignosulfonates, and trimethyl alkyl ammonium chloride (17−20°C) (Kaplin 1979). Biodegradation of nonionogenic surfactants can cause new problems.

Persistent products including nonylphenol are formed during the biological transformation and biological degradation of a number of nonionogenic surfactants (Granmo et al. 1989; Ekelund et al. 1990; Ahel et al. 1993; see Chapter 1). LC_{50} of nonylphenol (NP) for *Mytilus edulis* L. (semistatic and flowing test systems were used) was equal to 3 mg/l (96 h), 0.5 mg/l (360 h), or 0.14 mg/l (850 h) (Granmo et al. 1989). Such sublethal effects as a decrease in the strength of the byssus and a decrease in SFG (scope for growth) manifested themselves at concentrations of 0.056 mg/l (Granmo et al. 1989). Nonylphenol (0.01−10 µg/l; 24−48 h; at 25°C and 28°C) inhibited the settlement and colonization of the substrate by cypris-like larvae of *Balanus amphitrite* (Billinghurst et al. 1998).

In the treated wastewater, NP was found at concentrations ranging from 2 to 4,000 µg/l, i.e. up to 4 mg/l (Giger et al. 1981; Etnier 1985, cited from Ekelund et al. 1990). Up to 38,000 µg/l of nonylphenol per kilogram of dry weight was found in macrophytic algae *Cladophora glomerata* from Chriesbach Creek in the Glatt Valley in Northern Switzerland, while the mean concentration of nonylphenol in the water of this creek was seemingly insignificant (3.9 µg/l) (Ahel et al. 1993). The concentration of another product of degradation of nonionogenic surfactants, nonylphenol monoethoxylate NP1EO) in the same algae from the Glatt River was even higher reaching 80,000 µg/kg of dry weight at a concentration of this chemical in the water of the river equal to 16 µg/l (Ahel et al. 1993). The concentration of another product of degradation of nonionogenic surfactants, NP2EO (nonylphenol diethoxylate) in the latter case was equal to 29,000 µg/kg. The bioconcentration factors (BCF) for NP1EO were equal to 3,500−5,000, while for nonylphenol they were 6,600−10,000 (Ahel et al. 1993). The bioconcentration factor for *Mytilus edulis* L. was approximately equal to 3,400 (Ekelund et al. 1990), which is significantly greater than that reported earlier (see McLeese et al. 1981, cited in Ahel et al. 1993). The range of values of BCF is characterized by the following numbers: 10 (nonylphenol, *M. edulis*); 320 (nonylphenol, *M. edulis*); 100 (nonylphenol, *Cragon cragon*); 1250 (nonylphenol, *Gasterosteus aculeatus*); 280 (nonylphenol, *Salmo salar*); 271–344 (nonylphenol, *Pimephales promelas*); for some species of fish in Switzerland (Glatt River): 13–408 (nonylphenol), 3–300 (NP1EO), 3–326 (NP2EO) (Thiele et al. 1997). The total concentration of nonylphenol, NP1EO, and NP2EO in the edible parts (muscles) of fish *Barbus barbus* L. reached 5.8 mg/kg of dry weight; of wild ducks (*Anas boscas*), greater than 3.6 mg/kg (Ahel et al. 1993).

The following maximum concentrations of the degradation products and biological transformation products of nonionogenic surfactants were found in the aquatic systems: in the rivers of Japan up to 50–70 µg/l (NPEO 1982), in rivers of Switzerland up to 116 (NPEC 1987), in the rivers of Great Britain up to 180 (nonyl-phenol, 1995); in the bottom sediments, the concentrations were from one to three orders of magnitude higher and the maximum values were equal to (µg/kg of dry weight): in Canada up to 41,100 (nonylphenol, 1995) in Germany, Bavaria up to 10,000 (nonylphenol, 1997), in Spain up to 6,600, in Switzerland (the Glatt River), up to 13,100 (nonylphenol, 1990), in the U.S., Wisconsin up to 1,040 (nonylphenol, 1996), in Italy (Venetia lagoon) up to 6,600 (NP2EO 1990) (Thiele et al. 1997).

In active sludge (anaerobically stabilized) the content of the products of degrad-ation and biological transformation of nonionogenic surfactants reached (µg/kg of dry weight): in Toronto, Canada up to 470,000 (nonylphenol, 1995), in Germany up to 1,300,000 (nonylphenol, 1997), in Sweden up to 1,200,000 (nonylphenol, 1990), in Switzerland up to 410,000 (NP1EO 1997), and 2,530,000 (nonylphenol, 1984), in Great Britain up to 824,000 (nonylphenol, 1994), in Los Angeles, U.S. up to 370,000 (nonylphenol, 1994) (Thiele et al. 1997).

Thus, the hazard of nonionogenic surfactants for aquatic ecosystems is multi-faceted and it is important to accumulate new information about the biological effects of these chemicals. A comparison of the relative sensitivity of the organisms, which represent the different trophic levels of the ecosystem to anionic surfactants, is given in the summary Table 4.30.

When comparing the results of studies of the biological effects caused by the impact of a representative of nonionogenic surfactants, TX100, on various organisms in experimental conditions, the test systems could be ranked and the organisms arranged according to their tolerance. Under experimental conditions used, we have the following sequence according to the degree of increasing tolerance to the impact of TX100: *Thalassiosira pseudonana* < *Mytilus edulis* < *Hyphomonas* sp., *Synecho-coccus* sp. < *Fagopyrum esculentum*.

Comparatively more tolerant organisms can be used for the purposes of phyto-remediation.

The results of experiments on freshwater and marine mollusks support our hypothesis put forward in Chapter 3 in that nonionogenic surfactants can inhibit the filtration activity of mollusks. The sensitivity of molluskan filtration activity to non-ionogenic surfactants (consistent with the data of high sensitivity of other hydrobiont filter feeders to nonionogenic surfactants (Gillespie et al. 1996)) suggests a potential ecological hazard these chemicals pose in the sense that their biological effects can disturb the processes essential for water self-purification. Revealing analogies in the

Table 4.30 Qualitative comparison of the levels of general sensitivity to nonionogenic surfactants.

Organisms	Sensitivity estimate	References/comments
Bacteria *Hyphomonas* sp. MHS-3	L	Section 4.1; Vainer and Ostroumov, 1998
Bacteria *Hyphomonas* sp. VP-6	L	Section 4.1
Various heterotrophic bacteria	L	General estimate of nonionogenic surfactants as relatively low toxic chemicals (Stavskaya et al., 1988)
Cyanobacteria *Synechococcus*	L, S	Section 4.2.1; Waterbury and Ostroumov 1994
Green algae *Chlorella fusca* Shihers et Krauses	L, S	Wong, 1985
Diatoms *Thalassiosira pseudonana*	H	Section 4.2.2; Fischer, Maertz-Wente, and Ostroumov 1996
Seedlings of *Fagopyrum esculentum, Sinapis alba, Lepidium sativum* and others	L	Section 4.3.1; Comparatively low sensitivity makes possible application of these organisms for phytoremediation (Ostroumov 2000)
Freshwater mollusks *Unio* sp., other filter feeders (Simuliidae)	H	Section 4.3.2.1; Gillespie et al., 1996; Comparatively high sensitivity causes a hazard of distorting the balance in the trophic web (Ostroumov, 2000).
Marine mollusks *Mytilus edulis*	H	Section 4.3.2.2; Similarly a hazard can be caused of distorting the balance in the trophic web (Ostroumov et al., 1998).
Marine mollusks *Mercenaria mercenaria*	H	Section 4.3.2.3. A hazard of distorting the normal cycle of reproduction of the population.

Note: The general estimate of sensitivity to anionic surfactants (at inhibitory effects) is as follows: L is low, M is moderate, H is high, S is stimulating effect.

character of biological effects of anionic and nonionogenic surfactants put forward the importance of investigating the impact of cationic surfactants on the organisms. As a working hypothesis, it is reasonable to suggest that cationic surfactants would also inhibit water filtration by mollusks and removal of suspended matter from water by them. The results of experiments which verified this hypothesis are discussed in the next chapter.

5 Biological Activity of Waters Containing Cationic Surfactants

Cationic surfactants along with anionic and nonionogenic surfactants are one of the main classes of surfactants. The proportion of cationic surfactants in the total volume of synthetic surfactants used in the economy is approximately equal to 22% (for the U.S.; Ainsworth 1992). Cationic surfactants enter aquatic ecosystems together with polluted waters because they are widely used in many industries including petroleum industry, oil refining industry, petrochemical industry, and gas industry. Cationic surfactants (from the group of quaternary ammonium compounds (QAC)) are used in pest control in aquaculture for combating pathogenic organisms of fish. QAC are frequently added to water at concentrations of 1–2 mg/l (Austin 1985).

Cationic surfactants are the class of synthetic surfactants whose production is rapidly increasing at an annual rate of 5% (Dean 1985). Correspondingly, pollution of the environment with these synthetic surfactants including quaternary ammonium compounds increases. The biological effects of synthetic surfactants were studied, e.g., in Stavskaya et al. (1988) and Lewis (1991a,b), however, on a comparatively limited range of biological subjects.

Cationic surfactants have a negative impact on many investigated species of bacteria and fungi. The minimum bacteriostatic activity (MBA) of QAC was equal to the following values for the species (the range of MBA values for the set of 13 different QAC is given in parenthesis, in µg/ml) (Glusman et al. 1978, cited from Stavskaya et al. 1988): *Aspergillus niger* (2–1000); *Candida albicans* (1–1000); *Staphylococcus aureus* (1–1000); *Pseudomonas aeruginosa* (250–1000); *Escherichia coli* (16–1000).

Analysis of the literature demonstrates the following regularities (Stavskaya et al. 1988), which characterize the bactericidal activity of QAC: (1) bactericidal action depends on the length of hydrocarbon radicals. Maximum bactericidal action is observed for chemicals with 16–18 carbon atoms; (2) bactericidal action increases with the increase of electronic density of the nitrogen atom; (3) antimicrobial activity increases with the increase in the number of unsaturated links in radicals; (4)

bactericidal action increases if a broad fraction of long-chain amines is used for synthesis of QAC; (5) antimicrobial activity increases if benzyl radicals are substituted for alkyl radicals.

Additional discussion of antimicrobial activity of QAC and cationic surfactants in general is given in Stavskaya (1981) and Stavskaya et al. (1988).

Toxicity of some cationic surfactants for fish and invertebrates was also studied. LC_{50} (48–96 h) for fish is equal to 0.6–2.6 mg/l (Huber 1985).

LC_{50} (48 h) for invertebrates, e.g., *Daphnia magna*, is often equal approximately to 0.16–1.7 mg/l (Huber 1985) but can be significantly greater than 10 mg/l. For example, LC_{50} for the action of dodecyl trimethyl ammonium bromide on young larvae of *Chironomus riparius* was equal to 14.6 mg/l (48 h), and for the action of distearyl dimethyl ammonium bromide it was 11.3 mg/l (72 h) (Pittinger et al. 1989).

The values of NOEC (no observed effect concentration) in the chronic or partly chronic experiments with fish and invertebrates ranged from 0.05 to 0.5 mg/l (Huber 1985). The NOEC values in the investigation of the effect of dodecyl trimethyl ammonium chloride (DTMAC) on *Ceriodaphnia dubia* were equal to 0.17–0.35 mg/l (Lewis and Suprenant 1985; Woltering and Bishop 1985, cited from Versteeg et al. 1997). In some test systems the values of NOEC exceeded 10 mg/l (Pittinger et al. 1989). Pittinger with coauthors studied the effect of two cationic surfactants on the hatching of *Chironomus riparius* larvae. The values of NOEC were equal to 21.5 mg/l for distearyl dimethyl ammonium bromide and to 15.4 mg/l for dodecyl trimethyl ammonium bromide.

The values of LC_{50} (48 h, 25°C) were determined for four QAC using rotifers *Brachionus calyciflorus* (Versteeg et al. 1997). They were equal to (mg/l): 39 for octyl trimethyl ammonium chloride; 0.23 for dodecyl trimethyl ammonium bromide; 0.067 for hexadecyltrimethyl ammonium chloride; and 0.66 for alkyl dimethyl hydroxyethyl ammonium chloride. In the latter case, alkyl was presented by a mixture of dodecyl and tetradecyl.

The effect of lauryl trimethyl ammonium chloride (LTMAC) on bivalve mollusk *Corbicula fluminea* was investigated in riverine mesocosms by Belanger et al. (1993). The growth of mollusks was disturbed at LTMAC concentrations of 0.185 mg/l and greater. Mortality and cellulolytic activity did not change even at LTMAC concentrations below 1.153 mg/l. Additional discussion of the literature data about the biological effects of cationic surfactants is given in the concluding section of this chapter.

After the accident at the Chernobyl nuclear power station, investigations of synthetic surfactants, in particular, cationic surfactants were intensified in Ukraine. The content of cationic surfactants in Ukrainian reservoirs increased significantly. For example, in the water of the aquatic farm of Power Station 5, the concentrations of cationic surfactants were over the maximum permissible concentration (MPC, for fisheries) by a factor or 37-39, according to Davydov et al. (1997). According to that paper, the concentrations of cationic surfactants observed in this reservoir were equal to 0.45–0.47 mg/l. Such concentrations of cationic surfactants were observed not only in the water of the bay but also in fishing cribs. The authors stated that the installation for water purification used in the incubation workshop did not decrease the level of cationic surfactants in water significantly (Davydov et al. 1997).

5.1   Biological Effects of Ethonium

One of the main groups among the cationic surfactants belongs to quaternary ammonium compounds (QAC). Ethonium (or 1.2-[N,N-*bis*(dimethyl)-N,N-*bis*-decyl acetate] ethylene diammonium dichloride) remains poorly studied. Ethonium pollutes aquatic ecosystems entering into them together with sewage waters because it is used in many industries as emulsifier, water-repellent agent, softener of fabric, stabilizer of dispersions, and floating agent.

5.1.1   Effects on algae

The effect of aquatic medium containing cationic surfactant of this group (ethonium) on *Chlorella vulgaris* and *Monochrys lutheri* algae was studied. It was shown that already after 20 h, a significant decoloration of algal suspension was observed in the medium containing 0.067 mg/l of ethonium. A sharp decrease in the optical absorption of the algal suspension in the longwave maximum of chlorophyll absorption was a quantitative characteristic of algal degradation (Table 5.1). After 20–23 h, the optical absorption decreased approximately by a factor of 3. Interestingly, the effects of the water medium with ethonium were close for taxonometrically distant test species of different algal types (Ostroumov and Maksimov 1988).

We also studied the effects of ethonium on the seedlings of some plants.

5.1.2   Effects on the seedlings of angiosperm plants

In our work we have shown that ethonium inhibits the growth of the seedlings of the cucumber *Cucumis sativus*. We studied the effects of ethonium concentrations in the range from 0 (using a concentration of 0.03 mg/l and greater) up to 0.25 mg/l. At an ethonium concentration in the water medium equal to 0.25 mg/l, the mean length of the seedlings after 70 h of incubation was 57.6% of the control length, and after 85 h it was equal to 45.0%. If the concentration was decreased to 0.125 mg/ml, the mean length of the seedlings was equal to 93–95% of the control length (Table 5.2). Since 0.125 mg/ml is the same as 125 mg/l, the experiment demonstrates that a significant part of the seedlings endures the concentrations of this cationic surfactant of the order of 100 mg/l, which makes this organism promising for use in phytoremediation.

5.2   Biological Effects of Tetradecyl Trimethyl Ammonium Bromide (TDTMA)

Alkyl trimethyl ammonium chlorides and bromides pollute aquatic medium since they are widely used cationic surfactants. Their applications include antistatic agents

Table 5.1 Decrease of optical absorption (in the long-wave range) of the suspensions of *Chlorella vulgaris* and *Monochrysis lutheri* algae during their degradation under the action of cationic surfactant ethonium (1,2-[N,N-bisdimethyl-N,N′-bisdecylacetate ethylenediammonium] dichloride).

Incubation time, h	Absorption $A = I - T$, % at given wavelengths ($\lambda_{max} = \lambda$).		
	Control (no surfactant)	Experiment (surfactant, 0.067 mg/ml), $\lambda = 672$ nm	Experiment / Control · 100%
Chlorella vulgaris			
23	23.5	6.8	28.9
46	26.0	2.0	7.7
Monochrysis lutheri			
20	28.0	9.25	33.0
43	35.5	2.0	5.6

Note: For *Chlorella vulgaris*, $\lambda = 680$ nm; for *Monochrysis lutheri*, $\lambda = 681$ nm.

Table 5.2 Effect of cationic surfactant ethonium on the mean length of *Cucumis sativus* seedlings.

Experiment No	Concentration, mg/ml	Time, h	Length		SD	CL	CV, %	n
			mm	% of the control				
1	0	70	19.26	100	2.88	1.39	15.0	20
	0	95	39.26	100	6.99	3.37	17.8	20
	0.25	70	11.10	57.63	5.87	2.74	52.9	20
	0.25	95	17.67	45.01	8.58	4.01	48.6	20
2	0	72	22.78	100	7.40	3.68	32.5	18
	0	96	43.12	100	11.83	6.08	27.4	17
	0.125	72	21.68	95.17	7.41	3.57	34.2	19
	0.125	96	40.68	93.34	12.83	6.18	31.5	19
3	0	72	10.00	100	1.56	0.86	15.6	15
	0	96	21.87	100	3.56	1.97	16.3	15
	0.031	72	9.90	99.0	2.40	1.12	24.2	20
	0.031	96	21.85	100	5.94	2.78	27.2	20
	0.0625	72	10.00	100	2.13	0.99	21.3	20
	0.0625	96	20.90	95.56	2.27	2.46	10.9	20

Note: SD, standard deviation; CL, confidence limit; CV, coefficient of variation; *n*, number of seedlings.

in the textile industry, inhibitors of corrosion in acid media, softeners of fabric, de-emulsifiers in various industries including wood-chemical industry, floating agents, extracting agents in hydrometallurgical industry, additives to cleaning and washing compositions. We studied the effect of TDTMA on autotrophic and heterotrophic organisms including cyanobacteria, algae, angiosperm plants, annelids, pulmonate mollusks, and bivalve mollusks.

5.2.1 Effects on cyanobacteria and algae

We studied the effect of cationic surfactants and TDTMA on the cells of Cyano-phyceae blue-green algae *Nostoc muscorum*, cells of green Chlorophyceae algae *Bracteacoccus minor* and some other algal species. In the work coauthored with A.N. Tretyakova (Ostroumov and Tretyakova 1990) we demonstrated the following.

The effect of surfactant TDTMA at concentrations of 0.04 mg/ml caused the death of all Cyanophyceae strains studied. The results of optical microscopy and visual observations of the cultures 7 days after inoculation are given in Table 5.3. However, the death of Cyanophyceae algae and the appearance of white flakes were observed even earlier, in 3 days. The effect of TDTMA at a concentration of 0.001 mg/ml caused disturbances in the state of the cells, which could be found during microscope observation, while the green color of the culture (observed visually) remained. Thus, the general character of the effect of TDTMA on Cyanophyceae and Chlorophyceae freshwater green algae was the same; however, at one of the concentrations studied (0.004 mg/ml) Cyanophyceae algae were more sensitive.

Table 5.3 Effect of surfactant TDTMA on the state of 7-day old cultures of *Nostoc muscorum* and *Bracteacoccus minor*.

Concentration, mg/l	N. muscorum, strain 33	N. muscorum, strain 235	B. minor, strains 200 and 219
0	Culture of intensive green color; vegetative cells with heterocysts, no spores	Culture of intensive green color; dividing vegetative cells	Culture of intensive green color; many dividing vegetative cells
1	Culture of intensive green color; cells smaller than in the control; there are spores; degradation of filaments is observed	Culture of intensive green color; some filaments acquired a brown-yellow tint	Color of the culture is preserved; dividing cells are fewer; grains in cells
4	Algae perished; white flocs are seen	Suspension of green flocs with white margins	Suspension of weakly green color; intensive grains in cells; many cells whose contents detach from the walls and shrink
40	Algae perished	Algae perished; white flocs	Algae perished; white-color suspension

The negative effect of surfactant TDTMA on Cyanophyceae and Chlorophyceae algae was also revealed in the experiments with soil cultures (Table 5.4). In
a month after introducing a solution of TDTMA (0.1 mg/ml) in Petri dishes with the
culture, the total abundance of Cyanophyceae algae decreased by a factor of more
than 7 (from 1700 to 230 cells in 1 g of soil). The abundance of the Chlorophyceae
algae cells decreased to a lesser extent, approximately by 40% of the control. The
abundance of the diatom algae also decreased approximately by a factor of four (the
author thanks Professor E.A. Shtina for consultations and assistance in the work, and
A.N. Tretyakova for assistance).

Table 5.4   Effect of cationic surfactant TDTMA on the abundance of algal cells (and also
cyanobacteria) of soddy-podzolic soil (exposure, 1 month).

Variant	Abundance, thousand cells per 1 g soil		
	Cyanobacteria	Green algae	Diatomic algae
Control	1700 (1300 : 400)	10	52
Surfactant, 0.05 mg/ml	1010 (840 : 170)	14	32
Surfactant, 0.1 mg/ml	230 (160 : 70)	6	14

Note: Soil was placed into Petri dishes (30 g each dish), 10 ml each of surfactant solution or
distilled water (control) was added. For cyanobacteria, the numbers in brackets are ratios of
non-nitrogen-fixing bacteria and nitrogen fixers. A.N. Tretyakova participated in this work.

The results obtained on the soil and aquatic cultures agree in the following
aspects: (1) negative effect of surfactant TDTMA on all studied algae cultures was
found; (2) Cyanophyceae algae manifested a greater sensitivity to TDTMA than
Chlorophyceae algae.

The obtained results and the data of the previous experiments investigating the
effect of surfactant sodium dodecyl sulfate (SDS) on green algae *Scenedesmus
quadricauda* (e.g., Goryunova and Ostroumov 1986) demonstrated that the effect of
TDTMA on the growth of the soil algae studied is realized at smaller concentrations
than the effect of SDS on *Scenedesmus quadricauda*. Further investigations are
needed for more definite and specific conclusions.

The data we obtained supplement the previously published information about
the effect of other pollutants on soil algae, e.g., the effect of smoke gas pollution of
soil by emission of metallurgical plants, oil pollution and pollution with highly mine-
ralized drilling muds and water (e.g., Shtina 1985), pollution with mineral fertilizers
(Tretyakova 1987), and pollution with other chemicals.

Further work was continued using other species of organisms. The sensitivity
of other species of cyanobacteria to cationic surfactants at a concentration of the
order of 1 mg/l was demonstrated on *Synechocystis* sp. PCC 6803 (Ostroumov and
Kolotilova 1988), *Spirulina platensis* (Ostroumov et al. 1999a,b), and Chlorophyceae algae (Piskunkova and Ostroumov 1999).

Our results agree with the data by other authors, who demonstrated the sensitivity of algae to cationic surfactants including the sensitivity of *Chlorella vulgaris*

to catamine AB (alkyl methyl benzyl ammonium chloride) (Parshikova et al. 1994). Algistatic concentrations in a 5-day test in some works were equal to approximately 0.1–1 mg/l. Under the action of sublethal concentrations of cetyl trimethyl ammonium bromide (CTAB) at concentrations less than 14 μM or 5 mg/l on *Porphyridium purpureum* an increase in the concentration of saturated fatty acids was detected in mono- and digalactosyl diglycerides (Nyberg and Koskimies-Soininen 1984).

After 72 h of incubation in aquatic medium containing 12 μM (or 4.38 mg/l) of CTAB, all cells of chrysophytes *Poterioochromonas malhamensis* lost vitality. At concentrations of CTAB equal to 2–3 μM the growth of these algae was slowed down compared to the control (Röderer 1987).

According to the general plan of investigating the effect of cationic surfactants on various biological objects, part of the experiments was carried out on angiosperm plants.

5.2.2   Effects of TDTMA on the seedlings of angiosperm plants

No information is available in the literature on the effect of cationic surfactant-containing aquatic media on higher plants. As it was mentioned above, such works are important in the context of interest to the systems of biological testing alternative to biological testing on animals.

Our experiments showed that aquatic media containing cationic surfactant TDTMA inhibited the development of buckwheat *Fagopyrum esculentum* seedlings (Table 5.5). In 69 h of incubation at TDTMA concentrations equal to 0.05 mg/ml, the mean length of the seedlings was <50% of the control, and at concentration equal to 0.25 mg/ml it was <25% of the control (Ostroumov and Tretyakova 1990). Interestingly, the presence of TDTMA had only a slight effect on the germination of buckwheat seeds.

Table 5.5   Effect of surfactant TDTMA on the average length of *Fagopyrum esculentum* (cultivar Shatilovskaya 5) seedlings.

TDTMA, mg/ml	Time, h	Average length, mm	CL, mm	SD	CL margins of SD	n
0	45	8.33	1.92	4.56	3.68–6.04	24
	53	13.25	2.78	6.58	5.32–8.73	24
	69	26.25	5.26	12.45	10.07–46.51	24
0.05	45	6.19	0.92	2.28	1.86–2.98	26
	53	8.65	1.25	3.10	2.52–4.05	26
	69	11.58	1.57	3.88	3.16–5.08	26
0.25	45	4.62	0.54	1.33	1.08–4.74	26
	53	5.46	0.61	1.50	1.22–4.97	26
	69	6.04	1.70	1.73	141–2.27	26

Note: CL, confidence limit; SD, standard deviation; *n*, number of seedlings.

We also calculated the rates of elongation of the seedlings and the effect of TDTMA on the rate of their elongation (Table 5.6). It was found that the rate of elongation was smaller than in the control. Besides, the dynamics of the growth rate changed in time: in the control it increased during the experiment, in the presence of TDTMA it was smaller in the end of the experiment (53–69 h) than in the middle of the experiment (43–53 h).

Table 5.6   Effect of TDTMA-containing aqueous medium on the growth rate of *Fagopyrum esculentum*, excluding ungerminated seeds (in brackets, AAL increase rate).

TDTMA, mg/l	Growth rate of *Fagopyrum esculentum*, mm/h, in different periods of time		
	0–45 h	45–53 h	53–69 h
0	0.19 (0.12)	0.62 (0.41)	0.81 (0.55)
50	0.14 (0.10)	0.31 (0.22)	0.18 (0.13)
250	0.10 (0.08)	0.11 (0.08)	0.04 (0.03)

Note: AAL, apparent average length.
S.A. Ostroumov (1991) *Vodn. Resursy*, **2**: 112–116 (in Russian).

A comparison of the results of an experiment made on algae and seedlings indicates that the sensitivity of aquatic cultures of algae to TDTMA under experimental conditions is greater than the sensitivity of the seedlings. At the same time, the effects of TDTMA on soil cultures was realized at the concentrations of surfactants of the same order of magnitude as in the experiments with seedlings.

Different sensitivity of various biological test systems was many a time demonstrated by us in the biological testing of anionic, cationic, and nonionogenic surfactants. It gives additional confirmation to the principle that several test systems should be used in the study of the effects of xenobiotics and contaminants (Stroganov 1979; Burdin 1985; Cairns 1986; Filenko and Dmitrieva 1999). Therefore, it was interesting to include invertebrates into the range of the species under study. We used annelids (frequently used for biological testing), *Hirudo medicinalis*, and mollusks.

5.2.3   Effects on annelids

We studied the effect of TDTMA on *Hirudo medicinalis* (medicine leech). The experiments demonstrated the range of concentrations that cause a lethal effect (Tables 5.7, 5.8, and 5.9).

The observation of behavior reactions of the animals allows us to distinguish additional aspects of the biological activity of xenobiotics that pollute water. The changes in the behavior of leeches and other invertebrates in response to the action of xenobiotics were studied by B.A. Flerov (1989). In our experiments, the records of the behavior reactions allowed us to detect the effect of TDTMA at significantly lower concentrations than those needed for the traditional test based on mortality of the animals. For example, the behavior of leeches *Hirudo medicinalis* changed already at a TDTMA concentration of 1 mg/l, while the death of animals was detected at a significantly higher concentration equal to 10 mg/l. After 24 h of incubation at a concentration of 10 mg/l, 33% of the test animals perished. Moreover, the behavior reactions manifest themselves significantly earlier, almost a few minutes after the action of a cationic surfactant. (I am grateful to Professor B.A. Flerov and his colleagues, especially to L.N. Lapkina for recommendations and assistance in performing this series of experiments.)

We observed an increase in the mobility of the animals (a reaction, which is called in the laboratory headed by B.A. Flerov a change of the static state to dynamic

Table 5.7 Response of *Hirudo medicinalis* test organisms to the action of an aqueous medium with various concentrations of surfactant TDTMA (for details, see text and the next table).

TDTMA, mg/l	Intoxication phase	Response
1 (initial exposure)	Latent phase	No pronounced differences from the behavior of control animals are observed
1–5	Locomotor response to low concentration of xenobiotic	Increase in motor activity
5–10	Moderate intoxication	Anterior segments are bent under the ventral side of the body – a nonspecific symptom of toxication (NST) (can alternate with increased swimming activity)
10–50	Deep intoxication	Body is bent, flexion-extension movements, convulsions
250	Premortal state, acute intoxication	Animals perish after a short time

S.A. Ostroumov (1991) *Vodnye Resursy (Water Resources)*, **2**: 112–116 (in Russian).

Table 5.8 Effect of TDTMA-containing aqueous medium on the leech *Hirudo medicinalis*.

Concentration of TDTMA, mg/l	Organisms at different stages of experiment, h	
	23	31
0 (control)	100% of animals are alive, no NST	100% of animals are alive, no NST
5	100% of animals are alive, NST is observed in 67% of animals	100% of animals are alive, NST is observed in all animals
10	33% perished, NST is observed in the others	Number of perished animals did not increase, NST is still observed in all remaining animals
25	100% of animals perished	–
50	100% of animals perished (died in less than 1.5 h)	–

Note: NST, nonspecific syndrome of toxication.
S.A. Ostroumov (1991) *Vodnye Resursy (Water Resources)*, **2**: 112–116 (in Russian).

one, CSSD) at minimum concentrations tested in the experiment (1 and 5 mg/l) during a period from 10 to 60 min after the onset of incubation (Ostroumov 1991). Similar changes were also observed under the influence of other xenobiotics. At a sublethal concentration (one to two orders of magnitude lower than LC_{50}) the following chemicals caused the CSSD reaction: metal compounds (copper sulfate, mercury dichloride, potassium manganate, aluminum nitrate, potassium bromide, rivanol dye, phenol, carbamate pesticide Sevin (only 9 of the total 17 compounds studied) (Lapkina et al. 1987). Brilliant ethyl green dye and some pesticides (trichlorfon, metaphos, paraoxone, malathion) did not cause CSSD at such sublethal concentrations (Lapkina et al. 1987).

Table 5.9   Preliminary assessment of the median concentration of cationic surfactant TDTMA in an aqueous medium during the biotesting on *Hirudo medicinalis*.

Exposure time	Estimate of LC_{50}, mg/l
1.5 min	<250
1.5 h	<50
8 h	<25
24 h	10–25
48 h	<10

S.A. Ostroumov (1991) *Vodnye Resursy (Water Resources)*, **2**: 112–116 (in Russian).

Detection of the effect of low concentrations of xenobiotic during the recording of changes in the behavior (compared to mortality records) is characteristic of many chemicals and test organisms (Filenko 1988; Flerov 1989), but it is not an absolute rule, and new facts of this type should be considered useful. It was found in the experiments with *Hirudo medicinalis* that the concentration of many pesticides (DDVF, trichlorfon, paraoxone), at which an increase in the locomotor activity is observed, is significantly greater than LC_{50} (48 h) for the same pesticides (Flerov 1989, p. 38). Similarly, the concentration of a number of pesticides (dichlorodiphenyl trichloroethane, endrine, toxaphene, paratione, chlordane, pentachlorophenol) and other organic xenobiotics (arochlor 1254), which cause the avoidance reaction of some aquatic animals, exceeded the lethal concentrations several times and even by an order of magnitude (see Flerov 1989, pp. 60–61).

The effect of cationic surfactants on the behavior of leeches (CSSD) was discovered by us as a new fact, and cationic surfactants supplement the group of chemicals that are capable of causing CSSD. At the same time, not all xenobiotics automatically cause CSSD. The fact is interesting that some effects of sublethal concentrations of cationic surfactants on the behavior of leeches are manifested more clearly than the effects of sublethal concentrations of a number of pesticides.

The (discovered by us) fact of increased locomotor activity under the influence of low concentrations of cationic surfactants is of ecological importance. It was found that under the influence of cationic surfactants the animals, which were previously attached to the walls of the experimental beaker, detached and started to move in water. In conditions of a stream with floating water (river, creek) detachment from an immobile substrate is of vital importance, because if the velocity of the current is sufficient, the animal is transported downstream. Such behavior of annelids is adaptive, because the animals are transported away from the polluted part of the river and there is a possibility that they will arrive in a less polluted region. However, there is another aspect of this situation. Actually, the annelids emigrate from the ecosystem or the part of the ecosystem, which was previously their habitat. Of course, this is a loss of biological diversity for the ecosystem. In a certain sense we can state that from the point of the interests of the ecosystem a loss of organisms as a result of emigration is equal to their loss as a result of their death. Thus, while lethal effects were not observed at low concentrations, the overall result for the system can be equivalent to the death of the organisms.

These data are especially important because the medicinal leech is included into the Red Books and into the lists of endangered species and protected hydrobionts in many countries. In the economical aspect it is important that the leech occupies a place in the upper part of the ecological pyramid and is among the species regulating the species occupying lower positions. The ecological importance of such species–species interaction in aquatic systems was manifested especially clearly in the known experiments on manipulation of aquatic ecosystems by means of modifying the top of the ecological pyramid (Paine 1966).

Recently, the data about biological activity of aquatic solutions of cationic surfactants obtained additional importance in relation to the fact that modified methods of measuring the concentrations of cationic surfactants in aquatic objects indicated the presence of these chemicals in various reservoirs including the reservoirs in Ukraine (Kalenichenko 1996) at high concentrations (see above, Chapter 1).

5.2.4 Effects on mollusks

We studied the effects of cationic surfactant TDTMA on freshwater and marine mollusks.

5.2.4.1 Effects of TDTMA on freshwater mollusks

A. *Lymnaea stagnalis* (Gastropoda, Pulmonata). This species is a representative of an important group of mollusks with a wide range of food objects. These mollusks eat phytomass and dead parts of plants, thus performing an important sanitary function in water bodies. Assimilation of food of gastropods varies from 43 to 85% being on average equal to 67.6% (Monakov 1996). A significant part of the eaten food is excreted in the form of pellets, which rapidly deposit on the bottom of the water ecosystem (according to our data with a mean velocity of 0.82 cm/s, while the velocities in different measurements range between 0.6 and 1.6 cm/s). This accelerates the flux of chemical elements (C, N, P, and others) through aquatic ecosystems. The content of chemical elements in pellets (if the food was *Nuphar lutea*) was equal to: 69.74% of C, 2.3–2.9% of N, 0.4–0.5% of P, 1.1–1.7% of Si, 0.054–0.059% of Al (chemical analyses were performed by M.P. Kolesnikov

(Kolesnikov and Ostroumov 2000)). Taking into account the biomass of pulmonary mollusks in aquatic ecosystems, the flux of substance through the trophic web they represent can reach significant values up to 14.9 g/m^2 during 120 days of the most active vegetation season (Ostroumov 2000, Aquatic ecosystems and organisms - 2). Thus, the inhibition of feeding under the influence of synthetic surfactants denotes a hazard of decreasing the normal rate of migration of chemical elements through the ecosystem.

In our experiment, four beakers contained the tested solution of TDTMA and a total of 7.04 g of phytomass of *Nuphar lutea* (L.) Smith leaves (the total wet weight in four beakers) and 31 mollusks with a total weight of 55.72 g (wet weight). Four control beakers contained a total of 6.94 of phytomass and 20 mollusks with a total wet weight of 38.04 g. Concentrated solution of TDTMA was added to a final concentration of 2 mg/l. The incubation continued 45 h at a temperature of 22.5 ($\pm$1)°C. The values of eaten phytomass were obtained on the basis of the weight difference of phytomass at the beginning and end of incubation. The degree of inhibition of the feeding (normalized per unit weight of mollusks) exceeded 60% (Table 5.10). During the experiment, no mortality of mollusks was observed in the control and experimental beakers (Table 5.11).

Table 5.10 Effect of TDTMA on the trophic activity of *Lymnaea stagnalis* mollusks (see text for details).

TDTMA, mg/l	Phytomass (*Nuphar lutea*) eaten for 45 h, mg per 100 g mollusk weight (mg wet weight / 100 g wet weight)	Percent of the phytomass eaten by mollusks in the control	Extent of feeding rate inhibition, %
0	1970	100	0
2	680	34.5	65.5

B. *Unio pictorum* (common pearl shell). We studied the effect of TDTMA on water filtration by mollusks *Unio pictorum*. These mollusks are important as filter feeders participating in self purification of water and as components of ecosystems important for water transport through the ecosystems owing to the excretion of a significant part of filtered material in the form of faeces and pseudofaeces. According to incomplete data, the assimilation of food by freshwater bivalve mollusks is 40–47% (Monakov 1998).

In the experiment, eight mollusks *Unio pictorum* were placed in each of the beakers (1.5 liter) with settled tap water (STW). The mollusks were collected from sandy silt bottom of the Moskva River in the upper flows of the river upstream of Zvenigorod. Eight mollusks with a weight between 20.8 and 30.4 g (mean wet weight with the shell was 24.3 g) were in the beaker in variant A (control without cationic surfactant). Eight mollusks with a weight between 21.4 and 36.7 g (mean wet weight was 26.1 g) were in the beaker in variant B (with cationic surfactant tetradecyl trimethyl ammonium bromide (TDTMA), 2 mg/l). In both variants, some amount of the suspension of *Saccharomyces cerevisae* cells (SAF-moment preparation, S.I. Lesaffre, 59703 Marcq, France) was added beforehand into water. The final concentration of *S. cerevisae* on the basis of dry weight was 263 mg/l. In addition, an extra control was performed (variant C). In variant C, the beakers contained STW with the suspension of *S. cerevisae* but without any mollusks. TDTMA was not added in variant C. The beakers of all three variants were incubated at a temperature of 17°C. Aliquots were taken from the beakers and the optical density was measured at 500 nm (a Hitachi spectrophotometer 200-20, the optical path was equal to 10 mm). The measurement of optical density allowed us to characterize the concentration of cells in aquatic medium.

Table 5.11 Absence of a significant effect of TDTMA on mortality of *Lymnaea stagnalis* mollusks under experimental conditions during 2 days (the time period from the onset of experiment is given).

Time, days	TDTMA, mg/l	Number of mollusks (total for four beakers)	Survival, % of the initial number	Mortality, %
1	0	20	100	0
	2	31	100	0
2 (end of experiment with feeding)	0	20	100	0
2	2	31	100	0
3 (additional exposure after the main experiment was over)	0	20	100	0
3	2	28	90.3	9.7

The results of measurements indicated that in both variants A and B a decrease in the optical density was observed during the incubation (Table 5.12) as a result of water filtration by mollusks and removal of suspension cells from water by mollusks. A relatively stable level of the content of suspension cells was observed in variant C. It is clearly seen that in the presence of TDTMA the rate of withdrawal of cells from water was significantly less than in the control.

Thus, the results of the experiment indicate that cationic surfactant TDTMA significantly suppressed the removal of unicellular organisms from the water medium by mollusks, which indicates a drop in the water filtration rate under the influence of this chemical. Similar results were obtained during the study of the effect of Triton X-100 surfactant (5 mg/l) on the rate of the filtration removal of unicellular organisms from water (including *Synechocystis* sp. 6803, *Scenedesmus quadricauda*) by *Unio tumidus* mollusks.

5.2.4.2 Effects on marine mollusks

We studied the effect of TDTMA on water filtration by mollusks *Mytilus galloprovincialis* and *Crassostrea gigas*.

A. *Mytilus galloprovincialis* (Mediterranean–Black Sea mussel) juvenile mollusks were taken (see the section for methodology in Chapter 2), which settled

Table 5.12   Tetradecyl trimethyl ammonium bromide (TDTMA) inhibits the ability of *Unio pictorum* to filter cells of *Saccharomyces cerevisae* suspension from water.

Time from the onset of incubation, min	OD$_{500}$			
	Variant A (no TDTMA)	Variant B (with TDTMA)	Variant C (no mollusks, no TDTMA)	A/B, %
22	0.285	0.370	0.373	77.03
45	0.235	0.378	0.364	62.17
80	0.185	0.320	–	57.81
110	0.142	0.264	0.414	53.79

Note: Mollusks were placed in beakers with 1.5 l of settled tap water (STW). Variant A (control, no cationic surfactant), 8 mollusks of average weight 24.3 g (wet weight with shells); variant B (TDTMA, 2 mg/l), also 8 mollusks of average weight 26.1 g. In both variants, a cell suspension of *Saccharomyces cerevisiae* (263.1 mg/l) was preliminarily added to water. In additional control (variant C), the beakers contained STW with *S. cerevisiae* suspension, but without mollusks; no TDTMA was added. Temperature, 17°C. Optical density of aliquots from the beakers was measured at 500 nm. For details, see text and also *Food Industry at the Edge of the Third Millennium*, Moscow, 2000, pp. 251–254 (in Russian).

recently and started the attached period of life with a weight of approximately 1 mg. Approximately 500 animals were placed in one beaker. The total biomass of mollusks in one beaker was 0.5 g (wet weight with the shell). Mollusks were placed in beakers with 25 ml of seawater, and 25 ml of algal suspension *Monochrys lutheri* was added (12 million cells/ml) to the final concentration of 6 million cells/ml. The time of adding the algal suspension was considered as the beginning of incubation. The final volume of the incubation medium in each beaker was 50 ml. The incubation was performed at 25.8°C. The optical density of the incubation medium was measured at 650 nm (optical path was 10 mm) on a LOMO SF-26 spectrophotometer.

During the time of incubation (51 min), a decrease in the optical density of suspension occurred both in the control (without surfactant) and in the experiment (in the presence of TDTMA). A decrease in the optical density in the course of incubation indicated a decrease in the concentration of the algae due to their removal from water as a result of water filtration by mollusks (Table 5.13).

The experiments demonstrated an inhibitory effect of TDTMA at a surfactant concentration of 1 mg/l (Table 5.13). A three-fold difference between optical densities in the control and experiment was observed after 26 min of incubation, and a four-fold difference was observed after 50 min of incubation. An additional control was set, which demonstrated the absence of the effect of TDTMA (2 ml/l) on the optical density of *Monochrys lutheri* algae when they were incubated under the same conditions.

B. *Crassostrea gigas* (giant oyster). This Pacific species is very valuable for aquaculture. The conditions of the experiment are described in the notes to Table 5.14. High sensitivity of the organism and test reaction attract attention.

The efficiency of removal (EER) was greater than 200% after 5 min of exposure, more than 300% after 11 min, and more than 700% after 20 min. We emphasize that we used a comparatively low concentration of synthetic surfactants (0.5 ml/l). All organisms used in the experiment remained alive, i.e., the observed effect was sublethal.

Before, we showed negative sublethal effects of surfactants on water filtration by marine mollusks at the same concentrations of sodium dodecyl sulfate and Triton

Table 5.13 Effect of TDTMA on the filtration activity of *Mytilus galloprovincialis* Lam. mollusks.

Measurement period	Time from the onset of incubation, min	OD_{500}	
		Variant with TDTMA	Variant without TDTMA (control)
1	10	–	0.088
	11	0.135	–
2	14	–	0.061
	15	0.128	–
3	26	–	0.039
	27	0.125	–
4	41	–	0.030
	42	0.116	–
5	45	–	0.027
	46	0.114	–
6	50	–	0.023
	51	0.106	–

Note: Optical density of a suspension of *Monochrysis lutheri* algae during water filtration by mollusks is given. Concentration of TDTMA at the onset of filtration, 1 mg/l. See text for the other experimental conditions.

Table 5.14 Effect of TDTMA (0.5 ml/l) on water filtration by *C. gigas*.

Measurement No.	Incubation time, min	Optical density at 550 nm			B/A, %
		Variant A (no TDTMA)	Variant B (with TDTMA)	Variant C (only *S. cerevisiae*, no mollusks, no TDTMA)	
1	5	0.080	0.194	0.307	242.5
2	11	0.043	0.148	0.305	344.2
3	20	0.018	0.137	0.303	761. 1

Note: Experimental conditions: Ten mollusks each were placed in experimental beakers (total wet weight with shells: 47.3 g in beaker A; 55.2 g in beaker B). Age of the mollusks, 1 year. Incubation temperature, 27°C. Initial concentration of *S. cerevisiae*, 100 mg/l (dry weight). The volume of seawater during incubation, 500 ml.

X-100. The filtration activity of another important group of invertebrates, rotifers (*Brachionus angularis*, *B. plicatilis*), was also disturbed under the influence of surfactant TDTMA (0.5 ml/l) (Kartasheva and Ostroumov 2000). Taking into account the important role of filter feeders in aquatic ecosystems, these effects are important for a more complete estimate of the ecological hazards of chemicals. Therefore, it is reasonable to expand the amount of experimental data as a basis (in the study of mixed preparations containing surfactants) for more detailed analysis and generalizations.

5.3   Biological Effects of Benzethonium Chloride

In our joint research project with D. Galyama and other colleagues from Bratislava (Slovakia) we demonstrated the ability of cationic surfactant benzethonium chloride (benzyl diisobutyl phenoxyethoxyethyl dimethyl ammonium chloride) to suppress the vitality of cells of euglena (*Euglena gracilis* Klebs). The inhibitory activity of this cationic surfactant for euglenas was significantly greater than that of the two other synthetic detergents tested at the same time (Kristall and Lotos-Avtomat). The data of the experiments indicated that even the lowest of the tested concentrations (10 mg/l; 20 mg/l; and 100 mg/l) produced an inhibitory effect. It was shown that at all of these concentrations including 10 mg/l a complete suppression of growth occurred (Ostroumov 1991).

5.4   Other Data on the Biological Activities of Cationic Surfactants

High biological activity of cationic surfactants allowed the use of some of them as pesticides, and in particular as fungicides. For example, benzalconium bromide $C_{21}H_{38}BrN$, which is an alkyl ammonium bromide, is used as fungicide. In the paper by Steffann et al. (1988) the authors demonstrated the ability of this cationic surfactant to uncouple mitochondrial membranes of *Agaricus bisporus* fungi at a concentration of 6 µM (Steffann et al. 1988). This confirms the membranotropic character of the molecular mechanisms of action of cationic surfactants. (Indication of the ability for uncoupling the coupled membranes has something in common with a similar ability of anionic surfactant SDS – V.P. Skulachev, personal communication).

It was shown that the effect of catamine AB (alkyl methyl benzene ammonium chloride) on green algae *Chlorella vulgaris* is related to its effects on the photosynthetic apparatus of the cell (Parshikova et al. 1994).

Biological activity of aquatic media containing surfactants is not reduced in all cases to the traditionally expected toxicity. It is possible that the biological activity of surfactants is sometimes related to a certain degree to the ability of the molecules of surfactants to aggregate or integrate into supramolecular structures.

For example, among the cationic surfactants, the quaternary ammonium compound (QAC), dimethyl dioctadecyl ammonium bromide (DDA), has an interesting biological effect. In mammals, DDA intensifies immune responses including antineoplastic responses. DDA stimulates a number of activities in macrophages including fast adhesion to a substrate, phagocytosis, formation of interleukin-1, and lysosomal activity. The ability of DDA to induce peritoneal macrophages that possess cytotoxic activity and activity inhibiting the growth of some lines of cancerous cells was demonstrated (Prager et al. 1985).

Pinnaduwage and coauthors (1989) compared the toxicity of aquatic medium containing cetyl trimethyl ammonium bromide (CTAB) or CTAB together with liposomes prepared from phospholipid dioleyl phosphatidyl ethanol amine. It appeared that in the range of concentrations of CTAB from 2 to 20 µg/ml the aquatic medium with dissolved CTAB is more toxic than the medium containing CTAB at the same concentration and liposomes. Hence, the presence of phospholipids structures in aquatic medium can change the toxicity of dissolved surfactants.

Returning to the analysis of the information described in this chapter it is reasonable to review once again our results about the disturbances of filtration activity of mollusks under the influence of cationic surfactants. These results make it worth considering the data by other authors, who studied the effects of cationic surfactants on mollusks. It was found that colonization of a substrate by the larvae of *Corbicula fluminea* mollusks decreased at the concentration of lauryl trimethyl ammonium chloride (LTMAC) equal to 0.03 mg/l (Belanger et al. 1993). A significant role of these mollusks in ecosystems is characterized by the fact that populations of *Corbicula fluminea* in the Potomac River are capable of filtering the entire available water column during 1.5–2 days (Cohen et al., 1984, cited from Belanger et al. 1993).

Summarizing our new results and the data by other authors, we see that synthetic surfactants can affect several stages of the life cycle of mollusks: there are data about suppressing the growth of larvae, making disturbances of their orientation, producing a decrease of the ability to colonize a substrate, a decrease in the filtration activity of juvenile species settled on the substrate, a decrease in the filtration activity of adult species (sexually mature species), and a decrease in the growth of adult specimens. They create the conditions for summation of many biological effects, which are finally focused in one direction, to decrease the contribution of the mollusk population to the processes of water filtration in the ecosystem.

The ecological role and danger of cationic surfactants as pollutants increases because the process of their biological degradation occurs slower than that of many other xenobiotics. According to some data, the coefficient of biochemical oxidation in natural water at 20°C was equal to 0.02 day^{-1} for dimethyl benzyl octadecyl alkyl ammonium chloride, and from 0.002 to 0.05 day^{-1} for trimethyl alkyl ammonium chloride (Kaplin 1979). Additional information about biodegradation of synthetic surfactants is available in Swischer (1987) and Poremba et al. (1991).

Thus, we revealed and studied new biological effects and impacts of some cationic surfactants (such as quaternary ammonium compounds ethonium and tetradecyl trimethyl ammonium bromide) on various biological objects, which were not studied earlier (marine bacteria *Hyphomonas*, cyanobacteria, algae, plants,

Hirudo medicinalis). New negative biological effects were characterized quantitatively. They expand our scientific knowledge about the consequences of cationic surfactants' transfer in the environment and show new aspects of the ecological hazard they cause.

The summary table (Table 5.15) indicates that the overwhelming majority of the parameters we studied are highly sensitive to cationic surfactants. A comparison of the results of investigation of the biological effects of anionic, nonionogenic, and cationic synthetic surfactants on organisms demonstrates the following general regularities: all groups of synthetic surfactants inhibited the filtration activity of mollusks, which causes a danger of distorting the processes important for the self-purification potential of the aquatic ecosystems. On the other hand, relatively low sensitivity of angiosperm plants was revealed, which allows us to propose the use of these organisms for phytoremediation.

Table 5.15   Sensitivity of various organisms to cationic surfactants.

Organisms	Sensitivity	References/comments
Various heterotrophic bacteria	H	General assessment of cationic surfactants as substances with high antibacterial activity (Stavskaya et al. 1988)
Cyanobacteria	H	Kolotilova et al. 1998, Ostroumov and Kolotilova 1998, Ostroumov et al. 1999a,b
Chlorella vulgaris	H	Cationic surfactant catamin AB was more toxic than tested nonionogenic and anionic surfactants (Parshikova et al. 1994)
Seedlings of *Fagopyrum esculentum, Cucumis sativus,* and others	L or M	Comparatively low sensitivity made possible the use of these organisms for phytoremediation (Ostroumov 2000)
Annelids *Hirudo medicinalis*	H	Change of behavior can pose hazard of species emigration from ecosystems (Ostroumov 1991, 2000g)
Lymnaea stagnalis	H or M*	Decrease of trophic activity can introduce changes into the ecosystem (Ostroumov 2000a)
Freshwater mollusks *Unio* sp., rotifers	H	Comparatively high sensitivity poses hazard of disturbing the filtration activity and reducing suspension removal from water (Ostroumov 2000g, Kartasheva and Ostroumov 2000)
Freshwater mollusks *Corbicula fluminea*	H	Decrease of substrate colonization by molluskan larvae (at cationic surfactant of 0.043 mg/l) poses hazard of disturbing population reproduction (Belanger et al. 1993)
Marine mollusks *Mytilus galloprovincialis*	H	Comparatively high sensitivity poses hazard of disturbing the filtration activity and reducing suspension removal from water (Ostroumov 2000g)
Marine mollusks *Crassostrea gigas*	H	Comparatively high sensitivity poses hazard of disturbing the filtration activity

Note: L, low; M, moderate; H, high. *Preliminary estimate.

The use of vascular plants for biological testing allowed us to accumulate data (see above and also Ostroumov 1991a,b, 2000; Ostroumov and Semykina 1993), which can be used to establish the order of the representatives of different classes of synthetic surfactants in terms of biological activity they manifest. By showing increasing degrees of inhibitory action on *F. esculentum*, synthetic surfactants are positioned in the following order: polymeric surfactant CHMA (copolymer of hexene and maleic aldehyde) < anionic surfactant sodium dodecyl sulfate < noniono-genic surfactant Triton X-100 < cationic surfactant TDTMA.

The revealed biological effects of individual substances belonging to the group of synthetic surfactants (all three main classes, anionic, nonionogenic, and cationic surfactants) allow us to predict or put forward a working hypothesis that mixed preparations containing surfactants will also produce similar effects on test organisms. The next chapter gives the verification of this hypothesis.

6 Biological Effects of Surfactant-Containing Mixtures and Other Preparations

Along with the biological effects of individual chemicals, representatives of synthetic surfactants, it is also necessary to characterize the biological effects of mixtures that contain synthetic surfactants. One has to keep in mind permanently that real pollution of the environment is usually complex as mentioned in Patin (1979), Fedorov (1987), Filenko (1988), Venitsianov (1992), Losev et al. (1993), Bezel et al. (1994), and Krivolutsky (1994).

First of all, it is interesting to study the effects of those specific mixtures of chemicals (complex preparations and compositions) that include synthetic surfactants. Synthetic surfactants as part of these compositions are discharged to the environment. Synthetic detergents and foam detergents occupy important places among these mixtures, compositions, and preparations. According to estimates, 2 g of synthetic surfactants daily per person per day enters the drainage system in Russia, mainly owing to the use of synthetic and foam detergents (Akulova and Bushtuyeva 1986). In a number of countries this quantity is even greater: in Germany it exceeds 11 g (Steinberg et al. 1995, cited from Ostroumov 2000). Foam washing compositions pollute aquatic ecosystems in large degree as a result of their use to clean pipelines, tanks and vessels for storing oil products, tankers, etc.

Also, it is interesting to get comparative information about the sensitivity of the same biological tests to synthetic surfactants and other xenobiotics, which is necessary to compare the degree of the ecological hazard of synthetic surfactants and other chemicals polluting the environment, e.g., pesticides, which were studied earlier.

6.1 Impact of Aquatic Media With Surfactant-Containing Mixtures on Hydrobionts: Earlier Works

Synthetic surfactants frequently enter the environment as composite mixtures (Abramzon 1979; Lewis 1992). Aquatic media, in which surfactant-containing preparations are dissolved – such as oil dispersants (Nesterova 1980) and synthetic detergents – can disturb the vital activity of hydrobionts (Braginsky 1987, Flerov 1989).

The effect of oil dispersants on hydrobionts was analyzed in Fedulova et al. (1976), Nesterova (1980), Grozdov et al. (1981), Gapochka (1983, 1999), Gapochka et al. (1978, 1980), Gapochka and Karaush (1980), Bobra et al. (1989), Tukaj (1994), Burridge and Shir (1995), Singer et al. (1995), including the reviews (Patin 1997). Although the negative effect of dispersants has been demonstrated, in some works an opinion is put forward that modern dispersants in general are not highly toxic and present no hazard for the environment (see Chapter 1).

The effect of three dispersants used for combating oil spills on *Chlorella vulgaris*, *Gastrotricha infusorians*, *Stylonichia mytilus*, and *Daphnia magna* was studied in Grozdov et al. (1981). The least toxic chemical among the dispersants for all organisms was SN-79, and the most toxic was OM-6.

The sensitivity of several marine organisms to oil dispersant Corexit 9554 was studied by Singer and co-authors (Singer et al. 1995). The sensitivity of the species studied differed approximately by a factor of 20, as we can judge by the differences in the median effect (8–184 mg/l). The embryos of haliotis shell *Haliotis rufescens* were most sensitive. Moderate sensitivity was manifested by larvae of fish *Atherinops affinis* and zoospores of brown algae *Macrocystis pyrifera*. The least sensitive were mysids *Holmesimysis costata* (Singer et al. 1995). Germination of zygotes of brown algae *Phyllospora comosa* was studied under the influence of several dispersants (Burridge and Shir 1995). The strongest inhibition was observed under the influence of Corexit 9500 dispersant. Corexit 8667 and Corexit 7664 manifested a comparatively lower inhibition. Corexits also intensified the inhibitory effect of diesel oil (Burridge and Shir 1995).

The sensitivity of *D. magna* to five Corexit preparations was determined. Corexits 9527, 7664, 8667, 9660, and 9550 contain nonionogenic surfactants (Bobra et al. 1989). Corexit 9527 contains ionic synthetic surfactant along with nonionogenic surfactants. The value of LC_{50} for Corexits was sufficiently low (i.e., they manifested high toxicity) for daphnia *D. magna*. The LC_{50} value (48 h) ranged from 3 to 270 mg/l at a temperature of 5°C and from 0.5 mg/l to 88 mg/l at a temperature of 20°C. The strongest effect was observed for Corexit 8667, the least one was observed for Corexit 7664. The value of LC_{50} for a water-soluble fraction (WSF) of crude oil (Norman Wells crude oil, Esso Resource Canada Ltd.) was equal to 7 mg/l (at 20°C), i.e., the relative toxicity of this dispersant was approximately 200 times greater than the toxicity of a water-soluble fraction of oil. The values of LC_{50} (48 h, 20°C) for the mixture of three types of oil and five dispersants (Corexits 9527, 7664, 8667, 9660, and 9550; the volume ratio of dispersant to oil was 1:20) varied within a range of 1.1–5.2 mg/l. The values of LC_{50} for the mixture of Corexit 7664 and three types of oils were several times lower than LC_{50} of one dispersant applied without oil. It was also shown for all mixtures of Corexits and oil that their toxicity was higher than the toxicity of physical dispersions of oil without Corexits. Under conditions of the experiments, the toxicity of oil pollution increased when dispersants containing nonionogenic surfactants were added to the system (Bobra et al. 1989).

The study of the effect of some dispersants on *Scenedesmus quadricauda* Breb. did not reveal any significant effects at concentrations ranging from 0.01 to 0.1 g/l

(Fedulova et al. 1976). After 5–20 days, the Corexit dispersant at concentration equal to 0.1 mg/l increased the content of chlorophyll *a* (mg by 1 million of cells). Dispersant DN-74 increased the content of chlorophyll *a* by the 5th day, and later by the 10th and 15th days the inhibition appeared (Fedulova 1976). It is interesting to compare these results with a later work of the Polish investigator Z. Tukaj, who studied the effect of dispersants (as well as extracts and emulsions of diesel oil) on *S. quadricauda* and five other species of the *Scenedesmus* genus (Tukaj 1994). He demonstrated that the toxicity of emulsions is greater than the toxicity of emulsifiers (e.g., emulsifier DP-105) and that the sensitivity of the species decreases in the series *S. microspina* >> *S. obliquus* > *S. armatus* > *S. opoliensis* > *S. acutus* >> *S. quadricauda* G-15.

Oil emulsifier EPN-5 and oil dispersant DN-75 developed at the Institute of Oceanology RAS did not exert high toxicity for marine bacteria and other hydrobionts (Nesterova 1980). A concentration of EPN-5 of 2 mg/l was inert to daphnia. The effect of dispersant EPN-5 and other chemicals (Corexit 7664, diproxamine 157, berol) on cyanobacteria *Synechococcus aquaticus* and *Anabaena variables* was studied in works by Gapochka et al. (1980). It was shown that under certain experimental conditions toxicity of oil (arlanian, romashkian) and oil products (motor fuel, diesel fuel) decreases in the presence of dispersants (Gapochka et al. 1980).

An interesting research was performed at the University of Oregon (Corvallis) to study the effect of Corexit 9527 on glucose consumption by marine microorganisms (Griffiths et al. 1981). The inhibition of glucose consumption started at the concentration of glucose equal to 1 mg/l and greater. At 12 mg/l, the consumption of glucose decreased by 50%. This research was performed using 149 samples of seawater from a depth of 3 m and 95 samples of bottom sediments from depths ranging from 1 m to 2200 m collected at the coasts of Alaska. We think that these data are interesting in relation to the problem whether synthetic surfactants can distort the processes participating in self-purification of water (see Chapter 7 for details).

The other widely used classes of multicomponent preparations containing synthetic surfactants includes synthetic, foam, and liquid detergents. Their biological effects were investigated. On the one hand, they passed the tests on laboratory animals, and they are considered to satisfy the requirements (lack of clearly manifested toxicity, etc.) for chemicals that are in permanent contact with humans. On the other hand, certain data make us think about their ecological importance and potential hazard.

There are data on some of synthetic detergents and unfavorable effects on the laboratory animals and humans (Eskova-Soskovets et al. 1980; Ilyin 1980; Talakin et al. 1985). There are indications of allergic effects caused by synthetic detergents and synthetic surfactants (Eskova-Soskovets et al. 1980). Inhalation effect of detergents Lotos (18% of alkyl benzene sulfonate) and Era (8% of alkyl benzene sulfonate) for a month at a concentration of 50 mg/m^3 caused inhibition of nonspecific immune factors such as a decrease in the phagocytic activity and concentration of lysozyme in blood serum, and also affected lipid exchange processes and exchange of cyaloglycoproteins (Talakin et al. 1985).

It is important for understanding the phenomenon of complex pollution that a four- to five-fold decrease in LC$_{50}$ (i.e., an increase in toxicity) was demonstrated for a number of organic xenobiotics in experiments with laboratory mammals (white rats and mice, perorally) as a result of their interaction with synthetic surfactants (anionic azolate A, sulfonol NP-1, and nonionogenic surfactants OP-7 and OP-10) and redistribution in aquatic media containing surfactants at a level of only $1-10$ MPC (Ilyin 1986). Such a significant decrease in LD_{50} was observed for heptachlor, hexachloro-cyclohexane (HCCH) and gamma HCCH, carbophos, metaphos, trichlorine meta-phos, trichlorfon, granosan, cuprozan, cyneb, carbation, and aminophenol dyes (ortho-, meta- and paraisomers). In the presence of surfactants in the medium, such pesticides as HCCH (at a level of 1 MPC) caused significant changes in test mammals: a decrease in the number of erythrocytes, a decrease in the content of hemoglobin, a decrease in the activity of choline esterase or immune biological reactivity, and cardiotoxic action. In the presence of surfactant in the medium, a four-fold increase of mutagenic activity of HCCH and blastomogenic activity of 3,4-benzo-pyrene was observed. Virulence of pathogenic strains *Salmonella typhimurium* (5-fold), *S. typhi* (3.6-fold), *S. schottmulleri* (5.8-fold), Zone and Flexner pathogenic shigella (from 2.8 to 4-fold), pathogenic enterovirus LSc 2ab, ECHO Koksaki A-7 (from 7.3- to 93-fold) was increased in a medium with surfactants (Ilyin 1986). The author (Ilyin) associates these effects with the fact that traditional concepts about the uniform distribution of chemicals in the total volume of aquatic environment are not completely consistent.

In the presence of synthetic surfactants an adsorption film is formed at the water surface, at which the other contaminants (in the paper by Ilyin, pesticides and amino-phenol dye) and pathogenic microorganisms concentrate. It is possible that the presence of synthetic surfactants also increases the permeability of cell membranes for hydrophobic xenobiotics and pathogenic microorganisms, which in turn increases the observed virulence of the latter. This fact further emphasizes the hazards of complex pollution of the environment.

It is reasonable to remember the fact that some other forms of chemical pollution of aquatic medium also favor pathogenic microorganisms, e.g., pathogenic strains of mycobacteria grow faster in media containing oil hydrocarbons. They grow faster than in traditional media used for many years in medical microbiology for cultivation of, e.g., strains of tuberculosis pathogenic mycobacteria (Dr T.V. Koronelli, the report at a scientific seminar at the Department of Hydrobiology, MSU, Nov. 26, 1997).

Lewis (1992) analyzed the papers by English-writing authors from the viewpoint of whether antagonism or synergism takes place during the joint action of synthetic surfactants and other chemicals. According to his data (a total of 33 publications and 40 main combinations of chemicals were analyzed) antagonism was recorded in 5 cases, synergism was found in 23 cases, and in 12 cases synergism was not revealed (these cases were characterized by Lewis as non-synergetic (Lewis 1992). At the same time, it is noteworthy that that research was carried out based on the traditional approach without studying the surface film, as was done by Ilyin (1986; see above). Even using this analysis (it is possible that the probability of a sharp increase in the harmful effect, which was clearly brought out by Ilyin, could remain beyond the scope of the studies analyzed by Lewis), the number of synergism situations exceeded mere than four times the number of antagonism cases by a factor of four.

As was noted above, such preparations as surfactant-containing synthetic detergents enter environment in large amounts.

The behavior of medicine leech *Hirudo medicinalis*, guppy *Lebistes reticulates*, *Asellus aquaticus*, and branchiopods *Streptocephalus torvicomis* changed in aquatic medium with dissolved detergent Lotos-71 (Flerov 1989).

Synthetic detergents were lethal for daphnia (Braginsky et al. 1979). The value of LC_{50} was approximately 0.8–30 mg/l (Braginsky et al. 1979).

The effect of household detergent (2 and 4 mg/l; the latter corresponded to a concentration of LAS dodecyl benzene sulfonate of 0.8 mg/l) on the crustaceans *Tisbe holothuriae* in conditions of low and high population density was investigated (Faba and Crotti 1979). At low density of the population, the detergent caused a decrease in the mean number of nauplii hatched from eggs. At high density of population, the detergent caused an increase of this index, i.e., the detergent inhibited the mechanism that allowed the animals to feel overpopulation and react to it. It is possible that this mechanism is related to the excretion of a specific substance by the crustaceans. A suggestion was put forward that the detergent can interact with this substance or damage chemoreceptors of the animals (Faba and Crotti 1979).

Synthetic detergents inhibit the activity of some enzymes in fish gills. For example, 10–40 g of synthetic detergent Lotos inhibited succinate oxidase and cytochrome oxidase in tissues of minnow fish gills from Lake Baikal (Kolupayev and Putintseva 1983).

The authors compared the toxicity of some synthetic surfactant-containing preparations (both dispersants and synthetic detergents) to representatives of different links of the trophic web including Protococcaceae algae and Cladocera crustaceans, and concluded that synthetic surfactants were more toxic to representatives of higher levels of the trophic web (Grozdov et al. 1981). However, the range of biological effects of surfactant-containing preparations remained insufficiently complete.

We studied the effect of mixtures (several types of synthetic detergents) on autotrophic (flagellates, angiosperm plants) and heterotrophic organisms (mollusks).

6.2 New Results on the Impact of Surfactant-Containing Mixtures on Autotrophic Organisms

We studied the effect of some synthetic surfactant-containing preparations on different organisms. One should bear in mind that the content of synthetic surfactants in these preparations is always less than 100% and the usual content of synthetic surfactants in synthetic detergents does not exceed 15–20% (however, it may reach as high as 40%).

6.2.1 Effects of surfactant-containing preparations on phytoplankton

Part of our experiments was the study of the impact of synthetic detergents on the euglena culture and seedlings of plants. As was shown in our joint work with

Table 6.1 Effect of detergent Kristall on the culture *Euglena gracilis* Klebs var. Z. Prings-heim strain 1224-5/25 (abundance per 1 ml is given).

Detergent, g/l	Time, h			
	0	20.5	44.5	48
0	1.2×10^5	1.2×10^5	5.8×10^5	7.5×10^5
0.01	1.1×10^5	0.9×10^5	5.1×10^5	6.6×10^5
0.10	1.3×10^5	1.3×10^5	5.3×10^5	6.0×10^5
0.5	1.4×10^5	1.2×10^5	0.9×10^5	1.1×10^5

Detergent, g/l	Time, h			
	68	72	96	99.5
0	2.2×10^6	2.5×10^6	4.8×10^6	4.4×10^6
0.01	1.9×10^6	1.8×10^6	3.4×10^6	4.0×10^6
0.10	1.4×10^6	1.6×10^6	2.3×10^6	2.2×10^6
0.5	1.2×10^5	9.7×10^4	1.0×10^5	0.5×10^6

Note: Temperature, 26°C; illumination, 1500–2000 lux.
(Ostroumov and Vasternak (1991) *Vestn. Mosk. Universiteta, Ser. 16. Biol.*, **2**: 67–69 (in Russian).

K. Vasternak (Halle, Germany), inhibition or cessation of the growth of the culture of *Euglena gracilis* Klebs was observed in the aquatic medium containing the synthetic detergent Kristall (Table 6.1). During the first 48 h of the experiment with a detergent concentration equal to 0.1 mg/ml the effect was insignificant, but during the next stage of the experiment (68–99 h) the abundance of the remaining cells was noticeably lower than in the control. During the final stage of the experiment (72–99 h), at a detergent concentration of 0.01 mg/ml, we also observed a decrease in the abundance of the cells as compared to the control.

These results are in agreement with the overall results of another independent experiment carried out together with D. Galyama, I. Legotsky, and D. Slugen (Slovakian Polytechnic Institute, Bratislava, Slovakia). At a concentration of detergent Kristall equal to 0.3 mg/ml in aquatic medium the growth of euglena in the latter experiment was completely suppressed during 65 h, but by the end of the experiment (112 h) a slight increase in the density of the culture occurred (Table 6.2, see also Ostroumov 1991). At a concentration of detergent Kristall equal to 0.1 mg/ml the inhibitory effect was observed during both measurements (65 and 112 h). At a concentration of detergent Kristall equal to 0.2 mg/ml, a weak inhibitory effect was observed. Both independent experiments allow us to conclude that a significant or even major part of the range of inhibitory effect manifestation in the presence of detergent Kristall on euglenas falls between concentrations 0.01 and 0.5 mg/ml.

Aquatic medium with synthetic detergent Bio-S also suppressed the development of euglenas *E. gracilis*, as was revealed in a joint work with K. Vasternak

Table 6.2 Effects of detergents on the density of *Euglena gracilis* Klebs cultures.

Detergent, mg/ml	Experiment No.	Time, h	
		65	112
detergent Kristall			
0	1	0.40	0.45
	2	0.34	0.42
	3	0.38	0.44
		0.373 (mean)	0.436 (mean)
0.02	1	0.30	0.33
	2	0.30	0.34
		0.30 (mean)	0.335 (mean)
0.1	1	0.22	0.28
	2	0.20	0.18
		0.21 (mean)	0.23 (mean)
0.3	1	0.02	0.15
	2	0.002	0.16
		0.011 (mean)	0.155 (mean)
detergent Lotos-Avtomat			
0	1, 2, 3	0.37 (mean)	0.44 (mean)
0.02	1	0.39	0.53
	2	0.38	0.48
		0.385 (mean)	0.505 (mean)
0.1	1	0.31	0.36
	2	0.25	0.36
		0.28 (mean)	0.36 (mean)
0.3	1	0.28	0.26
	2	0.20	0.32
		0.24 (mean)	0.29 (mean)

Note: Density of the cultures is expressed in conventional units. D. Galyama, I. Legotsky, and D. Slugen participated in the experiments.

(M. Luther University, Halle, Germany). At a concentration of synthetic detergent equal to 0.5 mg/ml, the suppression was complete: the growth of the cells practically stopped (Table 6.3). During the final stage of the experiment (48–96 h), at a detergent concentration equal to 0.1 mg/ml the density of the culture was smaller than in the control. At a concentration of synthetic detergent equal to 0.01 mg/ml, no clear effects were detected. These data agree with the results of our joint work with the scientists from Slovakia on the effect of another detergent, Lotos-Avtomat, on euglenas. The data of the experiment indicate that detergent Lotos-Avtomat at a concentration of 0.02 mg/ml does not manifest any inhibitory effect. However, at a concentration of 0.1 mg/ml a slight inhibition was observed. At a concentration of synthetic detergent equal to 0.3 mg/ml, the inhibitory effect was more pronounced. Thus, both independent experiments indicate that the negative effect of synthetic detergent was realized in the range of concentrations from 0.1 to 0.5 mg/ml.

Experiments were carried out to study the effect of aquatic media containing liquid foam detergent on phytoflagellates (Ostroumov et al. 1990b). Preparation Kashtan (0.02 mg/ml) that contained surfactants caused degradation of cells of marine flagellates *Olisthodiscus luteus* N. Carter 1937 (Chromophyta, Class Raphidophyceae). After 4 h of incubation in the light, the optical density of the suspension of cells sharply decreased at all wavelengths. In particular, the ratios of optical density at 660 nm, 675 nm, and 690 nm to that at the same wavelengths in the control suspension (without synthetic detergent) were equal to 29.9, 30.1, and 22.8%, respectively.

The same preparation Kashtan caused destruction of cells of the photo-synthesizing bacteria *Rhodospirillum rubrum*. The experiments by the author demonstrated that after 3 h of incubation in the light a decrease in the optical density by more than 25% occurred over the entire part of the spectrum with a wavelength > 450 nm. After 21 h of incubation in the light, the spectrum lost its form and the former maxima of absorption at 515–520 nm and 550–555 nm were not detectable at all on the spectral curve.

The effect of surfactant-containing preparations on algae was studied by many authors. In the discussion we consider some literature data on the effect of mixed surfactant-containing preparations on phytoplankton.

Table 6.3   Effect of synthetic detergent Bio-S on the culture *Euglena gracilis* Klebs var. Z. Pringsheim strain 1224-5/25 (abundance per 1 ml is given).

Concentration, g/l	Time, h			
	0	20.5	44.5	48
0	1.0×10^5	1.6×10^5	3.7×10^5	7.7×10^5
0.01	1.0×10^5	1.2×10^5	5.4×10^5	6.8×10^5
0.10	6.0×10^5	1.6×10^5	5.3×10^5	5.6×10^5
0.5	1.2×10^5	1.8×10^5	1.3×10^5	8.9×10^5

Concentration, g/l	Time, h			
	68	72	96	99.5
0	2.2×10^6	1.9×10^6	4.4×10^6	4.1×10^6
0.01	1.7×10^6	2.4×10^6	3.5×10^6	4.0×10^6
0.10	1.2×10^6	1.7×10^6	3.6×10^6	3.2×10^6
0.5	1.0×10^5	1.0×10^5	9.6×10^4	1.2×10^5

Note: Temperature, 26°C; illumination, 1500–2000 lux.

The effect of synthetic detergents on phytoplankton is ambiguous. They differ in their ability to suppress algae. It was shown by the example of *Chlorella vulgaris* that synthetic detergent MK-1 (Novost, contains 40% of alkyl sulfates) was somewhat less harmful to algae than synthetic detergent MK-2 (Kristall, contains 20% of sulfonol) (Apasheva et al. 1976) and MK-3 (Lotos, 25% of sulfonol). After one day of the effect on algae at a high concentration (1 g/l), no significant decrease in the number of living cells was observed. After 3 days, a 40% decrease in their number was observed (Apasheva et al. 1976). This work also showed that MK-2 and MK-3 at the same concentrations caused a greater decrease in the number of living algal cells.

A sufficiently high concentration of preparation Kristall (140 mg/l) is required to affect *Gymnodinum kovalevskii* so that a 100% loss of mobility would be observed (Aizdaicher 1999). This testifies for a sufficiently high endurance of the cells of this species.

The ability of synthetic detergents to destruct the photosynthesis processes, primary production, and phytoplankton was shown in Braginsky et al. (1987).

The effect of surfactant-containing preparations on phytoplankton can also lead to a stimulation of the growth of organisms (possibly owing to the fact that many of these preparations contain phosphates). It was shown in our joint work with N.N. Kolotilova that in some experimental conditions synthetic detergents can stimulate the growth of cyanobacteria and green algae. Such a stimulation was shown for *Synechocystis* sp. PCC 6803, *Synechococcus elongates* (*Anacystis nidulans*), and *Scenedesmus quadricauda* (Kolotilova and Ostroumov 2000; Ostroumov and Kolotilova 2000). These results have something in common with the fact that during the effect of synthetic detergent preparations containing surfactants on some species of marine phytoplankton, stimulation of the growth of *Dunaliella tertiolecta* and *Platymonas* sp. was demonstrated at concentrations 1–10 mg/l (Aizdaicher et al. 1999).

6.2.2   Effects of mixed preparations on angiosperm plants

6.2.2.1   Powder synthetic detergents

The growth of rice seedlings *Oryza sativa* (cultivar Kuban-3) was inhibited in aquatic medium containing synthetic detergent Kristall (0.5 mg/ml). For example, after 26 h of incubation in water with Kristall the apparent average length (AAL) of the seedlings was 4.5 mm compared to 8 mm in the control. After 50.5 h of incubation, similar values were equal to 19.7 and 13.3 mm, respectively (N.F. Viktorova participated in this work).

Seedlings of buckwheat *Fagopyrum esculentum* (cultivar Shatilovskaya 5) were even more sensitive. In the medium containing 0.5 mg/ml of detergent Kristall the AAL of the seedlings decreased almost ten times. At a detergent concentration of 0.1 mg/ml (and greater), AAL decreased by more than 50% with respect to the control. The germination of the buckwheat seeds sharply decreased in the presence of synthetic detergent (Table 6.4) (see also Ostroumov 1991). Based on the comparison of this conclusion with the results of the experiments with seedlings of rice and buckwheat, we can suggest that the ranges of negatively affecting concentrations of detergent Kristall for euglenas and buckwheat seedlings are close. Hence, under certain conditions the seedlings of buckwheat can be used instead of euglenas for biological testing of polluted waters, which is interesting based on the fact that experiments with seedlings are simpler and less expensive.

Table 6.4 AAL of the roots of *Fagopyrum esculentum* (cultivar Shatilovskaya 5) seedlings in an aqueous medium containing detergent Kristall (Ostroumov (1991) *Khim. Tekhn. Vody*, Vol. 13, No. 3 (in Russian)).

Time, h	Detergent, mg/ml	AAL, mm	SD	SE	CV, %	M
28.5	0	4.2	5.26	0.38	125	14
	0.05	3.6	3.76	0.63	104	14
	0.1	1.3	2.73	0.46	210	25
	0.2	1.2	2.34	0.39	195	26
	0.4	0.8	1.80	0.30	225	27
42	0	11.9	15.18	2.53	128	14
	0.05	10.8	12.53	2.09	116	13
	0.1	3.3	7.0	1.17	212	25
	0.2	3.3	7.16	1.19	217	26
	0.4	2.4	5.38	0.90	224	26
49	0	16.4	19.27	3.21	118	14
	0.05	16.0	18.43	3.07	115	13
	0.1	4.5	9.12	1.52	203	25
	0.2	4.6	9.75	1.63	212	26
	0.4	3.6	7.45	1.24	207	26
54	0	19.8	22.72	3.79	115	14
	0.05	19.2	21.16	3.53	110	13
	0.1	5,4	11.01	1.84	204	25
	0.2	5.6	11.51	1.92	206	26
	0.4	4.3	9.28	1.55	216	26
65	0	29.8	31.36	5.23	105	14
	0.05	28.8	30.25	5.04	105	13
	0.1	7.2	14.67	2.45	204	25
	0.2	8.9	17.69	2.95	199	26
	0.4	6.1	13.21	2.00	217	26

Note: Each Petri dish contained 18 seeds and 10 ml test solution or water from Mozhaiskoe Reservoir (control); incubation temperature, 27.1°C; AAL, apparent average length; M, number of ungerminated seeds; SD, standard deviation; SE, standard error; CV, coefficient of variation; $n = 36$.

The information given above was related to powder detergents. Recently, the use of liquid detergents has been increasing. The effect of aquatic media polluted with them on biological objects was previously characterized insufficiently. Some of our experiments were carried out to fill the gap in this knowledge.

6.2.2.2 Liquid synthetic detergents

Liquid detergent Vilva at a concentration of 0.125 mg/l decreased the AAL of the buckwheat roots more than by 50% (Tables 6.5 and 6.6). The detergent also

Table 6.5 Effect of detergent Vilva on AAL of *Fagopyrum esculentum* (cultivar Shatilov-skaya 5) roots. Experiment 1.

Statistical parameters	Detergent, ml/l	
	0 (control, STW)	0.25
AAL, mm	11.73	0.87
Standard deviation	11.83	3.84
Standard error	2.16	0.70
Coefficient of variation, %	100.85	441.38
Germination of seeds, %	63.3	13.3
Sampling size	30	30

Note: Incubation time, 25 h, $t = 26°C$.

Table 6.6 Effect of detergent Vilva on AAL of *Fagopyrum esculentum* (cultivar Shatilovskaya 5) roots. Experiment 2.

Statistical parameters	Detergent, ml/l					
	0 (control, STW)		0.125		0.25	
	20 h	44.5 h	20 h	44.5 h	20 h	44.5 h
AAL, mm	7.23	24.3	3.00	8.60	0.17	0.43
Standard deviation	8.25	24.69	5.93	15.50	0.91	2.19
Standard error	1.51	4.51	1.08	2.83	0.17	0.40
Coefficient of variation, %	114.1	101.6	197.7	180.2	535.3	509.3
Germination of seeds, %	60	63.3	30	36.7	3.3	6.7
Sampling size	30	30	30	30	30	30

Note: Each Petri dish contained 10 seeds and 10 ml of test solution, $t = 26°C$.

decreased the AAL of the roots of another test subject, the rice seedling (Tables 6.7–6.10). However, the effect of the liquid detergent on the rice roots was less pronounced than its influence on buckwheat. The inhibition of AAL at a detergent concentration of 0.125 mg/l was less than 50%. EC_{50} for rice was within 0.15–0.20

Table 6.7   Effect of detergent Vilva on AAL of *Oryza sativa* (cultivar Kuban-3) roots. Experiment 1.

Statistical parameters	Detergent, ml/l		
	0 (control, STW)	0.125	0.25
AAL, mm	14.40	9.75	5.05
Standard deviation	7.61	6.62	3.84
Standard error	1.70	1.48	0.87
Coefficient of variation, %	52.85	67.90	77.03
Germination of seeds, %	100	90	95
Sampling size	20	20	20

Note: Each Petri dish contained 10 seeds and 10 ml of test solution, incubation time, 71.5 h; $t = 26°C$.

Table 6.8   Effect of detergent Vilva on AAL of *Oryza sativa* (cultivar Kuban-3) roots. Experiment 2 (temperature was decreased to 23°C).

Statistical parameters	Detergent, ml/l					
	0 (control, STW)		0.125		0.25	
	72 h	96 h	72 h	96 h	72 h	96 h
AAL, mm	4.00	14.80	3.85	14.60	2.30	10.65
Standard deviation	3.63	8.17	3.28	8.14	1.81	6.51
Standard error	0.81	1.82	0.73	1.82	0.40	1.46
Coefficient of variation, %	90.75	55.20	85.19	55.75	78.70	61.13
Germination of seeds, %	85	90	85	85	80	90
Sampling size	20	20	20	20	20	20

Note: Each Petri dish contained 10 seeds and 10 ml of test solution.

mg/l (at 26°C). The experiments showed that a decrease in the temperature of incubation caused a drop of sensitivity of the biological test (Tables 6.7 and 6.8).

The author also performed biological testing with preparation Kashtan in experiments with rice seedlings (Table 6.11) and *Cucumis sativus* (cultivar Nerosimy) (Table 6.12). This preparation inhibited the elongation of rice seedlings stronger than Vilva. These experiments are described in greater detail in Kartsev et al. (1990).

Biological activity of aquatic media containing preparation Kashtan was estimated in experiments with water lettuce (*Pistia stratiotes*). This species of aquatic macrophytes is a dangerous aquatic weed that grows rapidly in many aquatic ecosystems and causes problems for water management. At the same time, there are plans and suggestions to use this species and some other macrophytes for phytoremediation of polluted and eutrophic reservoirs. Thus, it is interesting to reveal the tolerance limits of this species to various contaminants.

Table 6.9 Effect of detergent Vilva on AAL of *Oryza sativa* (cultivar Kuban-3) roots. Experiment 3.

Statistical parameters	Detergent, ml/l							
	0 (control, STW)		0.15		0.20		0.25	
	72.5 h	95.5 h	72.5 h	95.5 h	72.5 h	95.5 h	72.5 h	95.5 h
AAL, mm	11.1	28.6	5.3	17.3	5.5	16.9	3.7	11.2
Standard deviation	7.3	14.4	5.1	11.9	4.2	9.8	3.1	8.1
Standard error	1.3	2.6	0.9	2.2	0.8	1.8	0.6	1.5
Coefficient of variation, %	65.6	50.4	95.5	68.9	76.3	57.7	83.5	71.8
Germination of seeds, %	83.3	90	66.6	83.3	80	83.3	86.7	90
Sampling size	30	30	30	30	30	30	30	30

Note: $t = 26°C$.

Table 6.10 Effect of detergent Vilva on AAL of *Oryza sativa* (cultivar Kuban-3) rice roots. Experiment 4 (content of detergent in water was increased).

Statistical parameters	Detergent, ml/l			
	0 (control, STW)	0.25	0.5	1.0
AAL, mm	12.85	3.20	1.80	0.65
Standard deviation	7.82	2.09	1.28	0.81
Standard error	1.75	0.47	0.29	0.18
Coefficient of variation, %	60.9	65.3	71.1	124.6
Germination of seeds, %	85	80	75	45
Sampling size	20	20	20	20

Note: Incubation time, 70.5 h; $t = 26°C$.

The experiments showed that after 10 days of cultivating young plants of water lettuce in aquatic medium containing detergent Kashtan (0.06 ml/l) the roots of hydrophytes were shorter than in the control. In addition, anomalous development of lateral branches of roots was observed. At the lower ends of the main vertical roots, the lateral branches were longer than in the upper parts of the roots. After 10 days of cultivation in aquatic medium containing a double concentration of detergent Kashtan (0.125 ml/l), a portion of the plants perished. Conversely, when the concentration of the preparation was decreased to 0.03 ml/l, no significant differences from the control were observed (Table 6.13; Ostroumov 1990 a).

Table 6.11   Effect of detergent Kashtan on the length of roots of *Oryza sativa* rice seedlings.

Sampling characteristics	Content of detergent in solution, ml/l			
	0	0.125	0.25	0.5
Sampling size	10	10	10	10
Mean, mm	33.6	19.5	14.3	7.5
Standard deviation	11.35	5.7	6.68	2.27
Standard error	3.59	1.80	2.18	0.72
Asymmetry	−0.806	0.358	−0.50	0.28
Standardized asymmetry	−1.041	0.462	−0.65	0.36
Excess	0.867	−0.252	1.72	−1.47
Standardized excess	0.560	−0.163	1.11	−0.95

Note: Cultivar Ognevsky, 7 days after soaking the seeds; incubation temperature, 21°C.

Table 6.12   Effect of synthetic detergent Kashtan on the mean length of *Cucumis sativus* (cultivar Nerosimy) seedlings.

Concen-tration, μl/ml	Length of seedlings		CL, %	SD	CL margins of SD	CV, %	*n*
	X, mm	with respect to the control, %					
0	32.35	100	5.68	12.16	9.66–16.67	37.59	20
2	4.25	13.14	1.82	3.89	3.09–5.33	91.53	20

Note: *X*, mean length of seedlings; CL, confidence limit; SD, standard deviation; CV, coefficient of variation; *n*, number of seedlings. Incubation time, 123 h.

Table 6.13   Effect of synthetic detergent Kashtan on *Pistia stratiotes*.

Content of detergent in the onset of experiment, ml/l	Observation time after the onset of experiment	
	10 days	45 days
0 (control)	Leaves of green color; normal growth and development	Leaves of green color; large rosettes; root system, much longer than 5 cm
0.03	No significant differences from the control were observed	No significant differences from the control were observed
0.06	Roots are shorter than in the control; abnormal development of lateral branches (lower parts of the roots are longer than upper parts)	No significant differences from the control were observed
0.125	Some plants perish; roots break off; leaves become yellow	Some plants perish; new young plants appear (rosettes, smaller than in the control by a factor of ~2); root system, less than 5 cm long

This experiment is interesting not only in that it gives complete information about the biological activity of the preparation containing surfactants and indicates the possibility of applying this new method of biological testing using pistia, but also because it establishes a concentration range within which this preparation containing surfactants can manifest a herbicide effect against this weed in aquatic systems. On the other hand, we revealed a range of concentrations that are likely to be endured by the plant in the case of using this species for phytoremediation.

The results obtained (in the experiments with seedlings) demonstrate that the application of waters polluted with synthetic detergents including liquid detergents above some critical values for watering agricultural lands can have negative consequences.

Heterotrophic hydrobionts were used for testing along with the autotrophic organisms.

6.3 New Results on the Impact of Surfactant-Containing Mixtures on Heterotrophic Organisms

We studied the effects of surfactant-containing preparations on freshwater and marine mollusks.

6.3.1 Freshwater mollusks

Unio tumidus Philipsson, 1788. In a typical experiment, 20 animals were used, 10 of which were subjected to the effect of synthetic detergent (variant A) and 10 others were used as controls (variant B). The total wet weight (with shells) of 10 mollusks in variant A was 205.7 g; in variant B it was 204.0 g.

Synthetic detergent was added 5 min before the experiment. The concentration of detergent given in the table and mentioned in the text in the discussion of the data (the same as in all other experiments) was the initial concentration at the moment when detergent was added to the beaker. Four variants were performed simultaneously (see Table 6.14). In variants A and B, mollusks were placed in the beakers. No mollusks were placed in the beakers in variants C and D, but the same suspension of cells with a volume of 0.5 l was added similar to variants A and B. Variants C and D were used as controls to determine the density of cell suspension at complete absence of biological filtration of water (because mollusks were not placed in these beakers). In the beaker of variant C, detergent was added at the same concentration as in the experiment (variant A).

Filtration was measured on the basis of decreasing the optical density of the incubation medium caused by the removal of cells (*S. cerevisiae* kindly given to us by N.N. Kolotilova) from the water column and correspondingly by a decrease in their concentration in water. (We repeatedly performed the comparison of the experiments, in which cells of different organisms were used as filtered suspension. On the basis of these experiments we made a conclusion about possible application of

S. cerevisiae due to its methodological convenience.) The experiments showed that at the initial concentration of detergent equal to 50 mg/l the rate of water filtration by *Unio tumidus* decreased significantly (Table 6.14). For example, after 1 h of filtration the optical density of the incubation medium decreased in the experiment (in the presence of detergent, variant A) only up to 0.534, while in the control (variant B) it decreased to 0.286. The optical density in the experiment (in the presence of detergent) exceeded significantly the density of the control and was equal to 186.7% of the latter. After 60 min of filtration, a decrease in the optical density compared to the initial level (0.649 units of optical density) was equal to 0.363 units in the control and to 0.115 units in the experiment. A decrease in the optical density in the control occurred 3.16 times faster than in the experiment, in which the decrease in the optical density (related to the removal of cells from water) was slowed down by the effect of detergent. In the absence of mollusks (variants C and D), in which water filtration was excluded, the optical density of the incubation medium was significantly greater than in the control with mollusks without detergent (variant B). Thus, under the influence of detergent (50 mg/l) the efficiency of water filtration by mollusks was 31.7% of the efficiency in the control, thus inhibition was 68.3%. Similar data were also obtained when the cells of another strain of *S. cerevisae* were used as suspension.

    The obtained data indicate that detergent decreases the rate of water filtration by freshwater mollusks and the efficiency of the removal of suspended particles by them. This result agrees with the data of the experiments on biological testing of anionic, cationic, and nonionogenic surfactants done with mollusks (Chapters 3–5) and confirms the hypothesis put forward in Chapter 5, which predicted this result.

Table 6.14 Temporal dynamics of the optical density (OD_{500}) of the incubation medium during the action of detergent (OMO, 50 mg/l) on the filtration of *S. cerevisiae* cells by mollusks *Unio tumidus*.

Time after addition of cells, min	(A) Experiment (+detergent)	(B) Control (with mollusks, −detergent)	(C) Control (without mollusks, +detergent)	(D) Control (without mollusks, −detergent)
20	0.599	0.534	0.640	0.657
30	0.617	0.526	0.676	0.674
40	0.542	0.402	0.636	0.652
50	0.552	0.329	0.644	0.651
60	0.534	0.286	0.616	0.660

Note: The data are averaged for three measurements. At the beginning of the experiment, 500 ml of *Saccharomyces cerevisiae* suspension was added to the beakers in all variants (A, B, C, and D); number of cells, 6.6×10^6 per 1 ml. Temperature, 19°C. In variants A and B, 10 mollusks each. Average weight of one mollusk (wet weight with shell): A, 20.6 g; B, 20.4 g.

6.3.2   Marine mollusks

The experiments with marine mollusks were carried out with the purpose to determine possible biological effects of two large groups of preparations: powder formulations of synthetic detergents and preparations containing liquid surfactants.

6.3.2.1 Powder formulations of synthetic detergents

Several types of detergents produced by different manufacturing companies were used for the study (see Chapter 2). The property of detergent Lotos-Extra to decrease the ability of mollusks to clarify water was revealed (the results of testing Avon Herbal Care are given for comparison). The difference between water turbidity in the experiment and control was recorded 7–9 min after the onset of the experiment (Table 6.15). For details, see *Food Industry at the Edge of the Third Millennium*, Moscow, 2000, pp. 248–251 (in Russian).

We studied the effect of detergent Losk-Universal (20 mg/l) on the changes in OD_{650} of *Pavlova lutheri* (Droop) Green (*Monochrys lutheri* Droop) algae during its filtration by juvenile mollusks. *Mytilus galloprovincialis* mussels were used in the experiments (juveniles, age 2 months, obtained from the Mariculture Department of IBSS) and algal suspension *Pavlova lutheri* (Droop) Green 1975 (*Monochrys lutheri* Droop 1953). Incubation was carried out at 27.8°C. During filtration of water, the mussels extracted algal cells from water, which led to a decrease in the optical density of algal suspension in the beaker with mussels. Algal suspension (50 ml) was introduced into the beakers at an initial concentration of cells approximately equal to 6×10^6 cells/ml. Optical density was measured simultaneously in three variants, two of which were controls. Mussels (0.5 g of wet weight with shells, a total of about 530 mollusks) were placed in the experimental beaker and suspension of algae and detergent Losk-Universal was added (20 mg/l). Control 1: the beaker contained mussels, algae without detergent. Control 2: the beaker contained only algae without mussels and without detergent. Optical density was measured at 650 nm (optical path was 10 mm) using an SF-26 spectrophotometer (LOMO).

Table 6.15 Surfactant-containing preparations AHC and LE decrease the ability of *Mytilus galloprovincialis* to clarify water.

Substances and concentrations, mg/l	Time interval (min) within which the difference between the control and experiment was recorded (see text)	Temperature, °C	*S. cerevisiae* suspension, g/l	Average weight (g, wet weight with shells) and/or average size (mm) of animals
AHC, 30	9–68	22.0	0.549	In experiment and control, 16 mollusks each, average weight 3.5
AHC, 60	18–30	22.0	1.081	
LE, 25	7–24	21.6	0.167	In experiment, 16 mollusks, average weight 4.0, average size 29; in control, 16 mollusks, average weight 4.0, average size 30
LE, 50	14–30	20.0	0.167	In experiment, 18 mollusks, average weight 3.45, average size 28; in control, 18 mollusks, average weight 3.51, average size 28

Note: AHC, Avon Herbal Care; LE, synthetic detergent Lotos-Extra. For details, see *Food Industry at the Edge of the Third Millennium*. Moscow: 2000, pp. 248–251 (in Russian).

Table 6.16   Effect of detergent Losk-Universal (20 mg/l) on the change of OD_{650} of a suspension of *Pavlova lutheri* (Droop) Green (*Monochrysis lutheri* Droop) algae in the course of its filtration by juvenile organisms of *Mytilus galloprovincialis*.

Measurement No.	Time from the onset of incubation, min	OD_{650}			EER, A/B ratio, %
		A Experiment (+detergent)	B Control 1 (with mussels, no detergent)	C Control 2 (no mussels, no detergent)	
1	19	–	0.075	–	193.3
	20	0.145	–	–	
	22	–	–	0.168	
2	34	–	0.031	–	480.6
	36	0.149	–	–	
	37	–	–	0.166	
3	43	–	0.024	–	595.8
	45	0.143	–	–	
	56	–	–	0.165	
4	52	–	0.005	–	2460.0
	53	0.123	–	–	
	54	–	–	0.155	

Note: EER, effect on the efficiency of removal of suspension from water.

Table 6.17   Effect of detergent Losk-Universal (7 mg/l) on the change of OD_{650} of a suspension of *Pavlova lutheri* (Droop) Green (*Monochrysis lutheri* Droop) algae in the course of its filtration by juvenile organisms of *Mytilus galloprovincialis*.

Measurement No.	Time from the onset of incubation, min	OD_{650}			A/B ratio, %
		A Experiment (+detergent)	B Control 1 (with mussels, no detergent)	C Control 2 (no mussels, no detergent)	
1	9	–	0.128	–	130.5
	10	0.167	–	–	
	13	–	–	0.173	
2	15	–	0.095	–	181.1
	17	0.172	–	–	
	18	–	–	0.166	
3	22	–	0.078	–	205.1
	23	0.160	–	–	
	24	–	–	0.165	

Table 6.17   (continued)

Measurement No.	Time from the onset of incubation, min	OD$_{650}$			A/B ratio, %
		A Experiment (+detergent)	B Control 1 (with mussels, no detergent)	C Control 2 (no mussels, no detergent)	
4	47	–	0.029	–	551.7
	48	0.160	–	–	
	49	–	–	0.167	

During the period of observation (more than 50 min), the OD of suspension in variant B (without detergent, i.e., normal rate of filtration) decreased from 0.168 (initial optical density in all variants of experiment) to 0.005 (see Table 6.16). Note that synthetic detergent Losk-Universal is absent in the medium. During the first period of filtration (19 min of the experiment) in the same variant with mollusks and without detergent Losk-Universal (control 1), the OD decreased to 0.075. If detergent Losk-Universal was present in the medium, the OD decreased more slowly. During 20 min of filtration, it decreased only to 0.145, which significantly differed from the control. In the course of further filtration, the difference between the experiments and controls became even greater. During the experiment, no increase in the OD was observed due to the growth of algae (see variant C, the column for control 2 in the table).

Thus, evidence was obtained of noticeable inhibition of water filtration under the influence of detergent Losk-Universal, as a result of which after 53 min the difference in OD of filtered algal suspension differed in control 1 and in experiment samples by a factor of 24.6. Similar data indicating the inhibition of filtration were obtained at the action of this detergent at a lower concentration (Table 6.17).

Thus, inhibition of the rate of water filtration by *M. galloprovincialis* juveniles was revealed in the concentration range of Losk-Universal from 7 to 20 mg/l.

In a similar experiment with detergent Tide-Lemon (50 mg/l) we demonstrated the inhibition of water filtration by juveniles of *M. galloprovincialis* under the influence of this detergent. The wet weight (with shell) of mussels in the beakers was: 0.10 g in beaker 3; 0.09 g in beaker 4, 0.11 g in beaker 5; 0.11 g in beaker 6. The number of mussels in beaker 5 was 118 species, i.e., the weight of an individual mussel was 0.93 mg. There were no mussels in beakers 1 and 2, because these beakers were used for controls. One can see that the value of EER (effect of the efficiency of removal) was approximately 207% (the measurements were made after the third period of filtration, after 97–101 min of exposure) (Table 6.17a).

The effect of detergent IXI Bio-Plus on the changes in *Mytilus galloprovincialis* led to a distinct inhibition of filtration. The value of inhibition expressed in EER (effect for suspension removal from water) depended on the concentration of detergent and regularly decreased when the latter decreased (Tables 6.18 and 6.19).

The inhibitory effect of synthetic detergents was also clearly shown in the experiments with another species of mollusks, oysters *Crassostrea gigas*.

Table 6.17a   Effect of detergent Tide-Lemon (50 mg/l) on the change of OD_{650} of a suspension of *Pavlova lutheri* (Droop) Green (*Monochrysis lutheri* Droop) algae in the course of its filtration by juveniles of *Mytilus galloprovincialis*.

Measurement No.	Time from the onset of incubation, min	OD_{650}						A/B ratio, %
		A Experiment (+detergent)		B Control 1 (with mussels, no detergent)		C Control 2 (no mussels, no detergent)		
		1	2	3	4	5	6	
1	44	–	–	0.12	0.10 (mean 0.11)	–	–	127.3
	47	–	–	–	–	0.15	0.15	
	49	0.134	0.145 (mean 0.14)	–	–	–	–	
2	75	–	–	0.072	0.070 (mean 0.071)	–	–	169.0
	78	–	–	–	–	0.14	0.14	
	80	0.12	0.12	–	–	–	–	
3	97	–	–	0.062	0.054 (mean 0.058)	–	–	206.9
	99	–	–	–	–	0.127	0.127	
	101	0.117	0.122 (mean 0.120)	–	–	–	–	

Note: Each variant was done in two repeats (beakers 1 and 2, variant A; beakers 3 and 4, variant B; beakers 5 and 6, variant C). In brackets, mean values for two repeats.

Detergent Deni-Avtomat exerted a very strong effect on the filtration activity of oysters. Under conditions of the experiment, the value of EER exceeded 150% after 2 min of exposure, and was 1070% after 10 min (the difference between the experiment and control was greater than 10 times), (Table 6.20). In the continuation of the experiment, the effects reached even more impressive values owing to almost complete water clarification in the control (in which the oysters filtered at a normal rate).

Detergent Lanza also inhibited the ability of *Crassostrea gigas* to clarify water (Table 6.21). The difference between the control and experiment characterizing the EER exceeded 150% during the entire period of measurements.

We used preparations of detergent Vesna-Delikat at a concentration one order of magnitude smaller than in the previous experiments (1 mg/l). It appeared that this concentration is sufficient to significantly decrease the filtration activity of oysters

Table 6.18   Effect of detergent IXI (20 mg/l) on the change of OD_{550} of a suspension of
S. cerevisiae in the course of its filtration by *Mytilus galloprovincialis*. Experiment 1.

Measurement No.	Time from the onset of incubation, min	OD_{550}			EER, A/B ratio, %
		A Experiment (+detergent)	B Control 1 (with mussels, no detergent)	C Control 2 (no mussels, no detergent)	
1	3	0.304	0.238	0.306	127.7
2	8	0.239	0.177	0.301	135.0
3	15	0.194	0.081	0.299	239.5
4	25	0.152	0.055	0.270	276.4

Note: Beakers A and B contained 16 mollusks each. Total weight (wet, with shells) of the mussels: A, 99.7 g; B, 101.5 g. Average weight of one species: A, 6.23 g; B, 6.34 g. Volume of water in the beakers: 500 ml. Temperature, 22.8°C. *S. cerevisiae*, 100 mg/l. EER = effect on the efficiency of removal of suspension from water.

Table 6.19   Effect of detergent IXI (20 mg/l) on the change of OD_{550} of a suspension of
S. cerevisiae in the course of its filtration by *Mytilus galloprovincialis*. Experiment 2
(concentration of detergent was decreased).

Measurement No.	Time from the onset of incubation, min	OD_{550}			EER, A/B ratio, %
		A Experiment (+detergent)	B Control 1 (with mussels, no detergent)	C Control 2 (no mussels, no detergent)	
1	3	0.310	0.269	0.352	115.2
2	8	0.192	0.129	0.342	148.8
3	15	0.119	0.083	0.342	143.4
4	25	0.071	0.045	0.331	157.8

Note: Total weight (wet, with shells) of the mussels: A, 96.2 g; B, 98.5 g. Average weight of one mussel: A, 6.01 g; B, 6.16 g. Volume of water in the beakers: 500 ml. Temperature, 22.3°C. *S. cerevisiae*, 100 mg/l. EER = effect on the efficiency of removal of suspension from water.

Table 6.20   Effect of detergent Deni-Avtomat (30 mg/l) on the change of OD_{550} of a
suspension of *S. cerevisiae* in the course of its filtration by *Crassostrea gigas*.

Measurement No.	Time from the onset of incubation, min	OD_{550}			EER, A/B ratio, %
		A Experiment (+detergent)	B Control 1 (with mollusks, no detergent)	C Control 2 (no mollusks, no detergent)	
1	2	0.262	0.171	0.325	153.2
2	10	0.151	0.014	0.305	1078.6

Table 6.20   (continued)

Measurement No.	Time from the onset of incubation, min	OD$_{550}$			EER, A/B ratio, %
		A Experiment (+detergent)	B Control 1 (with mollusks, no detergent)	C Control 2 (no mollusks, no detergent)	
3	20	0.116	0.002	0.307	5800.0
4	40	0.108	0.001	0.316	10800.0

Note: Beakers A and B contained 10 mollusks each. Total weight (wet, with shells) of the mollusks: A, 55.8 g; B, 48.0 g. Average weight of one species: A, 5.58 g; B, 4.8 g. Volume of water in the beakers: 500 ml. Temperature, 25.2°C. *S. cerevisiae*, 100 mg/l. EER = effect on the efficiency of removal of suspension from water.

Table 6.21   Effect of detergent Lanza (20 mg/l) on the change of OD$_{550}$ of a suspension of *S. cerevisiae* in the course of its filtration by *Crassostrea gigas*.

Measurement No.	Time from the onset of incubation, min	OD$_{550}$			EER, A/B ratio, %
		A Experiment (+detergent)	B Control 1 (mollusks, no detergent)	C Control 2 (no mollusks, no detergent)	
1	4	0.270	0.176	0.290	153.4
2	11	0.240	0.109	0.287	220.2
3	19	0.212	0.081	0.286	261.7
4	35	0.147	0.080	0.272	183.75

Note: Beakers A and B contained 10 mollusks each. Total weight (wet, with shells) of the mollusks: A, 52.6 g; B, 49.0 g. Average weight of one species: A, 5.26 g; B, 4.9 g. Volume of water in the beakers: 500 ml. Temperature, 23.4°C. *S. cerevisiae*, 100 mg/l. EER = effect on the efficiency of removal of suspension from water.

and the efficiency of suspension removal from water. The value of EER exceeded 170% after 22 min of exposure, and then rapidly reached 200% (Table 6.22).

Thus, all investigated powder detergents notably inhibited the filtration activity of marine mollusks. Seven preparations produced by different companies were studied, and all of them demonstrated their abilities to inhibit water filtration by mollusks (taking into account the experiments with freshwater mollusks, the number of preparations increases to eight). In the course of the experiment there was not a single case in which a preparation used for the tests did not produce any inhibitory effect. We note that the author was not attempting to reveal the minimum affecting concentrations but tried to establish the fact of inhibition. In further studies, a similar effect would be probably found at smaller concentrations of detergent. It was interesting to check whether liquid preparations containing surfactants manifested such action (foam and liquid detergents). Taking into account the experience of investigating other preparations we have adopted a supposition as a working hypothesis that liquid detergents (although they can have in their compositions components different from those in the detergent) can inhibit filtration activity of mollusks. The results of the experiments to test this hypothesis are given in the next section.

Table 6.22 Effect of detergent Vesna-Delikat (1 mg/l) on the change of OD_{550} of a suspension of *S. cerevisiae* in the course of its filtration by *Crassostrea gigas*.

Measurement No.	Time from the onset of incubation, min	OD_{550}			EER, A/B ratio, %
		A Experiment (+detergent)	B Control 1 (mollusks, no detergent)	C Control 2 (no mollusks, no detergent)	
1	7	0.196	0.162	0.237	121.0
2	22	0.097	0.055	0.174	176.4
3	27	0.078	0.039	0.176	200.0

Note: Beakers A and B contained 16 mollusks each. Total weight (wet, with shells) of the mollusks: A, 23.6 g; B, 23.5 g. Average weight of one mollusk: A, 1.475 g; B, 1.469 g. Volume of water in the beakers: 500 ml. Temperature, 23.4°C. *S. cerevisiae*, 60 mg/l. EER = effect on the efficiency of removal of suspension from water.

6.3.2.2 Liquid detergents

The effect of liquid detergent E (produced by Cussons International Ltd.) on two species of mollusks *M. galloprovincialis* (Table 6.23) and *C. gigas* (Table 6.24) was studied. Inhibition of the activity of oysters was manifested more clearly. The value of EER exceeded 200% after 19 min of exposure, and 300% after 26 min.

The effect of a Russian liquid detergent Mila was also studied in the experiments with both species of mollusks: in mussels (Table 6.25) and in oysters (Table 6.26). Unlike the previous mixed preparation the effect of this detergent was manifested more strongly in mussels than in oysters. After 30 min, the value of EER in mussels exceeded 230%, whereas in oysters (after 33 min) it was only 160%. However, the main result common for both preparations studied and for both species of hydrobionts was the fact that a comparatively low concentration (2 mg/l) caused a notable and sufficiently fast inhibitory effect in both organisms.

A liquid detergent of the third producer, Fairy (Procter&Gamble) also inhibited filtration of both species, and oysters appeared especially sensitive (Tables 6.27 and 6.28). As a result of the effect of the same detergent concentration (2 mg/l) on oysters, a more than 17-fold difference in optical densities in experiment and control samples was found after 27 min (Table 6.28). Similarly to the previous case, the author of these experiments put forward an objective to find and register the existence of the effect of inhibition and did not try to reveal the minimum acting concentration of liquid detergents. However, the value of the effect (more than 17-fold decrease in the efficiency of removing suspension from water) makes us think that even significantly lower concentrations of liquid detergent in further experiments could have some inhibitory effect on water filtration.

Table 6.23   Effect of synthetic detergent E (2 mg/l) on the change of OD_{550} of a suspension of *S. cerevisiae* in the course of its filtration by *M. galloprovincialis*.

Measurement No.	Time from the onset of incubation, min	OD_{550}			EER, A/B ratio, %
		A Experiment (+detergent)	B Control 1 (with mussels, no detergent)	C Control 2 (no mussels, no detergent)	
1	3	0.284	0.262	0.298	108.4
2	13	0.279	0.195	0.292	143.1
3	20	0.246	0.128	0.284	192.2
4	30	0.216	0.101	0.280	213.9

Note: Beakers A and B contained 50 mollusks each. Total weight (wet, with shells) of the mussels: A, 42.5 g; B, 42.4 g. Average weight of one species, 0.85 g. Volume of water in the beakers: 500 ml. Temperature, 23.0°C. *S. cerevisiae*, 100 mg/l. EER = effect on the efficiency of removal of suspension from water.

Table 6.24   Effect of synthetic detergent E (2 mg/l) on the change of OD_{550} of a suspension of *S. cerevisiae* in the course of its filtration by *Crassostrea gigas*.

Measurement No.	Time from the onset of incubation, min	OD_{550}			EER, A/B ratio, %
		A Experiment (+detergent)	B Control 1 (mollusks, no detergent)	C Control 2 (no mollusks, no detergent)	
1	3	0.277	0.268	0.310	103.4
2	9	0.269	0.197	0.292	136.5
3	19	0.223	0.104	0.295	214.4
4	26	0.183	0.060	0.298	305.0

Note: Beakers A and B contained 10 mollusks each. Total weight (wet, with shells) of the mollusks: A, 53.7 g; B, 49.5 g. Average weight of one mollusk: A, 5.37 g; B, 4.95 g. Volume of water in the beakers: 500 ml. Temperature, 23.4°C. *S. cerevisiae*, 100 mg/l. EER = effect on the efficiency of removal of suspension from water.

Table 6.25   Effect of synthetic detergent Mila (2 mg/l) on the change of OD_{550} of a suspension of *S. cerevisiae* in the course of its filtration by *M. galloprovincialis*.

Measurement No.	Time from the onset of incubation, min	OD_{550}			A/B ratio, %
		A Experiment (+detergent)	B Control 1 (with mussels, no detergent)	C Control 2 (no mussels, no detergent)	
1	3	0.280	0.270	0.323	103.7
2	10	0.241	0.185	–	130.3

Table 6.25 (continued)

| Measurement No. | Time from the onset of incubation, min | OD$_{550}$ | | | A/B ratio, % |
		A Experiment (+detergent)	B Control 1 (with mussels, no detergent)	C Control 2 (no mussels, no detergent)	
3	20	0.209	0.120	0.307	174.2
4	30	0.176	0.075	0.301	234.7

Note: Beakers A and B contained 24 mollusks each. Total weight (wet, with shells) of the mussels: A, 48.54 g; B, 47.32 g. Average weight of one species: A, 2.02 g; B, 1.97 g. Volume of water in the beakers: 500 ml. Temperature, 24.4°C. *S. cerevisiae*, 100 mg/l.

Table 6.26 Effect of synthetic detergent Mila (2 mg/l) on the change of OD$_{550}$ of a suspension of *S. cerevisiae* in the course of its filtration by *Crassostrea gigas*.

| Measurement No. | Time from the onset of incubation, min | OD$_{550}$ | | | A/B ratio, % |
		A Experiment (+detergent)	B Control 1 (mollusks, no detergent)	C Control 2 (no mollusks, no detergent)	
1	3	0.246	0.194	0.281	126.8
2	13	0.144	0.111	–	129.7
3	23	0.099	0.073	0.274	135.6
4	33	0.072	0.045	0.265	160.0

Note: Beakers A and B contained 10 mollusks each. Total weight (wet, with shells) of the mollusks: A, 51.8 g; B, 51.5 g. Average weight of one mollusk: A, 5.18 g; B, 5.15 g. Volume of water in the beakers: 500 ml. Temperature, 24.6°C. *S. cerevisiae*, 100 mg/l.

Table 6.27 Effect of synthetic detergent Fairy (2 mg/l) on the change of OD$_{550}$ of a suspension of *S. cerevisiae* in the course of its filtration by *M. galloprovincialis*.

| Measurement No. | Time from the onset of incubation, min | OD$_{550}$ | | | A/B ratio, % |
		A Experiment (+detergent)	B Control 1 (with mussels, no detergent)	C Control 2 (no mussels, no detergent)	
1	2	0.282	0.233	0.302	121.0
2	10	0.175	0.080	0.306	218.75
3	15	0.110	0.054	0.301	203.7
4	23	0.048	0.024	0.290	200.0

Note: Beakers A and B contained 20 mollusks each. Total weight (wet, with shells) of the mollusks: A, 158.4 g; B, 141.0 g. Average weight of one mollusk: A, 7.92 g; B, 7.05 g. Volume of water in the beakers: 500 ml. Temperature, 22.5°C. *S. cerevisiae*, 100 mg/l.

Table 6.28 Effect of foam detergent Fairy (2 mg/l) on the change of OD_{550} of a suspension of *S. cerevisiae* in the course of its filtration by *Crassostrea gigas*.

Measurement No.	Time from the onset of incubation, min	OD_{550}			EER, A/B ratio, %
		A Experiment (+detergent)	B Control 1 (mollusks, no detergent)	C Control 2 (no mollusks, no detergent)	
1	5	0.307	0.131	–	234.4
2	12	0.220	0.042	0.270	523.8
3	17	0.191	0.022	0.268	868.2
4	27	0.179	0.010	0.250	1790.0

Note: Beakers A and B contained 10 mollusks each. Total weight (wet, with shells) of the mollusks: A, 52.2 g; B, 50.4 g. Average weight of one mollusk: A, 5.22 g; B, 5.04 g. Volume of water in the beakers: 500 ml. Temperature, 24.0°C. *S. cerevisiae*, 100 mg/l.

Table 6.29 Effect of AHC (5 mg/l) on the change of OD_{650} of a suspension of *Pavlova lutheri* (Droop) Green (*Monochrysis lutheri* Droop) algae in the course of its filtration by juveniles of *Mytilus galloprovincialis*. Initial OD_{650}, 0.331.

Measurement No.	Time from the onset of incubation, min	OD_{650}			
		Variant 1 (control w/o AHC)	Variant 2 (experiment: (mussels + AHC)	Variant 3 (control with AHC, w/o mussels)	Variant 4 (control w/o AHC, w/o mussels)
1	5	0.293	–	–	–
	6	–	0.334	–	–
	7	–	–	0.307	–
	9	–	–	–	0.331
2	11	0.272	–	–	–
	12	–	0.289	–	–
	13	–	–	0.329	–
	14	–	–	–	0.325
3	16	0.246	–	–	–
	17	–	0.265	–	–
	18	–	–	0.314	–
	20	–	–	–	0.326
4	23	0.237	–	–	–
	26	–	0.262	–	–
	27	–	–	0.320	–
	28	–	–	–	0.332

Table 6.30 Effect of sublethal concentrations of AHC on *Mytilus galloprovincialis*.

Concentration of AHC, mg/l	Biological effects	Notes
5	Inhibition of filtration	5–26 min, 25.6°C; juveniles
25	No increase of mortality is observed	24 h and 48 h, 16°C; mature mussels; average length, 24.0 mm
30	Inhibition of filtration	9–68 min, 22°C; mature mussels; average wet weight of one mussel, 3.5 g
50	No increase of mortality is observed	24 h and 48 h, 16°C; mature mussels; average length, 24.3 mm
60	Inhibition of filtration	18–30 min, 22°C; mature mussels; average wet weight of one mussel, 3.5 g
100	No increase of mortality is observed	24 h and 48 h, 16°C; mature mussels; average length, 22.3 mm

Table 6.31 Summary of the new results on the effects of surfactant-containing detergents on the filtration and trophic activity of mollusks.

Type of detergent	Name of detergent	Organism	Example of effect (usually EER maximum; in brackets, concentrations in mg/l)
Powder	OMO	*Unio tumidus*	186.7 (50)
	Tide-Lemon	*Lymnaea stagnalis*	inhibition of trophic activity, 36% (75)
	Losk-Universal	*Mytilus galloprovincialis*	2460.0 (20)
	Losk-Universal	*Mytilus galloprovincialis*	551.7 (7)
	Tide-Lemon	*Mytilus galloprovincialis*	206.9 (50)
	IXI	*Mytilus galloprovincialis*	276.4 (50)
	IXI	*Mytilus galloprovincialis*	157.8 (10)
	Deni-Avtomat	*Crassostrea gigas*	10800.0 (30)
	Lanza	*Crastostrea gigas*	261.7 (20)
	Vesna-Delikat	*Crassostrea gigas*	200.0 (1)
Liquid	E	*Mytilus galloprovincialis*	213.9 (2)
	E	*Crassostrea gigas*	305.0 (2)
	Mila	*Mytilus galloprovincialis*	234.7 (2)
	Mila	*Crassostrea gigas*	160.0 (2)
	Fairy	*Mytilus galloprovincialis*	218.8 (2)
	Fairy	*Crassostrea gigas*	1790.0 (2)
	AHC	*Mytilus galloprovincialis*	114.0 (5)

The fourth surfactant-containing mixture from the class of liquid preparations tested on marine mollusks was AHC (Avon Herbal Care). This preparation (shampoo) directly interacts with the surface of the human body. No doubt, this preparation passed all necessary forms of sanitary hygienic testing. However, this did not guarantee its complete safety for hydrobionts. The experiments with mussels demonstrated that at a concentration equal to 5 mg/l, this preparation inhibited slightly the rate of filtration (Table 6.29). A notable effect was detected also at concentrations equal to 30 and 60 mg/l (Table 6.15). (We should note that during its application for hygienic purposes, the preparation contacts the surface of a human body in undiluted form, i.e., at concentrations one order of magnitude greater than in the experiments.) The tested concentrations and even a higher concentration of 100 mg/l did not cause death of mollusks (in 48 h) (Table 6.30), i.e., in standard tests to estimate the potential toxicity a conclusion was made that this preparation is not toxic

We note that the popularity and overall use of the preparations of the type studied here guarantee their continuous discharge into aquatic media together with domestic sewage. There are reasons to predict that in the future their discharge will continue not in lesser but possibly greater volumes than now. Correspondingly, the degree of the potentially hazardous anthropogenic stress caused by these preparations as a result of their effect on hydrobionts will only increase if the existing trends remain.

The negative impact of synthetic detergents (all eight preparations tested) and liquid detergents (all four preparations tested) on hydrobionts shown in all systems with mollusks (Table 6.31) casts a considerable doubt on the traditional perception of these detergents as ecologically safe chemicals or non-priority pollutants.

6.4 Assessment of the Biological Activities of Other Preparations and Samples

Along with the study of synthetic surfactants and preparations containing synthetic surfactants, we used some of the developed methods to estimate the biological activity of other preparations including pesticides and water samples from reservoirs taken in the period of algal blooms. Water solutions of pesticides were studied with the objective of comparing their effect on the organisms with similar effects caused by synthetic surfactants.

6.4.1 Pesticides

Chlorine organic pesticides lontrel and dinitroorthocresol (DNOC) were among the pesticides studied in our research.

6.4.1.1 Lontrel

In these experiments, ten seeds of *Cucumis sativus* were placed in Petri dishes and 10 ml of the tested solution or distilled water (in the control) was added. The experiment was carried out in the dark at room temperature.

After 95 h of incubation at a concentration equal to 0.02 mg/ml, the length of the seedlings was equal to (in mm, the confidence interval of the mean value is given in brackets): 39.26 mm (1.39) in the control; 23.60 (2.57) under the influence of lontrel. Thus, the length of the seedlings with respect to the control was 60.11%, and the degree of inhibition was approximately 40%.

It is interesting that simultaneously an experiment was carried out under the same conditions to study the effect of synthetic surfactant ethonium on seedlings. It was found that ethonium at a concentration of only one order of magnitude greater (0.25 mg/l) produces approximately the same inhibitory effect on the seedlings (or more precisely, slightly stronger). For example, after 95 h of incubation, the length of the seedlings of the same plant was equal to 17.67 (4.01) mm. The control for both experiments was the same, i.e., the length of the seedlings in the control was 39.26 mm (see above). The length of the seedlings in the variant with ethonium with respect to the control was 45.01%.

Hence, the biological activity of this synthetic surfactant is only one order of magnitude lower than the negative effect of pesticide (herbicide) lontrel.

In the next series of experiments, we studied the effect of lontrel on the seedlings of another plant, *Fagopyrum esculentum*. Similarly to the results obtained in the experiments with cucumbers, lontrel at a concentration equal to 0.02 mg/ml notably inhibited the growth of the buckwheat seedlings: their length was equal to 54% of the control after 47 h, and equal to 38% after 70 h of the experiment.

6.4.1.2   Dinitroorthocresol DNOC

One of the widely used pesticides is 2,4-dinitromethyl phenol, which is also known as dinitrocresol (DNOC), Arborol, Bryulex (more than 10 different synonyms). It is used as an aphicide, herbicide, desiccant, insecticide, and fungicide. The value of LD_{50} (for mice) is equal to 40–65 mg/l (Shamshurin and Krimer 1976).

We studied the effect of DNOC on some test objects, which allowed us to compare the action of this pesticide and surfactants. According to our data, a decrease in the mean length of *Sinapis alba* L. seedlings was observed under the influence of DNOC at concentrations ranging from 0.5 to 4 µg/ml. For example, after 48 h of growth at room temperature, the mean length of the seedlings in the control was 24.4 mm (confidence interval was 1.85 at the 0.05 level of significance). At the concentration of DNOC equal to 1 µg/ml, the mean length of the seedlings was equal to 16.9 mm (confidence interval was 1.16 mm). When the concentration of DNOC was increased to 4 µg/ml, the degree of inhibition of the seedlings correspondingly increased. The problems of the quantitative comparison of the results of biological testing of different chemicals on the same species of test organisms were considered in Ostroumov (1991a,b).

The results obtained in these experiments show that under conditions of the experiment, pesticides produced more clearly manifested effects than synthetic surfactants on the test objects we used in the experiments. Quantitative differences between the action of the pesticides and synthetic surfactants are of special interest. The difference in the biological activity (BA) between the representatives of different classes of xenobiotics was estimated as approximately one order of magnitude (the effects of synthetic surfactants were one order of magnitude weaker than those of pesticides). This difference should be related to the fact that the scale of the discharge of synthetic surfactants into ecosystems including aquatic ecosystems exceeds that of pesticides by one order of magnitude on average. Hence, the total ecological risk (including the product of BA by the amount or volume of the chemicals of a given class discharged to the ecosystem) of synthetic surfactants is the same as the risk of the pollution of ecosystems by pesticides, and in some cases it may even exceed it.

6.4.2 Water from man-made reservoirs

The objective of the study was to find whether some of the tested systems could serve to characterize the biological activity of waters from aquatic ecosystems.

A test with *Cucumis sativus* allowed us to distinguish the biological activity of water filtrate from a man-made reservoir in the period of active algal bloom. The mean length of the roots of the seedlings 173 h after the beginning of the experiment was 77% of the mean length of the seedling roots in the controls. Although the effect was not clearly manifested, it appeared statistically significant (the Student's mean difference was $P = 95\%$). The results of this experiment, which the author performed in the Mozhaiskoye Reservoir, were published in Kartsev et al. (1990). The dominating species at that period was *Aphanizomenon flos-aquae*.

It is interesting to compare the results of this test with the experiment carried out by L.A. Sirenko. It was found that the extracts of volatile substances emitted by blue-green algae (cyanobacteria) *Microcystis aeruginosa* (distillation with water vapor) inhibited the germination of radish seeds (Sirenko 1972). Extracts obtained from freshly collected vital colonies of *Microcystis* and algal mass in the state of fermentation manifested an inhibitory action. A slight inhibitory effect (83–92% of the control) was manifested by the freshly collected mass of *Microcystis* aggregated in lumps during the process of aging. It is interesting that volatile components of ether oils of *Aphanizomenon flos-aquae* similarly inhibited the seedlings of radish seeds (Sirenko 1972).

6.5 Some Concluding Remarks

The results of our experiments showed that the tested and developed methods for bio-assaying are applicable for estimating not only the biological activity of individual synthetic surfactants but also the activity of other preparations including mixed preparations containing synthetic surfactants and pesticides. We also demonstrated the applicability of these methods for characterizing natural waters of some types of ecosystems. It was shown for some specific synthetic surfactants and pesticides that their biological activity is only one order of magnitude weaker than the biological activity of pesticides. This points to a certain ecological risk, which can be caused by synthetic surfactants if they are transferred to ecosystems in mass amounts.

Another aspect of potential ecological hazard is worth mentioning. It is caused by the discharge of such preparations as synthetic detergents into water bodies and streams. We obtained the new data in the study of the effects of synthetic surfactants on some filter feeders (the effect of anionic and nonionogenic surfactants – see Chapters 3 and 4, respectively). The data lead to the conclusion that those chemicals are ecologically very hazardous. We think that if one would try to invent a means to disturb the aquatic ecosystems by misanthropic action, there could be no better tool than synthetic detergents. Indeed, these detergents contain synthetic surfactants, which can distort the living process of filter feeders. This was shown in model experiments using some species of freshwater and marine mollusks, and rotifers.

These results are in accord with the analysis of literature data, see Chapter 7 and our publications (Ostroumov et al. 1997, 1998, 2002, 2004, 2005a,b). Inhibition of filtration and a decrease in the removal of phytoplankton from aquatic medium can facilitate a disturbance of the balance between the factors regulating the abundance of phytoplankton. This may lead to the development of anomalously high concentrations of phytoplankton cells.

In small concentrations, synthetic detergents do not inhibit the development of phytoplankton (Braginsky et al. 1987; and our data on the effect of synthetic detergents on euglenas). In addition to this, many types of detergents contain dissolved compounds of phosphorus (see Chapter 2) frequently including 30–40% of sodium phosphates (Bock and Stache 1982, p. 173) or other phosphorus-containing compounds, which facilitates the improvement of the conditions for the growth of phytoplankton and a shift of the mentioned imbalance to increased development of algae. Eutrophication was observed in many reservoirs caused by the transfer of phosphorus-containing detergents with sewage waters (see, e.g., Godfrey 1982).

Additional evidence of the importance and necessity of normal functioning of the organisms of filter feeders and the flux of organic carbon to the bottom of the reservoir, which they catalyze, is given by the following numbers. According to estimates, the mean entry of P to sewage water is approximately 2 g of P per one citizen per day. According to the estimates of D. Uhlman, 89 g of oxygen is required to oxidize organic matter coming to the reservoir from one citizen (see Frantsev 1972). This value is significantly exceeded by the additional amount of oxygen (320 g) required to oxidize organic matter and algae growing in reservoirs owing to the transfer of 2 g of P (Frantsev 1972). Hence, a disturbance of normal water filtration and removal of algae and additional organic matter from the water column becomes especially dangerous in conditions of complex pollution including the transfer of excess P into the reservoirs.

Under certain conditions, addition of synthetic detergent into the growth medium can stimulate the growth of cyanobacterial cultures and Chlorophyceae algae (Kolotilova and Ostroumov 2000; Ostroumov and Kolotilova 2000; similar results were independently obtained in seawater, Aizdaicher et al. 1999), which can facilitate the imbalance in the interaction of organisms of neighboring trophic levels in the ecosystem.

It is interesting that increased transfer of nutrients from the coasts of Ozerninskoye Reservoir (Moscow Region) occupied by agricultural lands was accompanied with a decrease in the benthos biomass by 60%, a decrease of its production by 30%, and a decrease in the diversity of filter feeders to zero value (Solovyev and Pastukhova 1981). In the latter work, the authors observed that diversity of filter feeders decreased to a greater degree compared to other groups of benthos, swallowers, and predators.

Summation of the results of investigating the effects of surfactant-containing preparations on organisms confirms the regularities, which were found for specific classes of synthetic surfactants (Table 6.32). The role of nutrient transfer in the pollution of reservoirs has been for a long time the focus of attention in the analysis

Table 6.32   Sensitivity of various organisms to synthetic detergents and other surfactant-containing preparations.

Organisms	Sensitivity	References/comments
Euglena sp.	L	Ostroumov and Vasternak 1991, Vasternak and Ostroumov 1990, Ostroumov et al. 1998
Phytoplankton	L, M	Braginsky et al. 1983, 1987, etc.
Synechocystis sp. PCC 6803; *Scenedesmus quadricauda*; *Synechocystis elongatus*	L, S	Kolotilova and Ostroumov 2000, Ostroumov and Kolotilova 2000
Marine phytoplankton	S, L	Synthetic detergents stimulated microalgal growth (Aizdaicher et al. 1999)
Seedlings of *Fagopyrum esculentum, Oryza sativa, Cucumis sativus,* etc.	L, M	Comparatively low sensitivity enables the use of these organisms for phytoremediation (Ostroumov et al. 2000g)
Lymnaea stagnalis	M (estimate)	Decrease of the trophic activity may introduce changes to the ecosystem (Ostroumov 2000a)
Freshwater mollusks *Unio* sp.	M, H	There is danger of an inhibition of the filtration activity and a decrease of suspension removal from water (Ostroumov 2000b–d)
Marine mollusks *Mytilus galloprovincialis*	H	Comparatively low sensitivity poses hazard of disturbing the filtration activity and decreasing suspension removal from water (Ostroumov 2000b–d)
Marine mollusks *Crassostrea gigas*	H	Comparatively high sensitivity poses hazard of inhibition of the filtration activity (new results)

Note: Sensitivity in inhibition: L, low; M, moderate; H, high; S, stimulation.

of eutrophication, but our new data allow us to see a new ecological hazard caused by pollution by synthetic detergents, which contain both nutrients and synthetic surfactants.

By using vascular plants for testing, we accumulated information on the biological effects of different chemicals representing different classes of synthetic surfactants (see above and also Ostroumov 1991a,b; Ostroumov and Khoroshilov 1992;

Ostroumov and Semykina 1993). Summarizing what was said in Chapters 3–5 and the results described in this chapter we can state that synthetic surfactants can be arranged in the following order according to the degree of their inhibitory action on *F. esculentum*: polymeric surfactant copolymer of hexene and maleic aldehyde (CHMA) < anionic surfactant sodium dodecyl sulfate < detergent Vilva < nonionogenic surfactant Triton X-100 < cationic surfactant tetradecyl trimethyl ammonium bromide.

New facts and regularities presented in Chapters 3–6 make their generalized analysis reasonable in respect of the potential ecological hazard for hydrobionts caused by pollution of water bodies and streams. This problem is discussed in Chapter 7.

The author thanks former students I.A. Pavlova, V.S. Khoroshilov, and N.F. Viktorova for participating in some experiments. The impact of liquid foam detergent Vilva on plants was studied with V.S. Khoroshilov, a student of the Department of Hydrobiology. Thanks are also due to N.N. Kolotilova for assistance.

7 Biological Effects of Synthetic Surfactants and Participation of Hydrobionts in Water Purification

Experience indicates the negative impacts of surfactants and surfactant-containing preparations on representatives of the major functional ecosystemic blocks, including both autotrophic and heterotrophic organisms (see above). The author realizes that interpretation of the results obtained requires great caution and there are many factors that can affect the manifestation of biological effects of synthetic surfactants in more complex systems. Therefore, direct transfer of the laboratory results to natural ecosystems is inappropriate. The need for caution in interpreting experimental results was emphasized by N.S. Stroganov (1976, 1979, 1981), V.I. Lukyanenko (1967, 1983), S.A. Patin (1979), B.A. Flerov (1989), and other investigators when they analyzed the problems of using the results of laboratory experiments for understanding the situation in ecosystems under conditions of anthropogenic pollution.

It seems interesting to discuss the results obtained, touching upon important issues related to the functioning of ecosystems in conditions of anthropogenic impacts: (1) issues of self-purification of aquatic systems; (2) some applied problems related to self-purification of water; (3) problems of assessing the potential ecological hazard of anthropogenic impact on aquatic biota.

7.1 Self-Purification of Water and the Role of Hydrobionts in Aquatic Ecosystems

There are several definitions for *self-purification of water* in scientific literature. According to one of the definitions, self-purification is "the entire complex of biological, physical, and chemical processes that determine the ability of water bodies to get rid of pollution introduced by sewage waters and formed due to the vital activities of aboriginal organisms" (Metelev et al. 1971). "Self-purification of water in reservoirs is purification of water as a result of natural biological and physico-chemical processes, transformation of organic and partially inorganic substances" (Filenko 1988). The works by Skadovsky (1955), Drachev (1964), Vinberg and coauthors (Vinberg 1973, 1980), Konstantinov (1979), Vavilin (1983), Polikarpov (Polikarpov and Egorov 1986), and many others played an important role in explaining the processes of self-purification.

Natural ecosystems function under far from ideal conditions, being subject to the impacts of anthropogenic chemical pollution. All in all, the treatment facilities in Russia purify about 28 km^3 of water per year; herewith, approximately 97 km^3 are taken from the surface aquatic systems and more than 76 km^3 is annually discharged into natural ecosystems (the data of 1989) (Yakovlev et al. 1992; Ostroumov 1999e), i.e., only ~37% is subject to purification. The necessity of conserving and, if possible, increasing the self-purification potential of Russian aquatic ecosystems is indicated by the fact that inspection of public water supply sources revealed discrepancies of their quality with respect to the normal conditions (by the chemical indices) in five regions of European Russia (Kaluga Region, Nizhny Novgorod Region, and Saratov Region; Kalmykia and Mordovia); and in 14 more territories in the Volga River Basin. The frequency of such cases exceeds the average level for the Russian Federation by 27.7% (Elpiner 1999).

Complete purification is not always achieved after polluted waters have passed through water treatment facilities (Pupyrev 1992; Otstavnova and Kurmakayev 1997). Of the 184 purification systems in Moscow checked at their discharge points to the surface water systems, 88 (more than 30%) did not satisfy the requirements for the discharge of sewage waters into the environment (Otstavnova and Kurmakayev 1997). Thus, the essential function of ultimate water purification lies with natural ecosystems.

The processes of self-purification of aquatic ecosystems are important not only from the viewpoint of maintaining water quality as a resource for water consumption but also from the point of view of maintaining normal habitats needed for preserving the biodiversity.

7.1.1 Involvement of various hydrobionts in self-purification processes

Self-purification of an aquatic medium includes:

(1) physical and physico-chemical processes (see, e.g., Spellman 1996), among them
   (1.1) dilution,
   (1.2) washout of pollutants to the coast,
   (1.3) washout to adjacent aquatic objects,
   (1.4) sorption of pollutants by suspended particles followed by sedimentation,
   (1.5) sorption of pollutants by benthic sediments,
   (1.6) evaporation of pollutants;

(2) chemical processes (see, e.g., Bogdashkina and Petrosyan 1988; Skurlatov 1988; Shtamm and Batovskaya 1988), among them

(2.1) hydrolysis of pollutants,

(2.2) photochemical transformations,

(2.3) redox-catalytic transformations,

(2.4) transformations involving free radicals,

(2.5) binding of pollutants to dissolved organic matter (DOM) leading to a decrease in toxicity of pollutants,

(2.6) chemical oxidation of pollutants involving oxygen;

(3) biological processes (see, e.g., Voskresensky 1948; Vinberg 1973, 1980; Sushchenya 1975; Skarlato 1976; Vavilin and Vasilyev 1979; Alimov and Finogenova 1976; Konstantinov 1977, 1979; Alimov 1981; Kokin 1981; Vavilin 1983, 1986; Gutelmakher 1986; Vavilin et al. 1993; Koronelli 1996; Sadchikov 1997; Monakov 1998; Newell 1998; Ostroumov 1998, 2005a,b; Ostroumov and Fedorov 1999; Strayer et al. 1999), among them

(3.1) sorption and accumulation of pollutants and nutrients by hydrobionts,

(3.2) biotransformation (redox reactions, degradation, conjugation), mineralization of organic matter,

(3.3) extracellular enzymatic transformation of pollutants,

(3.4) removal of suspended particles and pollutants from the water column as a result of water filtration by hydrobionts,

(3.5) removal of pollutants from the water column as a result of sorption by pellets excreted by hydrobionts,

(3.6) assimilation of biogens (nutrients) by the benthos, which leads to preventing or slowing down the transfer of biogens and pollutants from benthic sediments into water,

(3.7) biotransformation and sorption of pollutants in soil during watering of lands with polluted waters,

(3.8) regulatory effects on other components of the water self-purification system, including the impact on organisms.

The list is incomplete; the phenomena are interrelated, and particular processes can be singled out only arbitrarily, for the purpose of their analysis and study.

It should be emphasized that the items on the above list formally categorized as physical or chemical processes in fact also involve biological factors that are considered in detail in Ostroumov (2000b–d).

Virtually all groups of hydrobionts are involved in self-purification. The role of microorganisms is certainly significant. The proportion of bacterial plankton in the transformation of organic matter increases from eutrophic to oligotrophic aquatic objects: the share of bacteria in eutrophic, mesotrophic, and oligotrophic lakes is equal to 55, 65, and 85%, respectively (Adamenko 1985). The role of microorganisms was analyzed in detail in the literature (Koronelli 1982, 1996; MacBerthouex and Rudd 1977; Vavilin 1983, 1986; Mishustina and Baturina 1984; Mishustina

1993; Swisher, 1987; Vavilin et al. 1993; Golovleva 1997; Varfolomeyev et al. 1997) and is not considered in detail here.

The rates of decay of pollutants are formed with the participation of almost all components of the ecosystem and are considered integral characteristics of the ecosystem (Gladyshev et al. 1996).

In some works, the contribution of hydrobionts to self-purification is considered as a permanent factor that does not depend on the harmful effects of an ecosystem's pollutants on the organisms (Lech and Vodicnik 1985). We can hardly agree with this. Under present-day conditions, the realization of the biological factors of self-purification in the ecosystems is affected by many factors, including pollution of water media. The composition of ecosystems in polluted reservoirs is formed under the influence of pollution. In 1908, R. Kolkwitz and M. Marsson developed a scale for estimating the degree of pollution of water bodies on this basis according to the presence of different organisms in them.

The experiments to characterize the biological effects of anionic, nonionogenic, cationic surfactants, and detergents on hydrobionts (Chapters 3, 4, 5, and 6, respectively) were carried out using the hydrobionts participating in self-purification of the water bodies and streams (Table 7.1). (Discussion in more detail is given in Ostroumov and Fedorov (1999) and Ostroumov (2000b–d).)

It was shown by the example of surfactants that under certain conditions they exert different effects on hydrobionts (which lead to inhibition of growth, changes in behavior, etc.). This can affect the processes, which with the participation of these hydrobionts lead to self-purification of water. Plankton and benthic organisms are important from this point of view.

One of the main factors of self-purification of aquatic ecosystems is the composition and abundance of plankton organisms. Bacterial plankton, phytoplankton and zooplankton participate in many processes leading to self-purification of water (a list of processes is given above).

The filtration activity of plankton was studied and estimated by many authors (Bogorov 1969; Kryuchkova 1972; Sushchenya 1975; Gutelmakher 1986; Gilyarov 1987). According to some estimates, rotifers can filter the volume of water in which they are located up to 7.7 times a day (Kryuchkova 1972). Plankton crustaceans filter daily some amount of water (from 5 to 90% of the volume) depending on the specific water body and its type (Gutelmakher 1986). The lower estimate of filtration activity (5%) denotes that the entire volume of the water column is filtered only by the crustaceans within 20 days, which coincides with the estimate given by V.G. Bogorov (1969) for the upper part of the World Ocean (0–500 m).

Plankton is the object of direct and indirect effects of pollutants. The direct effects of pollutants on phytoplankton were considered in the works by Braginsky et al. (1987), Ivanov (1976a), Lewis (1990), Poremba et al. (1991) and in many other works including our studies (see above), in which we distinguished the impacts of surfactants on *S. quadricauda* (impact of SDS), on *M. lutheri* (effect of cationic surfactant ethonium) (Ostroumov and Maksimov 1988), on marine cyanobacteria *Synechococcus* (effect of nonionogenic surfactant TX100) (Waterbury and Ostroumov 1994) on euglenas *E. gracilis* (impact of synthetic detergents Bio-S and Kristall) (Vasternak and Ostroumov 1990; Ostroumov and Vasternak 1991) on

Table 7.1 New results on the impacts of contaminants (synthetic surfactants, etc.) on representatives of the major blocks of the ecosystems involved in self-purification processes (examples).

Feeding types	Groups of organisms	Species (examples)	Substances and types of effects (examples)	Involvement of the given group of organisms in water self-purification
Auto-trophic and mixo-trophic	Cyano-bacteria	*Synechococcus* WH7805, WH8103	TX100 (i, l)	+ (2.6, 3.1, 3.2)
	Diatoms	*Thalassiosira pseudonana*	TX100 (i, l)	+ (2.6, 3.1)
	Flagellata	*Euglena*	synth. detergent	+ (2.6, 3.1, 3.2)
	Plantae vasculares	*Pistia stratiotes, Sinapis alba, Fagopyrum esculentum, Lepidium sativum, Oryza sativa*	sulfonol, SDS, TX100, TDTMA, DNOC, lontrel, synth. detergent (s, i, l)	+ (2.6, 3.1, 3.7)
Hetero-trophic	Heterotrophic bacteria	*Hyphomonas* sp. MHS-3, VP-6	TX100, TDTMA (i)	+ (3.2)
	Pulmonata	*Lymnaea stagnalis*	TDTMA (s)	+ (3.2)
	Bivalvia	*Mytilus edulis, M. galloprovincialis, Crassostrea gigas, Unio tumidus, U. pictorum*	SDS, TX100, TDTMA, synth. detergent (s)	+ (3.4)
	Annelida	*Hirudo medicinalis*	TDTMA (b, s, l)	+ (3.8)

Note: i, inhibition of growth of cultures or organisms; s, sublethal effects; b, change of behavior; l, lethal effect. See text for explanations.

The right column gives the numbers of the corresponding processes important for water self-purification (according to the numeration of listing of the information given in the text of this chapter; see above).

Dunaliella asymmetrica (effect of sulfonol) (Ostroumov et al. 1990), and on marine diatoms *T. pseudonana* (effect of nonionogenic surfactant TX100) (Chapter 4).

Indirect effect of pollutants on phytoplankton can be realized due to the fact that the abundance of phytoplankton depends on many abiotic and biotic factors. Among them, the rate of grazing by invertebrate filter feeders (Gutelmakher 1986; Vinogradov and Shushkina 1987; Gilyarov 1987), including benthic filter feeders (Voskresensky 1948; Alimov and Finogenova 1976; Kondratyev 1977; Vinberg 1980; Alimov 1981; Monakov 1998; Strayer et al. 1999). The role of benthic filter feeders in self-purification is discussed in the next section.

7.1.2   The role of benthic filter feeders

The role of benthic filter feeders should be discussed in greater detail. Their role in self-purification is related to the fact they exert a conditioning effect on water quality by removing suspensions of different nature from water. The assimilation of benthic filter feeders varies within broad limits, e.g., being 40–70% for some freshwater mollusks (Monakov 1998). The other part of filtered organic material is excreted and is transferred to the bottom sediments as pellets. This makes benthic filter feeders participants of significant geochemical fluxes related to the removal of suspended matter from water.

The extraordinary role of benthic filter feeders is illustrated by the following examples: zebra mussels *Dreissena polymorpha* in the western part of Lake Erie (up to 50,000 mollusks per 1 m^2) consumed 2–4 times more phytoplankton daily than its observed biomass per 1 m^2 (Stoeckmann and Garton 1995). According to another estimate, the populations of *Dreissena polymorpha* filter every day 70–125% of the water column (in the summer months, according to the data for the aquatic eco-systems of North America, see Strayer et al. 1999). Bivalve mollusks belonging to the family Corbiculidae in North American freshwater ecosystems filter 0.3–10 m^3 water daily per 1 m^2 at a population density of 1–30 g/m^2 (dry weight without shells) (Strayer et al. 1999). Mollusks from Red Lake (*Unio tumidus, U. pictorum, Anodonta complanata*) filtered in summer 123–174 g of suspended organic matter in a water column over 1 m^2 bottom (Kuzmenko 1976). Sponges in Lake Baikal filter a water layer 12 m thick approximately in 1.2 days (Savarese et al. 1997). Mollusks of the Dnieper– Boug Estuary (the Black Sea) filter the volume of the estuary more than 16 times during the vegetation season (Alekseyenko and Aleksandrova 1995). The bottom water layer 3 m thick is filtered by the mussels of the Black Sea at certain sites in approximately 30 h (Zaika et al. 1990). Oysters *Crassostrea virginica* before their intensive catches in Chesapeake Bay (volume 71.7× 10^9 m^3) filtered the entire volume of the bay in 3.3 days, 30% of carbon in filtered seston having been excreted as compact biological deposits becoming available for the benthic trophic web (Newell 1998).

The role of other invertebrate filter feeders is significant. The rate of filtration by ascidians *Styla clava* was 0.38 ml/s in the experiment (the mean weight of an animal was 179 mg dry weight); by polychaetes *Sabella penicillus*, 2.17 ml/s (the mean weight of an animal, 65 mg dry weight; temperature during the filtration measurements, 20°C) (Riisgard and Larsen 1995, cited from Ostroumov 2000). The total water filtration by invertebrates (mollusks, ascidians, polychaetes), was estimated as 1–10 m^3 m^{-2} day^{-1} (Riisgard et al. 1998, cited from Ostroumov 2000).

Benthic filter feeders can contribute to the regulation of processes related to eutrophication and blooming of toxic plankton species (Officer et al. 1982; Buskey et al. 1997).

The processes and phenomena occurring in the ecosystem as a result of water filtration are important for water self-purification and regulation of the functioning of the ecosystem; the processes and phenomena include:

• adsorbed pollutants are sedimented together with the suspended particulate matter;
• turbidity of water is decreasing and the conditions are improved for visual light and UV radiation penetration as well as for the effects they induce on hydrobionts and organic matter;

- the content of fine dispersed suspended matter is decreasing in water, which is favorable for the increase in the economical value (fishing) of the water bodies; in the opposite case, if the content of suspension in water is increasing, the rate of filtration of biological filter feeders is decreasing (see Sushchenya 1975; Alimov 1981, and others);
- mixing of water is intensified, which results in aeration of water, and affects phytoplankton and zooplankton; higher concentrations of phytoplankton, decrease in the concentration of nutrients, and zooplankton are observed in permanently mixed aquatic bodies and streams;
- aeration of water and conditions for oxygen consumption are improved, which facilitates oxidizing of organic matter;
- species composition and abundance of species of algobacterial community are regulated and this affects the rate of generation and destruction of hydrogen peroxide and the rate of self-purification by the free radical-dependent mechanism;
- the components of dissolved organic matter (DOM) are excreted;
- sedimentation of organic matter is accelerated due to the consumption of phytoplankton and bacterial plankton by benthic filter feeders, excretion of pellets of faeces and pseudofaeces by filter feeders;
- active growth and functional activity of mollusk filter feeders facilitates the development and functioning of heterotrophic bacteria in the lower zone of the ecosystem (see above for details in Chapter 3, the section about SDS and also Ostroumov 2000b–d).

There are data in the literature which help explain the ecological role of the processes related to water filtration: the processes of formation of certain transparency of water, which is important for the penetration of UV radiation and realization of the biological effects related to UV radiation in the water column (Reitner et al. 1997; Santas et al. 1997); the processes that decrease the amount of suspended matter (Zak 1960), negatively affecting many hydrobionts by decreasing their filtration activity (Sushchenya 1975; Alimov 1981; Mitin and Voskresensky 1982; Mitin 1984; Gorbunova 1988; Yamasu and Mizofuchi 1989) and others. The excess of suspended matter in water can increase the toxicity of pollutants. For example, in the presence of benthic clay (50 mg/l; particles smaller than 2 μm), the toxicity of herbicide glyphosate to *Daphnia pulex* increased more than two times; in the presence of benthic clay the value of EC_{50} (48 h, 15°C) was equal to 3.2 mg/l, while in the medium without suspended benthic clay it was 7.9 mg/l (Hartman and Martin 1984).

Thus, the problem of the degree to which filtration activity of hydrobionts can be inhibited under the influence of anthropogenic factors including chemical pollution is ecologically important.

7.1.3 Effects of pollutants on filter feeders

As we repeatedly noted, our experiments demonstrated inhibition of the filtration activity by hydrobionts under the influence of anionic surfactants (Chapter 3), nonionogenic surfactants (Chapter 4), cationic surfactants (Chapter 5), synthetic and liquid detergents (Chapter 6).

Pollutants can affect the rate of water filtration. We demonstrated this by the example of anionic, nonionogenic, and cationic surfactants (see Chapters 3, 4, and 5; Table 7.1), all of which inhibited the rate of removal of phytoplankton cells from the ecosystem. The experiments showed that synthetic surfactants inhibited water filtration by *M. edulis*, *M. galloprovincialis*, *C. gigas*, *U. tumidus*, and *U. pictorum* (see above, and also Ostroumov et al. 1997, 1998; Ostroumov 1998, 2000d, 2005a,b). The effects of synthetic surfactants were shown to be statistically significant (Chapters 3, 4, and also Ostroumov 2000d).

The results of our experiments agree with those obtained in studies of the impacts of pollutants on other mollusk species (e.g., Mitin 1984; Stuijfzand et al. 1995). Various pollutants cause an increase in the period of the time when the mollusk valves are closed (see, e.g., Karpenko et al. 1983 and also Tyurin 1985; cited from Flerov 1989). In our experiments, we also observed the closure of valves of *M. edulis*, but at a surfactant concentration (SDS, 20 mg/l) significantly greater than those much lower concentrations (1–2 mg/l) sufficient to inhibit the filtration rate. Mercury compounds (methyl mercury acetate at 0.4–2.8 mg/l) decreased the grazing of diatomic algae by *M. edulis* (Dorn 1976; cited from Flerov 1989).

While studying the effect of DDT on the Black Sea mussel, a weakening of water filtration was observed by Zaitsev and Golovenko (1981). The rate of water filtration by mussels *M. edulis* was inhibited by pesticides lindane, endrin, carbaryl, dichlorvos, flucitrinate, permethrin (Donkin et al. 1997); tributyltin, and dibutyltin (Widows and Page 1993). Inhibition of biological filtration of water by bivalve mollusks under the influence of pollutants was also demonstrated by other authors (Mitin 1984; Smaal and Widdows 1994).

According to many observations, pollution of water leads to the elimination of filter feeder organisms from the composition of macrozoobenthos in the polluted parts of rivers (e.g., Zhadin 1964; Kumsare et al. 1972, Zinchenko and Rozenberg 1996; and others) and storage reservoirs (Dyga and Lubyanov 1972), which finally decreases the filtration activity of benthic communities. Long-term investigations in the Upper Volga River demonstrated that in aquatic systems with unsatisfactory ecological states (strong toxic pollution or high load of organic matter on the reservoir) filter feeders (mollusks, Bryozoa, sponges) are almost absent in the composition of zooperiplankton (Skalskaya and Flerov 1999). The biomass of filter feeders sharply decreased in the aquatic ecosystems of Fennoscandia after the increase in the concentration of phosphorus (P_{total}) in water, a decrease in pH, and during toxification (near the sources of pollution with heavy metals) (Yakovlev 2000).

There are indications of the impacts of pollutants on the filtration activity of plankton (Day and Kaushik 1987; Filenko 1988; Matorin et al. 1989, 1990).

The rate of water filtration and feeding of freshwater crustacean *Daphnia magna* on the cells of *Chlamydomonas reinhardii* was sensitive to pyrethroid phenvalerate (Day and Kaushik 1987). The inhibition of the rate of grazing of *Daphnia magna* on the cells of chlorella under the influence of herbicide Saturn (0.001–0.1 mg/l), insecticides DDT (0.1–1 mg/l), and metaphos (2 mg/l), as well as under the influence of copper sulfate were demonstrated by D.N. Matorin and coauthors (Matorin et al.

1989, 1990) using the method of delayed fluorescence tested in studies of the biological effects of many pollutants (Matorin 1993) including synthetic surfactants (Parshikova et al. 1994). The effect of many pollutants on the rate of grazing of daphnia was demonstrated at lower concentrations of pollutants than the one at which the inhibition of algae was pronounced (Matorin 1998).

An inhibitory effect of synthetic surfactant TDTMA on the filtration activity of two rotifer species *Brachionus angularis* and *Brachionus plicatilis* was shown in Kartasheva and Ostroumov (2000).

Transfer of biogens into the water medium is an additional hazard that can lead to a decrease in the filtration activity of hydrobionts. Nutrients stimulate the development and increase in the biomass of phytoplankton. It was found for several groups of filter feeders that an increase in the concentration of nutrition particles (phytoplankton and bacterial plankton are examples) caused a decrease in the filtration rate (e.g., Sushchenya 1975, Alimov 1981).

As mentioned above, the mechanisms of self-purification of water include the processes that occur with the participation of heterotrophic bacteria, cyanobacteria, algae, flagellates, and filter feeders. The changes in the abundance, rates of growth, and nutrition of hydrobionts (Monakov 1998), and rates of excretion of pellets of faeces and pseudofaeces (Palaski and Booth 1995), and changes in the ratio of species in the composition of aquatic biocenoses under the influence of synthetic surfactants cause consequences for the processes of self-purification. The latter require trophic activity (Monakov 1998) of bivalve and pulmonal mollusks that form significant amounts of pellets. The pellets rapidly sediment to the bottom under gravity forces, thus contributing to the removal of organic matter from the pelagial. This organic matter is consumed by organisms as nutrition (Newell 1988; Alekseyenko and Aleksandrova 1995; Strayer et al. 1994, 1999; Ogilvie and Mitchell 1995; Newell and Ott 1999). An increase in the rate of sedimentation of pellets compared with the rate of sedimentation of individual cells of phytoplankton and their fragments was proved (Lisitzin 1983; Emelyanov 1998).

The differential biological effects of anthropogenic chemicals on the organisms of different ecological groups is especially important under conditions of complex pollution of aquatic media (Filenko 1988) including pollution with synthetic detergents, when nutrients are transported to water together with synthetic surfactants. Under certain conditions, synthetic detergents (containing surfactants and phosphorus compounds) can stimulate the growth of algae. For example, detergent Tide-Lemon at concentrations of 1–100 mg/l stimulated the growth of *Synechocystis* sp. PCC 6803 (Kolotilova and Ostroumov 2000). Similar data were independently obtained in experiments with marine microalgae (Aizdaicher et al. 1999). A potentially hazardous situation is when the growth of phytoplankton organisms is stimulated owing to the uptake of biogens, while the filtration activity leading to the removal of phytoplankton from the water column is inhibited under the influence of surfactants. Since a stable population of algae is possible only at a balance of factors leading to an increase in its abundance, and factors causing a decrease of the latter (these factors include grazing the algae by consumers including benthic filter feeders), there is a danger of imbalance between the processes determining the state of phytoplankton in polluted water, which would facilitate algal blooming.

Taking into account the diversity of biological effects produced by synthetic surfactants during their effect on representatives of all major groups of hydrobionts, we arrive at a deeper understanding of the fact that aquatic biota (including micro- and macroorganisms) are labile and vulnerable components in the system of water self-purification. Bivalve mollusks, which filter water (Ostroumov 1998, 1999a,b, 2000, 2005a,b), are vulnerable parts of the system.

7.2 Water Purification and Some Applied Problems

Many methods of using organisms for biotechnological purification of polluted waters (MacBerthouex and Rudd 1977; Stavskaya et al. 1988, 1989; Vasilyev and Vasilyev 1993; Zhmur 1997) and ecosystems or their components (Koronelli 1982, 1996; McCutcheon et al. 1995; Golovleva 1997; Varfolomeyev et al. 1997) are being developed and tested. These processes are modeled in order to optimize the control of the water purification processes (Vavilin 1983; 1986; Vavilin and Vasilyev 1979; Vavilin et al. 1993).

Many of the main groups of hydrobionts are either used or suggested to be used in biotechnological processes. The range of the species used in artificial ecosystems is widening. Since many of the artificial ecosystems are designed for cleaning polluted waters and are frequently operated at the upper limits of admissible concentrations of pollutants, information about the tolerance limits of hydrobionts to all major pollutants including synthetic surfactants is important (Ostroumov 1999c). There are limits of the maximum content of synthetic surfactants in the waters discharged to waste water treatment systems, which vary from 20 to 50 mg/l (Stavskaya et al. 1988). These limits are established for synthetic surfactants in general without differentiation into individual components or classes of synthetic surfactants.

Our results obtained in experiments with a wide range of objects point to the heterogeneity of synthetic surfactants with respect to their biological effects on organisms. Using angiosperm plants used as biotests, the sequence of representatives of different classes of synthetic surfactants was determined in the series of increasing the biological activity they manifest. For example, synthetic surfactants are positioned in the following sequence, according to the increase in the degree of their inhibitory effect on *F. esculentum*: polymeric surfactant copolymer of hexene and maleic aldehyde < anionic surfactant sodium dodecyl sulfate < detergent Vilva < nonionogenic surfactant Triton X-100 < cationic surfactant TDTMA.

Hence, the need to take into account the heterogeneity of synthetic surfactants in further work on establishing the norms of the chemical compositions of waters to be purified. In practice, the situation becomes even more pressing because the content of synthetic surfactants in the sewage frequently exceeds the allowable norms indicated above and can reach 30 g/l. The role of synthetic surfactants is enhanced by the fact that the efficiency of cleaning water removing these chemicals is often 48–80% on average, while in the winter period it is only 20% (see Boichenko and Grigoryev 1991). Some individual types of synthetic surfactants (e.g., noniono-genic surfactants belonging to the class of alkyl phenol derivatives) belong to hard-to-decompose xenobiotics and the percentage of their removal from water is even lower.

As there is a severe deficit of water resources in some regions, soils are irrigated or planned to be irrigated with polluted waters (Izmerov et al. 1978; Agricultural... 1989; Kaliev 1996). During irrigation with waters containing synthetic surfactants, they are accumulated in soils and in plants (Ignatova 1978; Mudryi 1990). Our experiments carried out with several species of plants demonstrated the general regularity for many species (*S. alba* L., *F. esculentum* Moench, *L. sativum* L., *O. sativa* L., *C. sativa* (L.) Crantz, *T. aestivum* L., and others) that synthetic surfactants at concentrations significantly smaller than their maximum content revealed in sewage waters inhibit the rate of elongation of plant seedlings (see above, Chapters 3–6). We also demonstrated that at concentrations lower than those that caused significant inhibition of the elongation of seedlings, synthetic surfactants damaged the formation of root hairs by the plant rhizoderm. This affects ecological interactions in the

Table 7.2 Impacts of some synthetic surfactants and surfactant-containing preparations on the hydrobiont functions important for water self-purification (by the example of mollusks and rotifers).

No.	Substances	Concentrations, mg/l	Species	Effects*
1	TX100	5	*Unio tumidus*	1
2	TX100	1	*U. tumidus*	1
3	TDTMA	1–2	*U. pictorum*	1
4	TDTMA	1–2	*U. pictorum*	1
5	TDTMA	2	*Lymnaea stagnalis*	2
6	Tide-Lemon	75	*L. stagnalis*	2
7	Tide-Lemon, Losk-Universal, Lotos-Extra	6.7–50	*Mytilus galloprovincialis*	1
8	Avon Herbal Care	5–60	*M. galloprovincialis*	1
9	SDS	1.7	*M. galloprovincialis*	1
10	TDTMA	1	*M. galloprovincialis*	1
11	SDS	1–5	*M. edulis*	1
12	TX100	0.5–5	*M. edulis*	1
13	SDS	0.5 and higher	*Crassostrea gigas*	1
14	TDTMA	0.5	*C. gigas*	1
15	Synthetic detergents (four preparations)	2 and higher	*C. gigas*	1
16	Liquid detergents (two preparations)	2	*C. gigas*	1
17	TDTMA	0.5	*Brachionus angularis, B. plicatilis*	1

*1, decrease of water filtration rate; 2, decrease of trophic activity and pellet formation rate. For details, see Ostroumov (2000d), Kartasheva and Ostroumov (2000). P. Donkin participated in experiments 11 and 12; N.N. Kartasheva, in experiments 2 and 4; N.V. Kartasheva, in experiment 17.

plant–soil system. In this relation, of interest are the results of our experiment that demonstrated a decrease in the abundance of cyanobacteria (including nitrogen-fixing bacteria) in soil during the effect of water solutions of synthetic surfactants (Ostroumov and Tretyakova 1990). Thus, the results of our investigations of the effect of synthetic surfactants on the plants and soil cyanobacteria significantly supplement the literature data and indicate the possibility of distortions in the structural functional parameters of the ecosystems, which affects their water purification potential and ecological capacity of agricultural lands with respect to their ability to serve for utilization of polluted waters.

For practical solution of the problems of cleaning, rehabilitation, and restoration of contaminated natural aquatic ecosystems, of great importance are the approaches related to bio- and phytoremediation.

Numerous biotechnological schemes exist for cleaning polluted waters using various organisms (MacBerthouex and Rudd, 1977). Among them are those that use immobilized microorganisms (Stavskaya et al. 1988, 1989; Ostroumov and Samoilenko 1990), floating or submerged macrophytes, e.g., *Potamogeton, Myriophyllium* sp., *Lemna* sp., *Sagittaria* sp., *Elodea canadenis, Ceratophyllum demersum* (McCutchon 1997), various species of algae (Schnoor et al. 1995), artificial ecosystems imitating aquatic ecosystems (constructed wetlands) (Medina and McCutcheon 1996), and schemes with application of staged or conveyor purification with participation of several organisms. Application of plants in many cases is based on the fact that certain concentrations of pollutants can be "assimilated by plants without any serious consequences" (Kozhova and Timofeyeva 1983). "By establishing metabolic concentrations [of pollutants, S.O.], it is possible to mobilize internal resources of hydrophytes to detoxicate pollutants and use this possibility for developing new methods for cleaning sewage sinks and elucidating the role of flora in the processes of water self-purification" (Kozhova and Timofeyeva 1983).

Use of artificial ecosystems and phytoremediation allowed the cleaning of waters from some very specific organic chemicals, e.g., trinitrotoluene and other chemicals of military purposes (Medina and McCutcheon 1996). The operational costs of degrading pollutants using artificial ecosystems and phytoremediation are significantly lower by at least 30–36% compared to other methods of degrading xenobiotics (Medina and McCutcheon 1996).

Newly obtained information about the sensitivity and tolerance of a number of plant species to synthetic surfactants obtained in our experiments appear useful for further work on phytoremediation of polluted waters and ecosystems.

7.3   Anthropogenic Impact on Hydrobionts: Assessment of the Ecological Hazards

Diversity and ecological importance of the biological (including sublethal) effects of synthetic surfactants that we revealed and studied enables us to distinguish insufficient adequacy of some of the existing criteria for estimating the ecological hazards of chemicals (Maki and Bishop 1985; Feijtel et al. 1997). An example of a traditional approach is the calculation of the EC_{50}/CSW ratio recommended in the U.S. as a first guideline, where CSW is the expected concentration of a pollutant in sewage water. The value of EC_{50} was taken equal to LC_{50} (96 or 48 h) (Maki and Bishop 1985). This ratio can prove sufficiently high, and, based on that, a chemical is concluded to be relatively safe ecologically. However, this approach is insufficient for such conclusions since some unfavorable effects on organisms and ecosystems can occur at significantly lower concentrations than LC_{50} (96 or 48 h).

An objective estimate of the ecological hazard of a chemical should necessarily consider sublethal effects (this opinion is supported by Flerov (1989)) and the impacts of the chemical on the capability of ecosystems self-purification (Filenko 1988; Index… 1999; Ostroumov 2000c, 2002, 2004). The latter should be understood in a broader sense than the processes performed by microorganisms.

A broader system of criteria is required, which would take into account the diverse (including sublethal) effects of pollutants (Patin 1979; Lukyanenko 1983; Simakov 1986; Ramade 1987; Filenko 1988; Flerov 1989; Moiseyenko 1999). The situation is complicated by the fact that "there are no universal criteria for estimating all forms of effect" (Moiseyenko 1999). Taking into account the results obtained, the author believes it reasonable to suggest for consideration and possible application a four-stage concept of systemic level–block analysis of potential ecological hazard of anthropogenic impacts on biota, which includes the analysis of anthropogenic disturbances:

- at the level of individual and population changes,
- at the level of aggregated parameters (e.g., such as aggregated characteristics of total productivity and biomass of groups of organisms),
- at the level of integrity and stability of the ecosystem, and
- at the level of the contribution of the ecosystem to biospheric processes (Table 7.3).

The foundations and development of the notions important for this concept are described in Yablokov and Ostroumov (1983, 1985), Yablokov and Ostroumov

Table 7.3. The concept of level–block analysis (Ostroumov 2000b,d) of the potential ecological hazard of the anthropogenic impacts on the biota (taking into account publications by many authors, including Yablokov and Ostroumov 1983, 1985; Ostroumov 1984, 1986a; Abakumov and Maksimov 1988; Stavskaya et al. 1988; Filenko 1988; Flerov 1989; Bezel et al. 1994; Krivolutsky 1994; Kasumyan 1995; Ostroumov and Fedorov 1999; Donkin et al. 1997; taking into account new experimental data by the author).

No.	Disturbance levels	Examples of disturbances and their consequences (some of them can be assigned to several levels)
1	Level of individual and population responses	Toxic effects on particular species (higher mortality rate, lower fertility, disturbed ontogenesis, pathologies, etc.), growth rate changes, variability change of morphological and functional indices, genotoxicity, change of behavior of organisms
2	Level of aggregated (supraorganismal) responses	Change of primary productivity, change of aggregated biomass, change of chlorophyll concentration in aqueous medium, change of dissolved oxygen concentration

Table 7.3 (continued)

No.	Disturbance levels	Examples of disturbances and their consequences (some of them can be assigned to several levels)
3	Level of stability and integrity of the ecosystems	Rearrangements and/or weakening of plankton–benthos relationship, disturbance of planktonic–benthic coupling, rearrangements and/or weakening of links in the trophic web
		Change of level of bacterial destruction, decrease of the rate of filtration and clarification rate / removal of particles from water, decline of water self-purification
		Decrease of regulatory effects as a result of extinction / escape from the ecosystem / trophic passivity of organisms of higher trophic levels (deregulation)
4	Level of the contribution of the ecosystems to the biospheric processes	Change of flows of C (e.g., sedimentation of pellets formed and excreted by filter feeders), N (e.g., nitrogen fixation), and other elements (e.g., S, P), and also energy fluxes

Table 7.4   Use of the level–block concept (Ostroumov 2000b,d) for classifying the types of potential ecological hazards of the impacts of substances on the organisms.

No.	Biological effects	Type of ecological hazard (see Table 7.3)	Sources of data
1	Surfactants (TX100, SDS, TDTMA), synthetic detergents suppress water filtration by *Mytilus edulis*, *M. galloprovincialis*, *Unio tumidus*	3, 4	New data by the author; Ostroumov et al. 1997a,b, 1998; Ostroumov 1998; Ostroumov and Fedorov 1999; Ostroumov 2000b
2	Pesticides decrease water filtration by *M. edulis*	3, 4	Donkin et al. 1997
3	TDTMA decreases the trophic activity of *Lymnaea stagnalis* and pellet formation rate	3, 4	Ostroumov 2000a; new data by the author
4	SDS, sulfonol, TX100, TDTMA, synthetic detergents, liquid detergents inhibit the growth rate of seedlings of *Sinapis alba*, *Fagopyrum esculentum* and other plants	1	This work; Ostroumov 1990, 1991a,b; Ostroumov et al. 1990; Ostroumov and Maksimov 1991; Ostroumov and Tretyakova 1990; Ostroumov and Golovko 1992; Ostroumov and Khoroshilov 1992
5	TDTMA induces behavioral changes in *Hirudo medicinalis*	1, 3	Ostroumov 1991a,b
6	TDTMA decreases the abundance of nitrogen-fixing cyanobacteria	4	Ostroumov and Tretyakova 1990
7	Cu (sulfate), Cd (sulfate), Pb (nitrate) inhibit water filtration by *M. galloprovincialis*	3, 4	Ostroumov 2004

(1991), Ostroumov (1999a), Ostroumov and Fedorov (1999), Ostroumov (2000b–d). Application of this concept as the basis of a classification scheme allows us to introduce an additional systematization and regularity of the effect on organisms. An example of such an application of this classification scheme for additional analysis and classification of new information on the biological effects of synthetic surfactants obtained in this work (Chapters 3–6) is given in Table 7.4.

This concept can be used in the solution of the problems related to the determination of specific numerical values of critical (ecologically admissible) loads on ecosystems, i.e., in the numerical determination of the "discharge of one or several contaminants into the habitat, which do not exert any harmful effects on the most sensitive components of the ecosystem (at the present level of knowledge)" (Moiseyenko 1999) (a concept with a similar sense is the concept of ecologically admissible effects (Izrael 1984)). Solving that group of the problems, the concept of level block analysis can be applied at the stage denoted as "diagnosis of the state of ecosystems and justification of the most informative criteria of the state of the organisms, populations, and communities" and in the final determination of "critical (admissible) loads, i.e. the volume of the transfer of contaminants to aquatic objects" (Moiseyenko 1999). The concepts developed by the author have something in common with the concept of assimilation capacity of the ecosystem developed in the case of marine ecosystem (Izrael et al. 1988a,b; Izrael and Tsyban 1983, 1989). Assimilation capacity of a marine ecosystem is an "integral function of its state, which reflects the ability of physical, chemical, and biological processes for removing pollutants and eliminating their effect on biota" (Izrael and Tsyban 1992). The "biotic component plays a dominating role in the fluxes of pollutants" (Izrael and Tsyban 1992, p. 20).

The obtained results and developed concepts can be used for perfecting the approach for estimating the potential hazard of chemicals in the ecological monitoring and forecasting. Comparatively high tolerance of angiosperm plants to synthetic surfactants revealed in our investigation can be used for phytoremediation purposes (Ostroumov 2000d).

Generalizing Remarks

Information on the biological effects of synthetic surfactants is required to better predict the ecological consequences of the entry of surfactants into aquatic ecosystems, to furnish a more comprehensive idea of potential hazards, and to execute more adequate ecological assessments of the state of ecosystems and the environment. Particular classes of surfactants need a differentiated approach in the assessment of their potential ecological hazards; special limits should be developed for each of them (including nonionogenic and cationic surfactants) in order to regulate their entry to waste water treatment plants.

Synthetic surfactants entering aquatic ecosystems can produce undesirable effects on the organisms and the structural/functional parameters of the ecosystems. The potential hazard of disturbing the processes of water filtration and self-purification of aquatic ecosystems as a result of pollution with surfactants is significant. Taking into account the results obtained, the author developed a concept that aquatic biota as an ecosystemic unit (including not only microbiota but also macrobiota) serve as a labile and vulnerable element in the system of water self-purification. Preventing the anthropogenic decrease of the self-purification potential of aquatic ecosystems is a necessary condition for the stable and sustainable use of the aquatic ecosystems' resources.

Some of the biological effects described or quantitatively characterized in the book are sublethal and subtoxic; others are attributed to behavioral changes. The results obtained demonstrate the potential ecological hazard of sublethal concentrations of synthetic surfactants and related physiological and behavioral responses to anthropogenic pollutants, which is consistent with the results of other studies of xenobiotics (Filenko 1988, 1989; Flerov 1989, Krishnakumar et al. 1990; Kasumyan 1995, 1997; Juchelka and Snell 1995; Stuijfzand et al. 1995).

A number of methods we studied and tested in the experiments with synthetic surfactants to assess the biological activities of substances (tests on seedlings) are alternative to the most frequently used methods for biotesting of toxicants on animals. Some of them have virtually not been used (tests of all classes of surfactants on seedlings; tests of nonionogenic and cationic surfactants and synthetic detergents on mollusks) to characterize the range of substances of interest. Existing biotesting methods were improved and new techniques were developed (the morphogenetic index, which characterizes the apparent average length; tests using the rhizoderm reaction; modifications of measurements of the effects on the filtration activity efficiency). This has greatly expanded the arsenal of methods and techniques for assessing the biological activity of the above classes of chemicals.

The experiments and their analysis emphasized the need for a conceptual approach to the assessment of the ecological hazard of chemicals, which would take into account the diversity of their biological effects. Along with the traditional assessment based on mortality for a certain period of time – a necessary component of total ecological hazard of compounds (Filenko 1988, 1989) – other approaches based on other types of effects on the organisms are useful (Ostroumov 1984; Yablokov and Ostroumov 1985; Flerov 1989; Kasumyan 1995). Anthropogenic induction of ecological imbalances was revealed to be a potential hazard (examples are different effects of surfactant TX100 on the representatives of adjacent links of the trophic chain – plankton (*Synechococcus* sp., *Hyphomonas* sp.) and benthic filter feeders (*Mytilus edulis, M. galloprovincialis, Crassostrea gigas, Unio* sp.). Surfactants affected both plankton organisms and benthic filter feeders that consume them as a nutrition resource. However, the filtration activity of the filter feeders proved more susceptible to the surfactant than the growth of the plankton organisms: at relatively low surfactant concentrations, the efficiency of filtration activity of the mollusks decreased significantly. The results of the experiments indicate a potential hazard when a decrease in the removal of plankton organisms from water by their consumers is not compensated for by an adequate decrease in the growth of plankton. Algae as a whole were noted "to be related to the group of hydrobionts most resistant to the toxic effect of pollutants" (Gapochka 1999). Moreover, under certain conditions surfactant-containing compounds (detergents that are mixtures of several chemicals) can stimulate the growth of algae (Aizdaicher et al. 1999; Kolotilova and Ostroumov 2000). Thus, the effects of a pollutant on organisms of the adjacent trophic levels may produce a potential hazard of an imbalance in the trophic chains.

From time to time, a necessity emerges to revise the system of priorities used as a basis for ranking the compounds by the degree of their ecological hazard and to develop a more adequate system of classification. When developing such a system, it is suggested to use approaches that analyze the potential hazards caused by pollutants to the processes of ecosystem self-purification, matter and energy transfer in the trophic chain, dynamic balance between interacting species, and information fluxes in and between ecosystems. The applicability of the generalized four-component concept of the level–block analysis of potential ecological hazard of anthropogenic effects on ecosystems was demonstrated when considering some biological effects of xenobiotics (Ostroumov 1999a, 2000b–d). Underestimation of sublethal effects and differential biological activity of xenobiotics (e.g., synthetic surfactants) relevant to their effect on various ecosystem types may prove a source of possible significant errors in predicting the consequences of anthropogenic stress on the ecosystems. This prompts new suggestions for the hydrobiological aspects of global changes (Vellinga and van Verseveld 2000); namely, the ecological mechanisms of the biogeochemical carbon fluxes (Kuznetsov 1980, 1993; Zavarzin 1984; Izrael and Tsyban 1989; Jonsson and Carman 1994; Kuznetsov et al. 1997; World Resources 1994), and absorption and storage of CO_2 and C_{org} by aquatic ecosystems. Studies of the potential hazards of anthropogenic effects should be focused on those aspects of the functioning of hydrobionts that are important for biogeochemistry, in particular, their filtration activity.

It has been noted that under certain conditions synthetic surfactants are more hazardous pollutants than believed earlier (Ostroumov 1991a,b). The discharge of synthetic surfactants into the environment is significant (Kouloheris 1989; Painter 1992) and continues to increase annually (depending on the class of surfactant) by about 2–5%. The range of biological effects caused by surfactants is wide and covers almost all major blocks and trophic levels in aquatic systems. It includes behavior disorders in organisms and disturbances of the processes that contribute to water self-purification. Many synthetic surfactants decompose very slowly as a result of micro-bial oxidation and biodegradation (Swisher 1987; Poremba et al. 1991). Not only certain surfactants but also some products of their biodegradation (as shown for non-ionogenic surfactants, such as alkyl phenols and their derivatives) are persistent and have high bioaccumulation coefficients. Along with other negative consequences, they exert an estrogenic effect on the biota (Huber 1985; Holt et al. 1992). All this indicates that under certain conditions synthetic surfactants can be more hazardous pollutants than considered earlier, which should be taken into account in improving and changing the system of nature-protection priorities.

In general, the results obtained and their analysis point to a significant potential hazard of the consequences of massive pollution of aquatic media with synthetic surf-actants. This information facilitates a more profound understanding of the processes of self-purification under conditions of anthropogenic impact (synthetic surfactants were found to be a potential hazard for the filtration activity of benthic mollusks) and contributes to developing the artificial ecosystems for bio- and phytoremediation.

New results make necessary a more adequate interpretation of some provisions of environmental law. For instance, the concept of ecological hazard or damage to the environment, ecosystems, and living resources. This concept will not be inter-preted completely enough if, along with other types of damage, it fails to include the disturbance of the ability of ecosystems for self-purification, including the ability for water filtration at a normal rate. Examples of the laws whose interpretation and im-plementation require exact rendering of the concept of ecological harm or damage to the ecosystems and living resources are the Russian Federal Laws "On Environ-mental Assessments" (1995), "On Animal World" (1995), "On the Continental Shelf of the Russian Federation" (1995), and some others.

Attention was drawn to the role of chemical and biochemical factors in stabili-zation and destabilization of the ecological balance (Ostroumov, 1986). Research into the biological effects of synthetic surfactants, including the processes important for water self-purification and maintenance of the stability of aquatic ecosystems provided new material for the analysis of the anthropogenic destabilization of the ecological balance.

The range of possible applications of the results obtained and concepts devel-oped includes the diagnostics of the state of the ecosystems, determination of critical (allowable) loads, ecological examination, monitoring, and prediction, required for the stable utilization of biological resources and sustainable development. The mate-rials of this work could also be used as the scientific basis for the prophylaxis of emergency situations due to massive pollution of the environment.

The ecological examination of the projects should include the following priority questions. In what way does an anthropogenic influence affect the self-purification potential of aquatic ecosystems, including the activity of organisms that are filter feeders? What are the sublethal effects of pollutants on organisms, including the manifestation of differential (unequal in character and expression) biological activity of compounds and how would this affect the ecological balance in an ecosystem? What concentrations of anionic, nonionogenic, and cationic synthetic surfactants can occur in a water ecosystem?

The new facts obtained and concepts developed indicate the necessity of greater attention to potential ecological hazard and damage to the environment due to the misuse of synthetic surfactants and pollution of aquatic ecosystem with them. Measures should be taken to enhance the control of this kind of pollution and decrease its level, to move synthetic surfactants to a higher rank in the system of nature-preservation priorities.

The revealed vulnerability of aquatic organisms, especially filter feeders to sub-lethal concentrations of pollutants, including synthetic surfactants, emphasizes the importance of preserving the normal level of functional activity of this group of hydrobionts in aquatic ecosystems under conditions of anthropogenic stress. The necessary prerequisite for maintaining the habitat of hydrobionts, and, hence, for preserving biodiversity is to provide conditions for a normal, sufficiently high level of functional activity of that block of the aquatic ecosystem that is represented by filter feeders.

Conclusions

1. Using autotrophic and heterotrophic hydrobionts and other organisms as test objects, we established and characterized some important biological effects caused by the impact of the aquatic environment containing synthetic surfactants. Thus, in studies of the effects of synthetic surfactants on autotrophic organisms, we established inhibition of growth of diatoms *Thalassiosira pseudonana* (Hustedt) Hasle et Heimdal and euglenas; disturbance of growth and development of angiosperm plants, including inhibition of elongation of plant seedlings (*Sinapis alba* L., *Fagopyrum esculentum* Moench, *Lepidium sativum* L., *Oryza sativa* L., and others), and growth of aquatic macrophytes (*Pistia stratiotes* L.). Disturbance of morphogenetic processes in the rhyzoderm leading to the formation of root hairs was found. The impact of synthetic surfactants on heterotrophic organisms was established to inhibit growth of marine bacteria (prosthecobacteria *Hyphomonas* sp.) and filtration activity of marine and freshwater mollusks (*Mytilus edulis* L., *M. galloprovincialis* Lamarck, *Crassostrea gigas* Thunberg, *Unio tumidus* Philipsson, *U. pictorum* L.); and to change the behavior of annelids *Hirudo medicinalis* L., and others.

2. As a result of action of synthetic surfactants (including anionic, nonionogenic and cationic) and surfactant-containing mixed preparations on water filtration by mollusks, the biological effects of these classes of substances, including a reduction of the removal of suspended particles and cells of unicellular organisms from water, can pose a potential ecological hazard to hydrobionts.

3. We established the order of organisms in the sequence of their increasing tolerance to a representative of nonionogenic surfactants, TX100. Under experimental conditions used, by their tolerance to the impact of TX100, the organisms are in the following order: *Thalassiosira pseudonana* < *Mytilus edulis* < *Hyphomonas* sp., *Synechococcus* sp. < *Fagopyrum esculentum*.

4. We revealed the order of the representatives of various classes of synthetic surfactants in the sequence of their increasing biological activity when used as a biotest of vascular plants. Thus, by the increase of the extent of inhibitory action on *F. esculentum*, under experimental conditions the synthetic surfactants are in the following sequence: polymeric surfactant CHMA < anionic surfactant SDS, foam detergent Vilva < nonionogenic surfactant Triton X-100 < cationic surfactant TDTMA.

5. To assess the potential ecological hazards of synthetic surfactants and other substances for hydrobionts, we suggest a conceptual approach based on a structured system of analysis of the potential hazard of substances, which includes the assessment of disturbances of the aquatic biota at four levels: (1) at the level of individual and population changes, (2) at the level of aggregated parameters, (3) at the level of

the integrity and stability of an ecosystem, and (4) at the level of the contribution of an ecosystem to the biospheric processes.

6. It is proposed to complement the system of priority test species and parameters for biotesting (to include the filtration activity of bivalve mollusks and other organisms) and improve the system of ranking of pollutants (to move higher the rank of synthetic surfactants priorities).

7. For assessing the biological activity of chemical substances, it is proposed to use an improved variant of the biotesting method by considering the new proposed and approbated morphogenetic index. The index integrates information on the germination of seeds and the rate of elongation of seedlings (an integral morphogenetic index – an apparent average length of seedlings, AAL). A new method of biotesting was developed, based on the first observed effect of inhibition of the formation of root hairs.

8. Based on the revelation and comparison of the tolerance of organisms of various taxa, it is proposed to use angiosperm plants for phytoremediation of the environment polluted with surfactants.

9. The potential ecological significance of the effects caused by synthetic surfactants and surfactant-containing mixed preparations on hydrobionts and a relation of these effects to the hazard of anthropogenic impacts on the processes important for water self-purification was theoretically and experimentally substantiated. Preservation of the self-purification potential of water bodies and streams is impossible without additional efforts to decrease the damage to hydrobionts and ecosystems due to the pollution of water with synthetic surfactants and surfactant-containing mixed preparations. It is proposed to take this conclusion into account when formulating the hydrobiological priorities for sustainable development, environmental assessments, preservation of biodiversity, and use of bioresources.

References

Abakumov, V.A. (1972) Significance of the temporal organization features in populations in the evolutionary process. Organization and evolution of the living. Leningrad: Nauka Publishers, pp. 37–45 (in Russian).

Abakumov, V.A. (1973) Temporal structure of a population and methods of predicting its abundance. Trans. All-Union Research Inst. of Marine Fisheries and Oceanography. Vol. 91, pp. 68–86 (in Russian).

Abakumov, V.A. (1983) Methods of hydrobiological analysis of surface waters and bottom sediments: a manual. Leningrad: Gidrometeoizdat Publishers, 240 pp. (in Russian).

Abakumov, V.A. (1984) The system for hydrobiological monitoring of the quality of natural waters in the USSR. Topical problems of the protection of the environment in the Soviet Union and the Federal Republic of Germany. Proc. of Symposium, Munich, pp. 491–529.

Abakumov, V.A. (1985) Regularities of the change of aqueous biocenoses under the action of anthropogenic factors. Integrated global monitoring of the World Ocean. Proc. First Int. Symposium, Leningrad: Gidrometeoizdat Publishers, pp. 273–283 (in Russian).

Abakumov, V.A. (1987a) Production aspects of the biomonitoring of fresh-water ecosystems. Trans. Zool. Inst. USSR Acad. Sci., Vol. 165, pp. 51–61 (in Russian).

Abakumov, V.A. (1987b) Anthropogenic changes of the environment and some evolutionary issues. Problems of ecological monitoring and modeling of ecosystems. Leningrad: Gidrometeoizdat Publishers, Vol. 10, pp. 22–35 (in Russian).

Abakumov, V.A. (1990) Formation of the biosphere as a planetary ecological system. Problems of ecological monitoring and modeling of ecosystems. Leningrad: Gidrometeoizdat Publishers, Vol. 13, pp. 25–43 (in Russian).

Abakumov, V.A. (1991) Ecological modifications and the development of biocenoses. Ecological modifications and criteria of ecological normalization. Proc. Int. Symposium. Leningrad: Gidrometeoizdat Publishers, pp. 18–40 (in Russian).

Abakumov, V.A. (1993) On the peculiarity of the biosphere thermodynamics. Problems of ecological monitoring and modeling of ecosystems. St. Petersburg: Gidrometeoizdat Publishers, Vol. 15, pp. 21–36 (in Russian).

Abakumov, V.A. and Maksimov, V.N. (1988) Ecological modulations as an index of the background state of an aqueous environment. Scientific bases of the biomonitoring of freshwater ecosystems. Proc. Soviet–French Symposium. Leningrad: Gidrometeoizdat Publishers, pp. 104–117 (in Russian).

Abakumov, V.A. and Sushchenya, L.M. (1991) Hydrobiological monitoring of freshwater ecosystems and ways for its improvement. Ecological modifications and criteria of ecological normalization. Proc. Int. Symposium, Leningrad, pp. 41–61 (in Russian).

Abakumov, V.A., Kurilova, Yu.V., and Semin, V.A. (1989) Cyclic variations of hydrometeorological characteristics and biomonitoring of anthropogenic changes. Problems of ecological monitoring and modeling of ecosystems. Leningrad: Gidrometeoizdat Publishers, Vol. 12, pp. 18–31 (in Russian).

Abakumov, V.A., Maksimov, V.N., and Ganshina, L.A. (1978) Ecological modulations as indicators of water quality change. Bases of water quality control by the hydrobiological indices. Proc. All-Union Conference, Moscow, November 1–3, Leningrad, pp. 117–136 (in Russian).

Abramzon, A.A. and Gayevoy, G.M. (eds.) (1979) Surfactants. Leningrad: Khimiya Publishers, 376 pp. (in Russian).

Adamenko, V.N. (1985) Climate and lakes. Leningrad: Gidrometeoizdat Publishers, 264 pp. (in Russian).

Agricultural usage of sewage waters (1989) Moscow: Rosagropromizdat Publishers, 223 pp. (in Russian).

Ahel, M., McEvoy, J., and Giger, W. (1993) Bioaccumulation of the lipophilic metabolites of nonionic surfactants in freshwater organisms. *Environm. Pollution*, **79(3)**: 243–248.

Ainsworth, S. (1992) Soaps and detergents. *Chem. Eng. News*, **70 (3)**: 27–63.

Aizdaicher, N.A. (1999) Response of the dinoflagellate *Gymnodinium kovalevskii* to the action of detergents. *Biol. Morya*, **25 (2)**: 87–88 (in Russian).

Aizdaicher, N.A., Malynova, S.I., and Khristoforova, N.K. (1999) Effect of detergents on the growth of microalgae. *Biol. Morya*, **25 (3)**: 234–238 (in Russian).

Akberali, H.M. and Trueman, E. (1985) Effects of environmental stress on marine bivalve molluscs. *Adv. Marine Biol.*, **22**: 101–198.

Akulova, K.I. and Bushtuyeva, K.A. (eds.) (1986) Communal hygiene. Moscow: Meditsyna Publishers, 608 pp. (in Russian).

Alabaster, J. and Lloyd, R. (1982) Water quality criteria for freshwater fish. London: Butterworth.

Alekseyenko, T.L. and Aleksandrova, N.G. (1995) The role of bivalve mollusks in mineralization and sedimentation of organic matter of the Dnieper–Bug estuary. *Gidrobiol. Zhurn.*, **31 (2)**: 17–22 (in Russian).

Alexander, J. (1993) In defense of garbage. Westport: Praeger, 341 pp.

Alimov, A.F. (1981) Functional ecology of freshwater bivalve mollusks (Trans. Zoological Institute, USSR Acad. Sci., Vol. 96). Leningrad: Nauka Publishers, 248 pp. (in Russian).

Alimov, A.F. and Finogenova, N.P. (1976) Quantitative assessment of the role of bottom-dwelling animal communities in self-purification of freshwater reservoirs. Hydrobiological bases of water self-purification. Leningrad: USSR Academy of Sciences Publishers, pp. 5–14 (in Russian).

A list of fishery standards: maximum permissible concentrations and reference safe levels of impact of harmful substances for water of aquatic objects of commercial fishing importance. Moscow: VNIRO Publishers, 304 pp. (in Russian).

A list of maximum permissible concentrations and reference safe levels of impact of harmful substances for waters of fishery reservoirs Committee of the Russian Federation on Fisheries. Moscow: Medinor, 220 pp. (in Russian).

Andronikova, I.N. (1976) Quantitative assessment of the involvement of zooplankton in self-purification processes based on the example of Lake Krasnoe. Hydrobiological bases of water self-purification. Leningrad: USSR Academy of

Sciences Publishers, pp. 30–35 (in Russian).

Anisova, S.N., Sokolova, S.A., Mineyeva, T.V., Lebedev, A.T., Polyakova, O.V., and Semenova, I.V. (1995) A list of maximum permissible concentrations and reference safe levels of impact of harmful substances for the waters of fishery reservoirs. Moscow: Medinor Publishers, 221 pp. (in Russian).

Apasheva, L.M., Budzhiashvili, D.M., Murza, L.I., Naidich, V.I., Bogdanov, G.N., and Emanuel, N.M. (1976) Changes of the paramagnetic properties of chlorella under the action of toxic chemical compounds. *Dokl. Akad. Nauk SSSR*, **228 (3)**: 723–725 (in Russian).

Aronson, R. and Precht, W. (1995) Landscape patterns of reef coral diversity: a test of the intermediate disturbance hypothesis. *J. Exp. Marine Biol. Ecol.*, **192**: 1–14.

Artyukhova, V.I. (1996) Integrated analysis of the action of the fungicide imazalyl sulfate on the culture *Scenedesmus quadricauda*. Ecological and physiological studies of the algae and their significance for the assessment of the status of natural waters: Brief report at the conference. Borok: December 3–5, Yaroslavl, pp. 163–164 (in Russian).

Artyukhova, V.I. and Dmitrieva, A.G. (1996) Bioassay of the quality of water in model ecosystems using algae. Ecological and physiological studies of the algae and their significance for the assessment of the status of natural waters: Brief report at the conference. Borok: December 3–5, Yaroslavl, pp. 115–117 (in Russian).

Artyukhova, V.I., Dmitrieva, A.G., Filenko, O.F., and Chao Izun (1997a) Changes in the culture growth dynamics and cell size of *Scenedesmus quadricauda* (Turp.) Breb. under the action of potassium bichromate. *Izv. RAN, Biol. Series*, **3**: 280–286 (in Russian).

Artyukhova, V.I., Dmitrieva, A.G., Filenko, O.F., and Chao Izun (1997b) Consequences of the action of potassium bichromate on the algal culture *Scenedesmus quadricauda* (Turp.) Breb. at toxic load changes. *Izv. RAN, Biol. Series*, **4**: 440–445 (in Russian).

Ashby, J. and Tennant, R. (1988) Chemical structure, *Salmonella* mutagenicity and extent of carcinogenicity as indicators of genotoxic carcinogenesis among 222 chemicals tested in rodents. *Mutan. Res.*, **204 (1)**: 17–115.

Augenfeed, J. (1980) Effects of Prudhoe Bay crude oil contamination on sediment working rates of *Abarenocola pacifica*. *Mar. Environ. Res.*, **3 (4)**: 307–313.

Austin, B. (1985) Antibiotic pollution from fish farms: effects on aquatic microflora. *Microbiol. Sci.*, **2 (4)**: 113–117.

Bagotsky, S.V. (1992) Mathematical models of multispecies algal complexes. Problems of ecological monitoring and modeling of ecosystems. St. Petersburg: Gidrometeoizdat Publishers, Vol. 14, pp. 173–187 (in Russian).

Bailey, R.E. (1996) Biological activity of polyoxyalkylene block copolymers in the environment. *In* Nonionic surfactants, Nace, V.M. (ed.). New York: Marcel Dekker, pp. 243–257.

Bakacs, T. (1980) Protection of the environment. Moscow: Meditsyna Publishers, 216 pp. (Russian translation of the Hungarian book: Bakacs, T. Kornyezetvedekem. Medicina: Budapest, 1977).

Barenboim, G.M. and Malenkov, A.G. (1986) Biologically active substances. Moscow: Nauka Publishers, 365 pp. (in Russian).

Batovskaya, L.O., Kozlova, N.B., Shtamm, E.V., and Skurlatov, Y.I. (1988) Role of microalgae in the regulation of hydrogen peroxide content in natural waters. *Dokl. Akad. Nauk SSSR*, **301 (6)**: 1513–1516 (in Russian).

Belanger, S., Davidson, D., Farris J., Reed, D., and Cherry, D. (1993) Effects of cationic surfactant exposure to a bivalve mollusc in stream mesocosms. *Environm. Toxicol. Chem.*, **12**: 1789–1802.

Belanger, S., Meiers, E., and Bausch, R. (1995a) Direct and indirect ecotoxicological effects of alkyl sulphate and alkyl ethoxysulfate on macroinvertebrates in stream mesocosms. *Aquatic Toxicol.*, **33 (1)**: 65–87.

Belanger, S., Rupe, K., and Baush, R. (1995b) Responses of invertebrates and fish to alkyl sulphate and alkyl ethoxylate sulfate anionic surfactants during chronic exposure. *Bull. Environm. Contam. Toxicol.*, **55 (5)**: 751–758.

Belevich, T.A., Ilyash, L.V., and Fedorov, V.D. (1997) Biomass dynamics and functional characteristics of the cultures of marine microalgae *Prorocentrum micans* and *Platymonas viridis* at the action of a low concentration of cadmium. *Vestn. MGU, Ser. 16*, **1**: 29–32 (in Russian).

Belousova, M.Y., Avgul, T.V., Safronova, N.S., Krasovsky, G.N., Zholdakova, Z.I., and Shlepnina, T.G. (1987) Basic properties of organic compounds normalized in waters. Moscow: Nauka Publishers, 104 pp. (in Russian).

Berth, P. and Jeschke, P. (1989) Consumption and fields of application of LAS. *Tenside Surfact. Deterg.*, **26**: 75–79.

Bezel, V.S. (1988). Problems of mammalian ecotoxicology. *In* Ecotoxicology and protection of the environment (Krivolutsky, D.A., Bocharov, B.V., et al., eds.). Riga: Institute of Biology, Latvian Acad. Sci., pp. 26–28 (in Russian).

Bezel, V.S., Bolshakov, V.N., and Vorobeichik, E.L. (1994) Populational ecotoxicology. Moscow: Nauka Publishers, 81 pp. (in Russian).

Beznosov, V.N., Plekhanova, I.O., Prokhorov, V.G., and Plekhanov, S.E. (1987) On accumulation of heavy metals by Black Sea mussels and oysters. Use and protection of USSR flora and fauna resources. Moscow: Russian Academy of Sciences Publishers, p. 54 (in Russian).

Biedlingmaier, S., Wanner, G., and Schmidt, A. (1987) A correlation between detergent tolerance and cell wall structure in green algae. *BioScience*, **42 (3)**: 245–250.

Billinghurst, Z., Clare, A., Fileman, T., Mcevoy, J., Readman, J., and Depledge, M. (1998) Inhibition of barnacle settlement by the environment oestrogen 4-nonylphenol and the natural oestrogen 17beta oestradiol. *Mar. Poll. Bull.*, **36 (10)**: 833–839.

Blondin, G., Knobeloch, L.M., Read, H., et al. (1987) Mammalian mitochondria as *in vitro* monitors of water quality. *Bull. Environ. Contam. Toxicol.*, **38 (3)**: 467–474.

Bobra, A., Shiu, W., Mackay, D., and Goodman, R. (1989) Acute toxicity of dispersed fresh and weathered crude oil and dispersants to *Daphnia magna*. *Chemosphere*, **19 (8/9)**: 1199–1222.

Bocharov, B.V. (1988) Chemical means of protection against biodamage as a source of pollution. Ecotoxicology and protection of the environment. Riga: Institute of Biology, Latvian Acad. Sci., pp. 33–35 (in Russian).

Bocharov, B.V. and Prokofyev, A.K. (1988) Ecotoxicological assessment of the danger of polluting marine environment by organotin compounds. Ecotoxicology and protection of the environment. Riga: Institute of Biology, Latvian Acad. Sci., pp. 38–40 (in Russian).

Bocharov, B.V., Anisimov, A.A., and Kryuchkov, A.A. (1985) Major means of protecting materials against microbial damage. Ecological bases of protection against biodamage. Moscow: Nauka Publishers, pp. 169–221 (in Russian).

Bocharov, B.V., Panteleyev, A.A., and Prokofyev, A.K. (1988) Pollution of the environment by chloroaromatic compounds. Ecotoxicology and protection of the environment. Riga: Institute of Biology, Latvian Acad. Sci., pp. 35–38 (in Russian).

Bock, K. and Stache, H. (1982) Surfactants. *In* The handbook of environmental chemistry. Berlin: Springer, Vol. 2, Part B, pp. 163–200.

Bodar, C., Donselaar, E.G. van, and Herwig, H. (1990) Cytopathological investigations of digestive tract and storage cells in *Daphnia magna* exposed to cadmium and tributyltin. *Aquat. Toxicol.*, **17 (4)**: 325–333.

Boethling, R.S. and Linch, D.G. (1992) Quaternary ammonium surfactants. *In* The handbook of environmental chemistry, Vol. 3, Part F. Berlin: Springer, pp. 145–178.

Bogdashkina, V.I. and Petrosyan, V.S. (1988) Ecological aspects of aqueous environment pollution with oil hydrocarbons, pesticides, and phenols. Ecological chemistry of the aqueous environment. Vol. 2, pp. 62–78 (in Russian).

Bogorov, V.G. (1969) Role of plankton in the exchange of substances in the ocean. *Okeanologiya*, **9 (1)**: 156–161 (in Russian).

Boichenko, V.K. and Grigoryev, V.T. (1991) On the method for the calculation of synthetic detergent entry into the Ivankovskoe water reservoir. *Vodn. Resursy*, **1**: 78–87 (in Russian).

Bolshakov, V.N. (1990) Preface to Introduction to problems of biological ecology by Telitchenko, M.M. and Ostroumov, S.A. Introduction to problems of biological ecology. Moscow: Nauka Publishers, pp. 3–4 (in Russian).

Bolshakov, V.N., Sadykov, O.F., Benenson, I.E., Korytin, N.S., and Kryamjinsky, F.V. (1987) Topical problems of population monitoring. Problems of ecological monitoring and modeling of ecosystems. Leningrad: Gidrometeoizdat Publishers, Vol. 10, pp. 47–63 (in Russian).

Braginsky, L.P. and Beskaravainaya, S.D. (1983) The oxygen method of studying primary production and destruction as a bioassay for the presence of toxicants. *In* Theoretical issues of bioassaying (Lukyanenko, V.I. (ed.)). Volgograd: Institute for the Biology of Internal Waters, USSR Acad. Sci., pp. 145–152 (in Russian).

Braginsky, L.P., Beskaravainaya, S.D., Burtnaya, I.L., et al. (1983) Toxicity for hydrobionts and degradation of synthetic surfactants in freshwaters. Kiev, 231 pp. The paper is deposited in VINITI, No 3247–83 (in Russian).

Braginsky, L.P., Burtnaya, I.L., and Shcherban, E.P. (1979) Toxicity of synthetic detergents for mass forms of freshwater invertebrates. Experimental studies of the effect of pollution on aquatic organisms. Apatity: Kola Branch of the USSR Acad. Sci., pp. 24–30 (in Russian).

Braginsky, L.P., Kalenichenko, K.P., and Shcherban, E.P. (1983) Planning toxicological assays. *In* Theoretical issues of bioassaying (Lukyanenko, V.I. (ed.)). Volgograd: Institute for the Biology of Internal Waters, USSR Acad. Sci., pp. 30–37 (in Russian).

Braginsky, L.P., Perevozchenko, I.I., Kalenichenko, K.P., and Pishcholka, U.K. (1980) Biological factors of degradation of pesticides and detergents (synthetic surfactants) in an aqueous environment. Self-purification and bioindication of polluted waters. Moscow: Nauka Publishers, pp. 193–196 (in Russian).

Braginsky, L.P., Velichko, I.M., and Shcherban, E.P. (1987) Freshwater plankton in a toxic environment. Kiev: Naukova Dumka, 179 pp. (in Russian).

Bressan, M., Brunetti R., Casellato, S., Fava, G., Giro, P., Marin, M., Negrisolo, P., Tallandini, L., Thomann, S., Tosoni, L., Turchetto, M., and Campesan, G. (1989) Effects of linear alkylbenzene sulfonate (LAS) on benthic organisms. *Tenside*

Surfactants Detergents, **26**: 148–158.

Bro-Rasmussen, F., Calow, P., Canton, J., Chambers, P., Silva-Fernandes, A., Hoffmann, L., Jouany, J., Klein, W., Persoone, G., Scoullos, M., Tarazona, J., and Vighi, M. (1994) EEC water quality objectives for chemicals dangerous to aquatic environments (List 1). *Rev. Environm. Contam. Toxicol.*, **137**: 83–110.

Bryzgalo, V.L., Korshun, A.M., Nikanorov, A.M., and Sokolova, L.P. (2000) Hydro-biological characteristics of the low Don River under conditions of a long-term anthropogenic effect. *Vodn. Resursy*, **27 (3)**: 357–363 (in Russian).

Budayeva, L.M. (1991) Biological monitoring of large Caucasus rivers. Problems of ecological monitoring and modeling of ecosystems. Leningrad: Gidrometeoizdat Publishers, Vol. 13, pp. 54–60 (in Russian).

Bulion, V.V. and Nikulina, V.N. (1976) The role of phytoplankton in self-purifica-tion processes in water streams. Hydrobiological bases of water self-purification. Leningrad, pp. 15–24 (in Russian).

Burdin, K.S. (1985) Bases of biological monitoring. Moscow: Moscow University Press, 158 pp. (in Russian).

Burdin, K.S. (1989) Introduction to biochemical ecology (a review). *Zh. Obshch. Biologii*, **50 (3)**: 429 (in Russian).

Burkovsky, I.V. (1984) Ecology of free-living infusoria. Moscow: Moscow University Press, 208 pp. (in Russian).

Burkovsky, I.V. (1992) Structural–functional organization and resistance of marine bottom-dwelling communities. Moscow: Moscow University Press, 208 pp. (in Russian).

Burkovsky, I.V., Kashunin, A.K., and Azovsky, A.I. (1999) White Sea microbenthos community as an indicator of the state of the aqueous environment. *Gidrobiol. Zhurn.*, **35 (5)**: 86–94 (in Russian).

Burridge, T., and Shir, M. (1995) The comparative effects of oil dispersants and oil/dispersant conjugates on germination of the marine macroalgae *Phyllospora comosa* (Fucales: Phaeophyta). *Mar. Pollut. Bull.*, **31 (4)**: 446–452.

Buskey, E., Mantagna, P., Amos, A., and Whitledge, T. (1997) Disruption of grazer populations as a contributing factor to the initiation of the Texas brown tide algal bloom. *Limnol. Oceanol.*, **42 (5)**: 1215–1222.

Cain, R. (1987) Biodegradation of anionic surfactants. Biochem. Soc. Trans. V. 15 (Supplement), pp. 7S–22S.

Cairns, J. (1986) The myth of the most sensitive species. *BioScience*, **36 (10)**: 670–672.

Cairns, J. and Niederlehner, B.R. (1987) Problems associated with selecting the most sensitive species for toxicity testing. *Hydrobiologia*, **153**: 87–94.

Cano, M., Dyer, S., and DeCarvalho, A. (1996) Effect of sediment organic carbon on the toxicity of a surfactant to *Hyalella azteca*. *Environ. Toxicol. Chem.*, **15 (8)**: 1411–1417.

Chan, G.M. (1994) Nutrition studies of Japanese scallop larvae and spat. *Izv. TINRO*, **113**: 18–25 (in Russian).

Chawla, G., Viswanathan, P.M., and Santha, D. (1986) Effect of linear alkylbenzene sulfonate on *Scenedesmus quadricauda* in culture. *Environ. Exp. Bot.*, **26 (1)**: 39–51.

Cherbadzhi, I.I. (1996) Effect of some drilling solutions' components on the primary production of phytoplankton. Methodology and protocol for assessment of the impact of marine oil-and-gas industry on Arctic environment. Abstracts of papers of an international workshop. Murmansk, March 10–22, pp. 82–83 (in Russian).

Cherepanov, S.K. (1995) Tracheophytes of Russia and CIS (within the limits of the former USSR). St. Petersburg: Mir i Semya Publishers, 991 pp. (in Russian).

Chernenkova, T.V. (1987) Peculiar features of germination and growth of pine and spruce seeds in soils with different degree of heavy-metal pollution. *In* Impact of industrial enterprises on the environment (Krivolutsky, D.A. (ed.). Moscow: Nauka Publishers, pp. 168–182 (in Russian).

Choucri, N. (ed.) (1993) Global Accord. Cambridge: MIT Press, 562 pp.

Concept of ecological safety of Russia. Ecological safety of Russia, Vol. 2, pp. 52–56 (in Russian).

Concept of the transition of the Russian Federation to sustainable development (approved by the decree of the President of the Russian Federation of 1 April 1996 No 440). Newest ecological laws. Moscow: Inst. International Law and Economics, pp. 360–368.

Connell, D. and Miller, G. (1984) Chemistry and ecotoxicology of pollution. New York: Wiley, 444 pp.

Cormack, R. (1962) The development of root hairs in angiosperms, II. *Bot. Rev.*, **28**: 446.

Cosovic, B. and Ciglenecki, I. (1997) Surface active substances in the Eastern Mediterranean. *Croat. Chem. Acta*, **70 (1)**: 361–371.

Criteria of assessing the ecological situation of territories for revealing zones of extreme ecological situations and zones of ecological disaster. Moscow (in Russian).

Danilov-Danilyan, V.I. (1995) State and problems of the protection of the environment in the Russian Federation. Scientific and technical aspects of environment protection, No 10, pp. 60–67 (in Russian).

Davies, I.J. (1990) Biomonitoring of freshwater ecosystems in Canada: A program of the Department of Fisheries and Oceans. *In* Problems of ecological monitoring and modeling of ecosystems. Leningrad: Gidrometeoizdat Publishers, Vol. 13, pp. 75–88 (in Russian).

Davies, M.S. (1991) Effects of toxic concentrations of metals on root growth and development. *In* Plant root growth: an ecological perspective, ed. D. Atkinson. Oxford: Blackwell Scientific Publ., pp. 211–227.

Davies, K.L., Davies, M.S., and Francis, D. (1991) The influence of an inhibitor of phytochelatin synthesis on root growth and root meristematic activity in *Festuca rubra* L. in response to zinc. *New Phytol.*, **118**: 565–570.

Davydov, O.N., Balakhnin, I.A., Kalenichenko, K.P., and Kurovskaya, L.Ya. (1997) Adsorption and desorption of cationic surfactants by Aerosyl preparation and its effect on the immune-physiological indices of carp blood. *Gidrobiol. Zhurn.*, **33 (2)**: 68–75. (in Russian).

Day, K. and Kaushik, N. (1987) Short-term exposure of zooplankton to the synthetic pyrethroid, fenvalerate, and its effects on rates of filtration and assimilation of the alga *Chlamydomonas reinhardii*. *Arch. Environ. Contam. Toxicol.*, **16**: 423–432.

Dean, J. (1985) Consumer and institutional surfactants. *Chemical Week. Special Advertising Section*, **136 (20)**: 3–26.

De Bruijn, J. and Struijs, J. (1997) Biodegradation in chemical substances policy. *In* S. Hales, T. Feijtel, H. King, K. Fox, and W. Verstraete (eds.) Biodegradation kinetics. Brussels: SETAC–Europe, pp. 33–45.

Decree of the President of the Russian Federation (1996) "On the concept of the transition of the Russian Federation to sustainable development" of April, 1, 1996, No 440). Newest ecological laws. Moscow: Institute of International Law and Economics, p. 359 (in Russian).

Denisenko, V.P. and Rudi, V.P. (1975) Relation of the biological activity and physico-chemical properties of *bis*-quaternary ammonium compounds. *In* Physiological role of surfactants. Chernovtsy, pp. 40–41(in Russian).

Denisova, A.I., Timchenko, V.M., Nakhshina, E.P., Novikov, B.I., Ryabov, A.K., and Base, Y.I. (1989) Hydrobiology and hydrochemistry of Dnieper and its reservoirs. Kiev: Naukova Dumka, 216 pp. (in Russian).

Devi, Y. and Devi, S. (1986) Effect of synthetic detergents on germination of fern spores. *Bull. Environ. Contam. Toxicol.*, **37 (6)**: 837–843.

Dmitrieva, A.G. (1976) Studies of the action of polymetallic ores and concentrates on the vital activity of the blue-green alga *Microcystis aeruginosa*. PhD thesis. Moscow, 160 pp. (in Russian).

Dmitrieva, A.G., Artyukhova, V.I., and Lysenko, N.L. (1996a) Model population of algae in ecological biomonitoring. *In* Ecological and physiological studies of algae and their significance for assessing the condition of natural waters. A short conference report, Borok, December 3–5, Yaroslavl, pp. 133–135 (in Russian).

Dmitrieva, A.G., Sokolova, S.A., Artyukhova, V.I., and Startseva, A.I. (1996b) On standardization of bioassaying methods using laboratory algal cultures. *In* Ecological and physiological studies of algae and their significance for assessing the condition of natural waters. A short conference report, Borok, December 3–5, Yaroslavl, pp. 135–136 (in Russian).

Dmitrieva, A.G., Veselova, T.V., and Veselovsky, V.A. (1989) Bioassaying of sewage waters and their components and bioindication of natural waters using luminescent methods (Filenko, O.F. (ed.)). Moscow: Moscow University Press, pp. 21–34 (in Russian).

Donkin, P. (1994) Quantitative structure–activity relationships. *In* P. Calow (ed.) Handbook of Ecotoxicology, Vol. 2. Oxford: Blackwell Scientific Publications, pp. 321–347.

Donkin, P. (1997) None of the priority pollutant lists (US, UK, EU etc) that I am familiar with include surfactants. Personal communication, 14 May (email).

Donkin, P. and Ostroumov, S.A. (1997) Ecological hazard of sodium dodecyl sulfate. *Toksicol. Vestnik*, **3**: 37 (in Russian).

Donkin, P., Widdows, J., Evans, S.V., and Brinsley, M.D. (1991) QSARs for the sublethal responses of marine mussels (*Mytilus edulis*). The science of the total environment, Vol. 109/110, pp. 461–476.

Donkin, P., Widdows, J., Evans, S.V., Staff, F., and Yan, T. (1997) Effects of neurotoxic pesticides on the feeding rate of marine mussels *Mytilus edulis*. *Pestic. Sci.*, **49**: 196–209.

Drachev, S.M. (1964) Control of river, lake and reservoir pollution with industrial and communal sewage. Moscow, Leningrad: Nauka Publishers, 274 pp. (in Russian).

Dubinin, N.P. (1988) Introduction to biochemical ecology (review of the book by S.A. Ostroumov). *Izv. AN SSSR, Ser. Biol.*, **1**: 158 (in Russian).

Dutka, B. and Gorrie, J. (1989) Assessment of toxicant activity in sediments by the ECHA Biocide Monitor. *Env. Pollution*, **57**: 1–7.

Dyga, A.K. and Lubyanov, I.P. (1972) Zebra mussels and their larvas as indicators of reservoir pollution. Theory and practice of biological self-purification of polluted waters. Moscow: Nauka Publishers, pp. 164–166 (in Russian).

Ecological modifications and criteria of ecological normalization (1991) Proc. Int. Symposium. Leningrad: Gidrometeoizdat Publishers, 384 pp. (in Russian).

Edgerton, L. (1991) The rising tide: global warming and world sea levels. Washington, DC: Island Press, 140 pp.

Eilenberg, H., Klinger, E., Przedeki, F., and Shechter, I. (1989) Inactivation and activation of various membranal enzymes of the cholesterol biosynthetic pathway by digitonin. *J. Lipid Res.*, **30 (8)**: 1127–1137.

Ekelund, R., Bergman, A., Granmo, A., and Berggren, M. (1990) Bioaccumulation of 4-nonylphenol in marine animals – a re-evaluation. *Environm. Pollution*, **64**: 107–120.

Elpiner, L.I. (1999) Quality of natural waters and the health of the population in Volga River basin. *Vodn. Resursy*, **26 (1)**: 50–70 (in Russian).

Emelyanov, E.M. (1998) Barrier zones in the ocean. Kaliningrad: Yantarny Skaz Publishers, 416 pp. (in Russian).

Environmental and human health aspects of commercially important surfactants (1982) *In* Solution behavior of surfactants, eds. A. Sivak, M. Goyer, J. Permak, and P. Thayer. N.Y.: Plenum Press, Vol. 1, 739 p.

EPA. Methods for measuring the acute toxicity of effluents and receiving waters to freshwater and marine organisms. Fourth edition. EPA/600/4 – 90/027. Office of Research and Development. Washington, DC (edited by C.I. Weber). 293 pp.

Eskova-Soskovets, L.B., Sautin, A.I., and Rusakov, N.V. (1980) On allergenic properties of some surfactants. *Gigiena Sanit.*, **2**: 14–17 (in Russian).

European market (1988) *JAOCS*, **65 (1)**: 165.

Faba, G. and Crotti, E. (1979) Effetto di un detersivo commerciale e di uno dei suoi componenti, LAS, sulla produzione di nauplii in *Tisbe holothuriae* (Copepoda, Harpacticoida) in condizioni di alto e basso affollamento. *Atti Acad. naz. Lincei. Rend. Cl. sci. fis., mat e natur.*, **66 (3)**: 223–231.

Facts and figures (statistical data about surfactants) (1992) *Chem. Eng. News*, **70 (26)**: 32–75.404. Fedulova, A.N., Khromov, V.M., and Maksimov, V.N. (1976) Effect of some detergents, used to control oil pollution, on protococcal algae. *Biol. Nauki*, **5**: 90–95 (in Russian).

Federal law "On the animal world" (passed by State Duma on 2/24/1995). Newest ecological laws of the Russian Federation. Moscow: IMPE, pp. 16–54 (in Russian).

Fedorov, V.D. (1970) Primary production as a function of the structure of a phytoplankton community. *Dokl. Akad. Nauk SSSR*, **192 (4)**: 901–904 (in Russian).

Fedorov, V.D. (1974) Stability of ecological systems and its measurement. *Izv. Akad. Nauk SSSR, Biol. Series*, **3**: 402–415 (in Russian).

Fedorov, V.D. (1977) The problem of assessing the norm and pathology of ecosystems. Experience of using various systems of biological indication of polluted waters. Scientific bases for monitoring the quality of waters by hydrobiological parameters. Proc. Soviet-American seminar. Leningrad: Gidrometeoizdat Publishers, pp. 6–12 (in Russian).

Fedorov, V.D. (1979) On methods of studying phytoplankton and its activity. Moscow: Moscow University Press, 167 pp. (in Russian).

Fedorov, V.D. (1980) Pollution of aquatic ecosystems (principles of studying and assessment of action). Self-purification and bioindication of polluted waters. Moscow: Nauka Publishers, pp. 21–38 (in Russian).

Fedorov, V.D. (1987) Topical and non-topical in hydrobiology. *Nauchn. Dokl. Vyssh. Shkoly, Biol. Nauki*, **8**: 6–26 (in Russian).

Fedorov, V.D. (1992) The cause of ecological crisis and way out of it: issues of strategy and tactics. *Biol. Nauki*, **8**: 27–31 (in Russian).

Fedorov, V.D., Ilyash, L.V., Smirnov, N.A., Sarukhan-Bek, K.K., and Radchenko, I.G. (1992) Ecology of White Sea plankton. II. Kinetics of consumption of various forms of carbon by phytoplankton. *Biol. Nauki*, **8**: 77–90 (in Russian).

Fedorov, V.D., Koltsova, T.I., Sarukhan-Bek, K.K., Smirnov, N.A., and Fedorov, V.V. (1988a) Ecology of White Sea plankton. I. Phytoplankton. *Vest. MGU, Ser. Biol.*, **4**: 25–31 (in Russian).

Fedorov, V.D., Smirnov, N.A., and Fedorov, V.V. (1988b) Some regularities of organic-matter production by phytoplankton. *Dokl. Akad. Nauk SSSR*, **299 (2)**: 506–508 (in Russian).

Fedorov, V.D., Smirnov, N.A., and Koltsova, T.I. (1982) Seasonal complexes of White Sea phytoplankton and the analysis of similarity indices. *Izv. Akad. Nauk SSSR, Ser. Biol.*, **5**: 715–721 (in Russian).

Fedulova, A.N., Khromov, V.M., and Maksimov, V.N. (1976) Effects of some detergents, used to control oil pollution, on protococcal algae. *Biol. Nauki*, **5**: 90–95 (in Russian).

Fendinger, N., Begley, W., McAvoy, D., and Eckhoff, W. (1995) Measurement of alkyl ethoxylate surfactants in natural waters. *Environm. Sci. Technol.*, **29 (4)**: 856–863.

Fendinger, N., Versteeg, D., Weeg, E., Dyer, S., and Rapaport, R. (1994) Environmental behavior and fate of anionic surfactants. Environmental chemistry of lakes and reservoirs. Washington, DC: ACS, pp. 527–557.

Filenko, O.F. (1985) Relationship of bioassaying with normalization and toxicological control of reservoir pollution. *Vodn. Resursy*, **3**: 130–134 (in Russian).

Filenko, O.F. (1986) Preface. Kotelevtsev, S.V. et al. Ecological and toxicological analysis based on biological membranes. Moscow: Moscow University Press, pp. 3–5 (in Russian).

Filenko, O.F. (1988) Aquatic toxicology. Chernogolovka, 156 pp. (in Russian).

Filenko, O.F. (1989) Introduction. Problems and methods of bioassaying of an aqueous medium. Methods of bioassaying the quality of an aqueous medium (Filenko, O.F. (ed.)). Moscow: Moscow University Press, pp. 3–9 (in Russian).

Filenko, O.F. (1990a) Some universal regularities of the action of chemical agents on aquatic organisms. DSc thesis. Moscow: 36 pp. (in Russian).

Filenko, O.F. (1990b) Some universal regularities of the action of chemical agents on aquatic organisms. DSc (Biology) paper. Moscow: 311 pp. (in Russian).

Filenko, O.F. and Dmitrieva, A.G. (1999) Bioassaying as a means of controlling the toxicity of polluted aqueous medium. *Pribory Systemy Upravl.*, **1**: 61–63 (in Russian).

Filenko, O.F. and Khobotyev, V.G. (1976) Metal pollution. *In* Results of science and technology. General ecology. Biocenology. Hydrobiology. Vol. 3, Aquatic toxicology. Moscow: VINITI, pp. 110–150 (in Russian).

Filenko, O.F. and Lazareva, V.V. (1989) Effects of toxic agents on general biological and cytogenetic indices in daphnia. *Gidrobiol. Zhurn.*, **25 (3)**: 56–59 (in Russian).

Filenko, O.F. and Parina, O.V. (1983) Organotin compounds and the regulation of phosphate uptake by carp organs and tissues. Reactions of hydrobionts to pollution. Moscow: Nauka Publishers, pp. 151–158 (in Russian).

Filenko, O.F. and Sokolova, S.A. (eds.) (1998) Recommendations for establishing ecological and fish-economy rates (ultimate admissible level and approximate safe level of impact) of pollutants for waters of reservoirs of commercial fishing importance. Moscow: VNIRO Publishers, 145 pp. (in Russian).

Filenko, O.F., Lazareva, V.V., and Isakova, E.F. (1989) The ratio of cytogenetic and general biological indices in daphnia in culture. *Gidrobiol. Zhurn.*, **25 (2)**: 39–42 (in Russian).

Filippenko, V.N. (1985) Differentiation of rhizoderm cells in gramineous plants (determination stages and interrelation with divisions). PhD thesis. Moscow: 22 pp. (in Russian).

Filippova, L.M., Insarov, G.E., Semevsky, F.N., and Semenov, S.M. (1978) Problems of ecological monitoring and modeling of ecosystems. Leningrad: Gidrometeoizdat Publishers, vol. 1, pp. 19–32 (in Russian).

Finogenova, N.P. and Alimov, L.F. (1976) Assessment of the extent of water pollution by the composition of benthic animals. Methods of biological assaying of freshwaters. Leningrad: Nauka Publishers, pp. 95–106 (in Russian).

Flerov, B.A. (1983) Bioassaying: terminology, problems, prospects. Theoretical problems of bioassaying. Volgograd, pp. 13–20 (in Russian).

Flerov, B.A. (1989) Ecological and physiological aspects of freshwater animal toxicology. Leningrad: Nauka Publishers, 144 pp. (in Russian).

Flerov, B.A. and Lapkina, L.N. (1976) Avoidance of the solutions of some toxic substances by medical leech. *Biology of Inner Waters: Information Bulletin*, **30**: 48–52 (in Russian).

Flerov, B.A., Kozlovskaya, V.I., and Nepomnyashchikh, V.A. (1988) Assessment of the ecological hazard of pollutants by the physiological, biochemical, and behavioral reactions of aquatic animals. Ecotoxicology and protection of the environment. (Krivolutsky, D.A., Bocharov, B.V., et al. (eds.)). Riga: Inst. of Biology, Latvian Acad. Sci., pp. 195–197 (in Russian).

Flerov, B.A., Lapkina, L.N., Zhmur, N.S., and Yakovleva, I.I. (1988) A method of bioassaying the toxicity of sewage containing metal ions (Cu^{2+}, Hg^{2+}, Cd^{2+}, Al^{2+}, etc.) by the change of static to dynamic state in medical leech. *In* Methods of water bioassaying. Chernogolovka, pp. 114–116 (in Russian).

Frantsev, A.V. (1972) Some problems of controlling the quality of water. Theory and practice of the biological self-purification of polluted waters (Telitchenko, M.M. (ed.)). Moscow: Nauka Publishers, pp. 24–28 (in Russian).

Fucuda, M., Fujitsu, M., and Ohby, K. (1987) Interaction between liposomal membrane and sodium dodecyl sulfate as studied by an ESR spin probe method. *J. Jap. Oil Chem. Soc.*, **36 (7)**: 469–473.

Fujita, I., Takeshige, K., and Minakami, S. (1987) Characterization of the NADH-dependent superoxide production activated by sodium dodecyl sulfate in a cell-free system of pig neutrophils. *Biochim. Biophys. Acta*, **931 (1)**: 41–48.

Ganitkevich, Y.V. (1975) The problem of surface phenomena in the organism and of the physiological role of surfactants. *In* The physiological role of surfactants. Chernovtsy, pp. 3–5 (in Russian).

Gapochka, L.D. (1981) On adaptation of algae. Moscow: Moscow University Press, 80 pp. (in Russian).

Gapochka, L.D. (1983) On the phenotypical adaptation of blue-green algae to dispersants. *In* The reaction of hydrobionts to pollution. Moscow: Nauka Publishers, pp. 122–128 (in Russian).

Gapochka, L.D. (1999) Populational aspects of the resistance of cyanobacteria and microalgae to the toxic factor. DSc thesis. Moscow: Moscow University Press, 64 pp. (in Russian).

Gapochka, L.D. and Karaush, G.A. (1980) Effect of dispersant EPN-5 on mixed culture of blue-green algae. *Nauchn. Dokl. Vyssh. Shkoly. Biol. Nauki*, **8**: 65–68

(in Russian).

Gapochka, L.D., Artyukhova, V.I., Lobacheva, G.V., and Lebedeva, T.E. (1980) Studies of adaptation of blue-green algae *Synechocystis aquatilis* Sanv. 428 and *Anacystis nidulans* to dispersant DN-75. *Vestn. Mosk. Universiteta, Ser. Biol.*, **2**: 30–38 (in Russian).

Gapochka, L.D., Brodsky, L.I., Kravchenko, M.E., and Fedorov, V.D. (1980) Combined effects of oil, petroleum products, and dispersants on blue-green algae *Synechocystis aquatilis* and *Anabaena variabilis*. *Gidrobiol. Zhurn.*, **16 (2)**: 105–110 (in Russian).

Gard-Terech, A. and Palla, J. (1986) Comparative kinetics study of the evolution of freshwater aquatic toxicity and biodegradability of linear and branched alkylbenzene sulfonates. *Ecotoxicol. Environm. Safety*, **12 (2)**: 127–140.

Gehm, H. and Bregman, J. (eds.) (1976) Handbook of water resources and pollution control. New York: Van Nostrand Reinhold Co., 840 pp.

Geletin, U.V., Zamolodchikov, D.G., Levich, A.P., et al. (1991) Assessment and prediction of the state of aquatic ecosystems by the method of ecological modifications. Ecological modifications and criteria of ecological normalization. Proc. Int. Symposium. Leningrad: Gidrometeoizdat Publishers, pp. 318–329 (in Russian).

Ghilarov, A.M. (1991) Species redundancy versus non-redundancy: is it worth further discussion? *Zhurn. Obshch. Biol.*, **58 (2)**: 100–105.

Gibson, C. (1984) Sinking rates of planktonic diatoms in an unstratified lake: a comparison of field and laboratory observations. *Freshwater Biol.*, **14 (6)**: 631–638.

Gillespie, W., Rodgers, J., and Crossland, N. (1996) Effects of a nonionic surfactant ($C_{14-15}AE-7$) on aquatic invertebrates in outdoor stream mesocosms. *Environ. Toxicol. Chem.*, **15 (8)**: 1418–1422.

Gilyarov, A.M. (1987) Dynamics of the abundance of freshwater planktonic crustaceans. Moscow: Nauka Publishers, 189 pp. (in Russian).

Gilyarov, M.S. (1985) Preface to Levels of protection of the living nature. Moscow: Nauka Publishers, pp. 3–4 (in Russian).

Gladyshev, M.I. (1999) Basics of the ecological biophysics of aquatic systems. Novosibirsk: Nauka Publishers, 113 pp. (in Russian).

Gladyshev, M.I., Gribovskaya, I.V., Kalacheva, G.S., and Sushchik, N.N. (1996) Experimental study of the rate of self-purification as an integral functional characteristic of various types of aquatic ecosystems. *Sib. Ekolog. Zhurn.*, **3 (5)**: 419–431 (in Russian).

Gledhill, W., Saeger, V., and Trehy, M. (1991) An aquatic environmental safety assessment of linear alkylbenzene. *Environ. Toxicol. Chem.*, **10 (10)**: 169–178.

Godfrey, P.J. (1982) The eutrophication of Cayuga Lake: a historical analysis of the phytoplankton's response to phosphate detergents. *Freshwater Biol.*, **12 (2)**: 149–166.

Gollerbakh, M.M. (1977) Chlorophyta. *In* Life of plants, Vol. 3 (Gollerbakh, M.M. (ed.)). Moscow: Prosveshchenie Publishers, p. 266 (in Russian).

Gollerbakh, M.M. and Shtina, E.A. (1969) Soil algae. Leningrad: Nauka Publishers, 228 pp. (in Russian).

Gollerbakh, M.M., Kosinskaya, E.K., and Polyansky, V.I. (1953) Blue-green algae (Identifier of freshwater algae of the USSR, Issue 2). Moscow: Sov. Nauka, 652 pp. (in Russian).

Golovleva, L. (1997) Microorganisms involved in the biodegradation of organic compounds. *In* J.R. Wild, S.D. Varfolomeyev, and A. Scozzafava (eds.) Perspectives in bioremediation: technologies for environmental improvement. (NATO ASI Series. 3. High Technology – Vol. 19). Kluwer Academic Publishers: Dordrecht, pp. 57–63.

Golubev, A.A., Lyublina, E.I., Tolokontsev, N.A., and Filov, V.A. (1973) Quantitative toxicology. Leningrad: Meditsyna Publishers, 288 pp. (in Russian).

Gonzales-Mazo, E. and Gomez-Parra, A. (1996) Monitoring anionic surfactants (LAS) and their intermediate degradation products in the marine environment. *Trends Anal. Chem.*, **15 (8)**: 375–380.

Gorbunova, A.V. (1988) Effects of suspended substances on planktonic filter feeders. Proc. State Research Inst. Lake River Fisheries, No 288, pp. 69–70 (in Russian).

Gordeyeva, L.M. and Kozlova, M.V. (1980) *Limax* amoebas in sewage waters at different stages of purification. *In* Self-purification and bioindication of polluted waters. Moscow: Nauka Publishers, pp. 155–158 (in Russian).

Gore, A. (1992) Earth in the balance. Boston: Houghton Mifflin Co., 408 pp.

Gorshkov, V.T. (1997) Inhibition of carbon-cycle global changes by marine biota. *Dokl. Akad. Nauk*, **353 (3)**: 390–393 (in Russian).

Goryunova, S.V. and Ostroumov, S.A. (1986) Impact of anionic detergent on green protococcal alga and seedlings of some angiosperms. *Nauchn. Dokl. Vyssh. Shkoly, Biol. Nauki*, **7**: 84–86 (in Russian).

GOST 17.1.2.04-77 (1987) Indices of composition and rules of taxation of fish water reservoirs. (Protection of the environment. Hydrosphere). Moscow: Gosstandart Publishers, 17 pp. (in Russian).

Granmo, A. (1972) Development and growth of eggs and larvae of *Mytilus edulis* exposed to linear dodecyl benzene sulphonate, LAS. *Marine Biol.*, **15**: 356–358.

Granmo, A., Ekelund, R., Magnusson, K., and Berggren, M. (1989) Lethal and sublethal toxicity of 4-nonylphenol to the common mussel (*Mytilus edulis* L.). *Environm. Pollution*, **59**: 115–127.

Greek, B. and Layman, P. (1989) Higher costs spur new detergent formulations. *Chem. Eng. News*, **67 (4)**: 29–49.

Griffiths, R., McNamara, T., Caldwell, B., and Morita, R. (1981) A field study on the acute effects of the dispersant corexit 9527 on glucose uptake by marine microorganisms. *Mar. Environ. Res.*, **5 (2)**: 83–91.

Grozdinsky, A.M. and Grozdinsky, D.M. (1973) A short reference book on plant physiology. Kiev: Naukova Dumka, 592 pp. (in Russian).

Grozdov, A.O., Pereladov, M.V., and Startseva, A.I. (1981) Bioassaying of surfactants. *In* Bioassaying of natural and sewage waters. Moscow, pp. 64 –69 (in Russian).

Guidelines on the bioassaying of sewage waters using medical leech. Moscow: Ministry of Water Facilities of the Russian Federation, 24 pp. (in Russian).

Guillard, R. and Ryther, J. (1962) Studies of marine planktonic diatoms. I. *Cyclotella nana* Hustedt and *Detonula confervacea* (Cleve) Gran. *Can. J. Microbiol.*, **8**: 229.

Gusev, M.V. (1988) Introduction to biochemical ecology (book review). *Fiziol. Rasten.*, **35 (2)**: 412–413 (in Russian).

Guskov, G.V., Gorshkova, E.F., Vinogradova, L.A., Parkhomchuk, T.K., and Kamenev, A.I. (1986) Experimental studies on the assessment of the effect of biologically purified sewage waters on the sanitary regime and reservoir self-purification processes. *Gigiena Sanit.*, **12**: 7–10 (in Russian).

Gutelmakher, B.L. (1986) Metabolism of plankton as a whole. Leningrad: Nauka Publishers, 156 pp. (in Russian).

Hansen, B., Fotel, F., Jensen, N., and Wittrup, L. (1997) Physiological effects of the detergent linear alkylbenzene sulphonate on blue mussel larvae (*Mytilus edulis*) in laboratory and mesocosm experiments. *Marine Biol.*, **128**: 627–637.

Hartman, W. and Martin, D. (1984) Effect of suspended bentonite clay on the acute toxicity of glyphosate to *Daphnia pulex* and *Lemna minor. Bull. Environ. Contam. Toxicol.*, **33 (3)**: 355–361.

Hill, I., Heimbach, F., Leeuwangh, P., and Matthiessen, P. (eds) (1994) Freshwater field tests for hazard assessment of chemicals. Boca Raton: Lewis Publishers, 561 pp.

Holcombe, G., Phipps, G., Sulaiman, A., et al. (1987) Simultaneous multiple species testing: Acute toxicity of 13 chemicals to 12 diverse freshwater amphibia, fish and invertebrate families. *Arch. Environ. Contam. Toxicol.*, **16 (6)**: 697–710.

Holt, M.S., Mitchell, G.C., and Watkinson, R.J. (1992) The environmental chemistry, fate and effects of nonionic surfactants. The handbook of environmental chemistry, Vol. 3, Part F. Berlin: Springer, pp. 89–144.

Holysh, M., Paaterson, S., and Mackay, D. (1986) Assessment of the environmental fate of linear alkylbenzenesulphonates. *Chemosphere*, **15 (1)**: 3–20.

Huber, L. (1985) Stand der Kenntnisse über das ökologische Verhalten von Tensiden. Schadstoff-belastung und Ökosystemschutz im aquatischen Bereich. Munchen: Oldenbourg Verlag, pp. 189–208.

Ilyichev, V.D., Bocharov, B.V., and Gorlenko, M.V. (1985) Ecological bases of the protection from biodamage. Moscow: Nauka Publishers, 264 pp. (in Russian).

Ilyin, I.E. (1980) Studies of the toxicity of surfactant transformation products formed in water chlorination. *Gigiena Sanit.*, **2**: 11–14 (in Russian).

Ilyin, I.E. (1986) Studies of the hazard of re-distribution of chemical and biological pollutants in an aqueous medium. *Gigiena Sanit.*, **6**: 8–11 (in Russian).

Industrial surfactants (1984) *Chemical Weeks*, **135 (24)**: 3–24.

Isomaa, B. and Hägelstrand, H. (1988) Effects of nonionic amphiphiles at sublytic concentrations on the erythrocyte membrane. *Cell Biochem. Function*, **6 (3)**: 183–190.

Ivanov, V.B. (1974) Cellular basis of plant growth. Moscow: Nauka Publishers, 222 pp. (in Russian).

Ivanov, V.B. (1982) Cellular basis of root growth. Sov. Sci. Rev. Ser. Biol., Overseas Publ., Vol. 2, pp. 365–392.

Ivanov, V.B. (1992) Analysis of different chemical actions on root growth. *In* Root ecology and its practical applications (eds. Kutschera, L., Hubl, E., Lichtenegger, E., Persson, H., and Sobotik, M.). Klagenfurt: Verein für Wurzelforschung, pp. 307–310.

Ivanova, M.B. (1976a) Effect of pollution on planktonic crustaceans and the possibility of using them to determine the extent of river pollution. Methods of freshwater biological assaying. Leningrad: USSR Academy of Sciences, pp. 68–80 (in Russian).

Ivanova, M.B. (1976b) Experience of assessing the involvement of planktonic animals in water self-purification processes (by the example of zooplankton of Izhora River banks). Hydrobiol. bases of water self-purification. Leningrad: USSR Academy of Sciences, pp. 36–42 (in Russian).

Izmerov, N.F., Kirillov, V.F., and Trakhtman, N.N. (1978) General and communal hygiene. Moscow: Meditsyna Publishers, 408 pp. (in Russian).

Izrael, Y.A. (1984) Ecology and monitoring of the environment. Moscow: Gidro-meteoizdat Publishers, 560 pp. (in Russian).

Izrael, Y.A. and Abakumov, V.A. (1991) On the ecological state of surface waters of the USSR and the criteria of ecological normalization. Ecological modifications and criteria of ecological normalization. Proc. Int. Symposium. Leningrad: Gidrometeoizdat Publishers, pp. 7–18 (in Russian).

Izrael, Y.A. and Tsyban, A.V. (1983) On the assimilation capacity of the World Ocean. Dokl. Akad. Nauk SSSR, **272 (3)**: 702–704 (in Russian).

Izrael, Y.A. and Tsyban, A.V. (1989) Anthropogenic ecology of the ocean. Leningrad: Gidrometeoizdat Publishers, 526 pp. (in Russian).

Izrael, Y.A. and Tsyban, A.V. (1992) Studies of the ecosystems of the Bering and Chukotka Seas. St. Petersburg: Gidrometeoizdat Publishers, 656 pp. (in Russian).

Izrael, Y.A., Semenov, S.M., and Khachaturov, M.A. (1992) Bioclimatology and topical problems of assessing the consequences of the global change of the climate for land ecosystems. In Problems of ecological monitoring and modeling of ecosystems. Leningrad: Gidrometeoizdat Publishers, Vol. 14, pp. 8–20 (in Russian).

Izrael, Y.A., Semenov, S.M., and Kunina, I.M. (1990) Ecological normalization: methodology and practice. In Problems of ecological monitoring and modeling of ecosystems. Leningrad: Gidrometeoizdat Publishers, Vol. 13, pp. 10–24 (in Russian).

Izrael, Y.A., Tsyban, A.V., Kudryavtsev, V.M., Shchuka, S.A., and Zhukova, A.I. (1995) Penetration of biologically active UV radiation and its impact on the essential biological processes in the Bering and Chukotka Seas. Meteorol. Gidrologiya, **10**: 13–28 (in Russian).

Izrael, Y.A., Tsyban, A.V., Ventzel, M.V., and Shigaev, V.V. (1988a) A generalized model for the assimilation capacity of the marine ecosystem. Dokl. Akad. Nauk SSSR, **380 (2)** (in Russian).

Izrael, Y.A., Tsyban, A.V., Ventzel, M.V., and Shigaev, V.V. (1988b) Scientific justification of ecological normalization of the anthropogenic impact on a marine exosystem (by example of the Baltic Sea). Okeanologiya, **28 (2)** (in Russian).

Jablokov, A.V. and Ostroumov, S.A. (1988) Omul si Natura. De la probleme la solutii. Ocrotirea Natura. Cluj – Napoca: Dacia, pp. 65–80.

Jablokov, A.V. and Ostroumov, S.A. (1991) Ochrana Zive Prirody. Praha: Academia, 345 pp.

Jonsson, P. and Carman, R. (1994) Changes in deposition of organic matter and nutrients in the Baltic Sea during the twentieth century. Mar. Pollut. Bull., **28 (7)**: 417–426.

Juchelka, C. and Snell, T. (1995) Rapid toxicity assessment using ingestion rate of cladocerans and ciliates. Arch. Environ. Contam. Toxicol., **28 (4)**: 508–512.

Kabanova, Y.G. and Nesterova, M.P. (1975) Effects of surfactants on phytoplankton production. Vodn. Resursy, **5**: 117–124 (in Russian).

Kalenichenko, K.P. (1996) Determination of cationic surfactants in natural waters. Gidrobiol. Zhurn., **32 (6)**: 70–76 (in Russian).

Kaliev, A.J. (1990) Assessment of the effect of long-term irrigation by gas-processing industry sewage waters on the environment. Ekologiya, **6**: 436–440 (in Russian).

Kamshilov, M.M. (1979) Evolution of the biosphere. Moscow: Nauka Publishers, 256 pp. (in Russian).

Kaplan, A.Y. (1987) Carp terminal brain rhythmic activity induced by adequate stimulation. Zhurn. Evolut. Biokhim. Fiziol., **23**: 492–496 (in Russian).

Kaplan, A.Y. (1988) Rhythmic activity of the olfactory bulb of carp under conditions of nembutal anesthesia. *Biol. Nauki*, **10**: 50–54 (in Russian).

Kaplin, V.T. (1979) Current state and main directions in studies of chemical substances transformation in natural waters. Materials of the 6th All-Union Symp. on Modern Problems of Reservoir Self-purification and Regulation of Water Quality, Section 2. Tallinn, Part 1, pp. 3–17 (in Russian).

Karpenko, A.A., Tyurin, A.N., and Morozov, A.V. (1983) Effect of sublethal effects of some environmental factors on the motor activity of bivalve mollusks and the possibility of using them for biological monitoring. Applied ethology: Materials of the 3rd All-Union Conf. on Animal Behavior. Moscow: USSR Academy of Sciences, pp. 144–146 (in Russian).

Kartasheva, N.V. and Ostroumov, S.A. (2000) Studies of the ability of surfactants to inhibit the filtration activity of rotifers. Food industry at the turn of the third millennium. Moscow: Moscow State Technol. Academy, pp. 245–247 (in Russian).

Kartsev, V.G., Ostroumov, S.A., and Pavlova, A.I. (1990) Use of seedlings of *Cucumis sativus* and other plants for bioassaying. Allelopathy and plant productivity. Kiev: Naukova Dumka, pp. 124–129 (in Russian).

Kasumyan, A.O. (1995) Olfactory and taste reception and behavior of fish: ethological–physiological and ontogenetic aspects. DSc thesis. Moscow: Moscow University Press, 46 pp. (in Russian).

Kasumyan, A.O. (1997) Gustatory reception and food behavior in fish. *Vopr. Ikhtiologii*, **37 (1)**: 78–93 (in Russian).

Kemp, R., Meredith, R., Gamble, S., and Frost, M. (1983) A rapid cell culture technique for assessing the toxicity of detergent-based products *in vitro* as a possible screen for eye irritancy *in vivo*. *Cytobios*, **36 (36)**: 153–159.

Keonjyan, V.P., Kudin, A.M., and Terekhin, Y.V. (eds.) (1990) Practical ecology of Black Sea marine areas. Kiev: Naukova Dumka, 252 pp. (in Russian).

Khanislamova, G.M., Kabirov, R.R., and Khazipova, R.H. (1988) Surfactants in land ecosystems. Ufa: Biological Research Center, Urals Division of the USSR Acad. Sci., 143 pp. (in Russian).

Khlebovich, T.V. (1976) Role of infusoria in water self-purification. *In* Hydrobiological bases of water self-purification. Leningrad: USSR Academy of Sciences, pp. 25—29 (in Russian).

Khristoforova, N.K. (1989) Bioindication and monitoring of marine water pollution by heavy metals. Leningrad: Nauka Publishers, 192 pp. (in Russian).

Khristoforova, N.K., Aizdaicher, N.A., and Berezovskaya, O.Y. (1966) Effects of copper ions and detergent on green microalgae *Dunaliella tertiolecta* and *Platymonas* sp. *Biol. Morya*, **22 (2)**: 114–119 (in Russian).

Khublaryan, M.G. (1992) Waters of Russia. *Biol. Nauki*, **8**: 31–34 (in Russian).

Kihlstrom, M. and Salminen, A. (1991) Selective effects of some anesthetics and detergents on lipid peroxidation of mouse heart homogenates. *Comp. Biochem. Physiol.*, **100B**: 789–793.

King, A., Lowe, K., and Milligan, B. (1988) Microbial cell responses to a non-ionic surfactant. *Biotechnology Lett.*, **10 (3)**: 177–180.

Klyuev, N.A. (1996) Control of superecotoxicants in environmental objects and sources of their pollution. *Zhurn. Anal. Khimii*, **51 (2)**: 163–172 (in Russian).

Kojova, A.M. and Timofeeva, S.S. (1983) Ecological and toxicological problems in the system of monitoring. *In* Theoretical issues of bioassaying (Lukyanenko, V.I. (ed.)). Volgograd: Institute for the Biology of Internal Waters, USSR Acad. Sci., pp. 165–169 (in Russian).

Kokin, K.A. (1981) On the filtering role of higher aquatic plants in self-purification of Moskva River. *Nauchn. Dokl. Vyssh. Shkoly, Biol. Nauki*, **4**: 104–108 (in Russian).

Kolesnikov, M.P. and Ostroumov, S.A. (2000) Biogeochemical flows (C, N, P, Si, Al) through *Lymnaea stagnalis* and the effects of a surfactant on them. *Ecol. Studies, Hazards, Solutions*, **3**: 14.

Kolotilova, N.N. and Ostroumov, S.A. (2000) Growth during the action of a surfactant-containing preparation. Problems of ecology and physiology of organisms. Moscow: Dialog-MGU, p. 66 (in Russian).

Kolotilova, N.N., Piskunkova, N.F., and Ostroumov, S.A. (1998) Effect of cationogenic synthetic surfactants on freshwater cyanobacteria and green algae. Modern problems of mycology, algology, and phytopathology. Moscow, pp. 337–338 (in Russian).

Kolupayev, B.O. and Putintseva, V.A. (1983) Activity of enzymes in fish gill tissues in solutions of Lotos detergent. Materials on comparative physiology and adaptation of animals to abiogenic factors of the external medium. Yaroslavl: Yaroslavl University Press, pp. 74–76 (in Russian).

Komarovsky, F.Y. (1975) Effect of some surfactants on carp yearlings under conditions of an acute experiment. Self-purification, bioproductivity, and protection of reservoirs of Ukraine. Kiev: Naukova Dumka, pp. 107–108 (in Russian).

Kondrasheva, N.Y. and Kobak, K.I. (1996) Possible changes of the localization of the natural zones of the Southern Hemisphere at the global warming of the climate. *In* Problems of ecological monitoring and modeling of ecosystems. Leningrad: Gidrometeoizdat Publishers, Vol. 16, pp. 90–99 (in Russian).

Kondratyev, G.P. (1977) Biofiltration. Volgograd reservoir (Konstantinov, A.S. (ed.)). Saratov: Saratov University Press, pp. 179–187 (in Russian).

Kondratyeva, E.N., Maksimova, I.V., and Samuilov, V.D. (1989) Phototrophic microorganisms. Moscow: Moscow University Press, 376 pp. (in Russian).

Konstantinov, A.S. (1979) General hydrobiology. Moscow: Vysshaya Shkola Publishers, 480 pp. (in Russian).

Konstantinov, A.S. (1977) Volgograd water reservoir. Saratov: Saratov University Press, 222 pp. (in Russian).

Korolev, A.A., Bogdanov, M.V., and Vitvitskaya, B.R. (1975) Hygienic assessment of surfactant-destruction products in ozonation of water. *Gigiena Sanit.*, **1**: 16–20 (in Russian).

Korol, V.M. (1989) Conducting toxicological studies on aquatic higher plants. Methods of bioassaying the quality of an aqueous medium (Filenko, O.F. (ed.)). Moscow: Moscow University Press, pp. 34–40 (in Russian).

Koronelli, T.V. (1982) Microbial degradation of hydrocarbons and its ecological consequences. *Biol. Nauki*, **3**: 5—13.(in Russian).

Koronelli, T.V (1996) Principles and methods of intensification of biological degradation of hydrocarbons in the environment. *Prikl. Biokhim. Mikrobiol.*, **32 (6)**: 579–585 (in Russian).

Korte, F., Bekhadir, M., Klein, W., Lay, J.P., Parlar, G., and Scheunert, I. (1997) Ecological chemistry (Korte, F. (ed.)). Moscow: Mir Publishers, 396 pp. (in Russian) [Korte, F., Bekhadir, M., Klein, W., Lay, J.P., Parlar, G., and Scheunert, I. (1992) Lehrbuch der ökologischen Chemie. Georg Thieme Verlag: Stuttgart.

Koskova, L.A. and Kozlovskaya, V.I. (1979) Toxicity of synthetic surfactants and detergents for aquatic animals. *Gidrobiol. Zhurn.*, **15 (1)**: 77–84 (in Russian).

Kostovetsky, Y.I., Rakhov, G.M., and Steinberg, E.I. (1975) Synthetic surfactants in city sewage waters and the efficiency of their purification by the biological station. *Gigiena Sanit.*, **2**: 95–96 (in Russian).

Kotelevtsev, S.V., Stepanova, L.I., and Glaser, V.M. (1994) *In* Bio-monitoring of coastal waters and estuaries (ed. Kramer, K.J.M.). Boca Raton: CRC Press, 227–246.

Kotelevtsev, S.V., Stepanova, L.I., and Glaser, V.M. (1997) Accumulation of mutagenic xenobiotics in water ecosystems with special reference to Lake Baikal. Biomarkers: a pragmatic basis for remediation of severe pollution in Eastern Europe. NATO Advanced Research Workshop (September 21–25, 1997), Cieszyn, p. 34.

Kotelevtsev, S.V., Stvolinsky, S.L., and Beim, A.M. (1986) Ecological and toxicological assay based on biological membranes. Moscow: Moscow University Press, 105 pp. (in Russian).

Kouloheris, A. (1989) Surfactants. *Chem. Engineering*, **96 (10)**: 130–136.

Koyama, T. (1993) Zoobenthos effects on the gaseous metabolism in lake sediments. *Int. Ver. Theor. Angew. Limnol.*, Stuttgart, pp. 827–831.

Krainyukova, A.N. (1988) Bioassaying in protection of waters from pollution. Methods of bioassaying of waters. Chernogolovka, pp. 4–14 (in Russian).

Kratasyuk, V.A., Kuznetsov, A.M., Rodicheva, E.K., Egorova, O.I., Abakumova, V.V., Gribovskaya, I.V., and Kalacheva, G.S. (1996) Problems and prospects of bioluminescent assaying in ecological monitoring. *Sibirsk. Ekolog. Zhurn.*, **3: (5)**: 397–403 (in Russian).

Krishnakumar, P., Asokan, P., and Pillai, V. (1990) Physiological and cellular responses to copper and mercury in the green mussel *Perna viridis* (Linnaeus). *Aquat. Toxicol.*, **18 (3)**: 163–174.

Krivolutsky, D.A. (ed.) (1988) Ecotoxicology and protection of the environment. Riga: Inst. of Biology, Latvian Acad. Sci., 236 pp. (in Russian).

Krivolutsky, D.A. (1994) Soil fauna in the ecotoxicological control. Moscow: Nauka Publishers, 272 pp. (in Russian).

Krivolutsky, D.A. and Pokarzhevsky, A.D. (1988) The microbial link in trophic chains. *Ekologiya*, **5**: 10–20 (in Russian).

Krivolutsky, D.A. and Pokarzhevsky, A.D. (1990) Introduction to biogeocenology. Moscow: Moscow University Press, 104 pp. (in Russian).

Kryuchkova, N.M. (1972) Zooplankton as an agent of self-purification of reservoirs. Theory and practice of biological self-purification of polluted waters. Moscow: Nauka Publishers, pp. 58–61 (in Russian).

Kuksa, V.I. (1994) Southern seas (Aral, Caspean, Azov, and Black) in conditions of anthropogenic stress. St. Petersburg: Gidrometeoizdat Publishers, 320 pp. (in Russian).

Kumsare, A.Y., Laganovskaya, R.U., Matisone, M.N., and Melberga, A.G. (1972) Self-purification factors and the use of biological assay in studies of Daugava River. Theory and practice of biological self-purification of polluted waters. Moscow: Nauka Publishers, pp. 167–169 (in Russian).

Kuzhinovsky, V.A. and Mitskevich, I.N. (1992). Microbial activity of Black Sea mussel collectors in water. *Gidrobiol. Zhurn.*, **28 (3)**: 50–53 (in Russian).

Kuzmenko, K.N. (1976) The role of bivalve mollusks in self-purification of Lake Krasnoye (Karelia isthmus). Methods of biological assay of freshwaters. (Skarlato, O.A. (ed.)). Leningrad: USSR Academy of Sciences, pp. 134–135 (in Russian).

Kuzmenko, M.I. (1996) Distribution of radionuclides in the shallow-water biotope ecosystem. *Gidrobiol. Zhurn.*, **32 (6)**: 42–51 (in Russian).

Kuznetsov, A.P. (1980) Ecology of the bottom communities of the World Ocean. Moscow: Nauka Publishers, 244 pp. (in Russian).

Kuznetsov, A.P. (1993) On photosynthesis, biotic balance, and trophic structure of marine bottom biota. *Izv. RAN, Ser. Biol.*, **2**: 287–304 (in Russian).

Kuznetsov, A.P. and Sagaidachnyi, A.Y. (1987) On the biogeochemical role of bivalve mollusks – filter feeders of the Sea of Okhotsk. *Dokl. Akad. Nauk SSSR*, **297 (3)**: 751–754 (in Russian).

Kuznetsov, A.P. and Trotsyuk, V.Y. (1995) On the scale of organic-matter dispersion in marine sediments. *Izv. RAN, Ser. Biol.*, **5**: 606–611 (in Russian).

Kuznetsov, A.P. and Trotsyuk, V.Y. (1997) On the disposal of organic matter in ocean and marine sediments. On the problem of the assessment of fluxes of carbon, carbon dioxide, oxygen, and oil-and-gas resources. *In* Composition and distribution of benthic invertebrates in seas of Russia and adjacent aquatories (Kuznetsov, A.P. and Zezina, O.N. (eds.)). Moscow: Shirshov Institute of Oceanology, pp. 6–11 (in Russian).

Kuznetsov, A.P., Geodekyan, A.A., and Marina, M.M. (1997) On the scale of organic-matter dispersion in marine sediments. *Izv. RAN, Ser. Biol.*, **1**: 59–63 (in Russian).

Lai, H., Mirsa, V., Viswanathan, P., and Krishna Murti, C. (1983) Comparative studies on ecotoxicology of synthetic detergents. *Ecotoxicol. Environm. Saf.*, **7 (6)**: 538–545.

Lapkina, L.N. and Flerov, B.A. (1979) Studies of acute poisoning of leeches by some toxic substances. Physiology and parasitology of freshwater animals. Leningrad: Gidrometeoizdat Publishers, pp. 50–59 (in Russian).

Lapkina, L.N., Flerov, B.A., Chalova, I.V., and Yakovlev, I.I. (1987) Use of behavioral reactions of leech young for bioassaying. Problems of comparative physiology and aquatic toxicology. Yaroslavl: Yaroslavl State University, pp. 11–17 (in Russian).

Lavrenko, E.M. (1984) Protection of the environment: problems and prospects (review of the book by A.V. Yablokov and S.A. Ostroumov). *Botan. Zhurn.*, **12**: 1706–1710 (in Russian).

Lavrentyev, P., Gardner, W., Cavaletto, J., and Beaver, J. (1995) Effects of the zebra mussel (*Dreissena polymorpha* Pallas) on protozoa and phytoplankton from Saginaw Bay, Lake Huron. *J. Great Lakes Res.*, **21**: 545–557.

Lay, J., Peichl, L., Klein, W., and Korte, F. (1985) Influence of benzene on the phytoplankton and on *Daphnia pulex* in compartments of an experimental pool. *Ecotoxicol. Environ. Safety*, **10**: 218–227.

Lay, J., Schauerte, W., Muller, A., Klein, W., and Korte, F. (1985) Long-term effects of 1,2,4-trichlorobenzene on freshwater plankton in an outdoor-model ecosystem. *Bull. Environ. Contam. Toxicol.*, **34**: 761–769.

Layton, D., Mallon, B., Rosenblatt, D., and Small, M. (1987) Deriving allowable daily intakes for systemic toxicants lacking chronic toxicity data. *Regul. Toxicol. Pharmacol.*, **7 (1)**: 96–112.

Lazzari, M. (ed.) (1994) Environmental viewpoints. Detroit: Gale Research Inc., Vol. 3, 528 pp.

Lebedeva, G.D. (1985) Ecology of periphytons in freshwaters. Ecological bases of protection from biodamage. Moscow: Nauka Publishers, pp. 78–85 (in Russian).

Lech, J. and Vodicnik, M. (1985) Biotransformation. *In* Fundamentals of aquatic toxicology (eds. Rand, G. and Petrocelli, S.). New York: Hemisphere Publ. Corporation, pp. 526–557.

Lesyuk, I.I., Kotsyumbas, I.Y., Komarinets, O.T., Reshetilo, S.G., Ugrin, A.T., and Kospok, O.T. (1983) Effects of nonionogenic surfactants on loach embryos and prelarvae. *Gidrobiol. Zhurn.,* **19 (4)**: 35–40 (in Russian).

Leland, H. and Kuwabara, J. (1985) Trace metals. *In* Fundamentals of aquatic toxicology (eds. Rand, G. and Petrocelli, S.). New York: Hemisphere Publ. Corporation, pp. 374–415.

Lenova, L.I., Borisova, E.V., Lukinov, D.I., and Vasser, S.P. (1989) Studies of the dynamics of the distribution of cells by size to characterize the state of the micro-algal population. *Dokl. Akad. Nauk Ukr. SSR, Ser. B.*, **5**: 69–71 (in Russian).

Lenova, L.I., Stavskaya, S.S., and Ratushnaya, M.Y. (1980) Effect of sodium dodecyl sulfate on unicellular green algae *Chlorella. Gidrobiol. Zhurn.,* **16 (3)**: 83–87 (in Russian).

Lewis, M.A. (1986) Comparison of the effects of surfactants on freshwater phyto-plankton communities in experimental enclosures and on algal population growth in the laboratory. *Environ. Toxicol. Chem.*, **5 (3)**: 319–332.

Lewis, M.A. (1991a) Chronic toxicities of surfactants and detergent builders to algae: a review and risk assessment. *Ecotoxicol. Environm. Safety*, **20 (2)**: 123–140.

Lewis, M.A. (1991b) Chronic and sublethal toxicities of surfactants to aquatic animals: a review and risk assessment. *Water Research*, **25 (1)**: 101–113.

Lindell, M. and Edling, H. (1996) Influence of light on bacterioplankton in a tropical lake. *Hydrobiologia*, **323 (1)**: 67–73.

Lipnitskaya, G.P., Parshikova, T.V., and Topalova, E.K. (1989) Effect of chlorine sodium dodecyl sulfate on the growth of chlorella and microcystis in culture *Gidrobiol. Zhurn.,* **25 (2)**: 63–66 (in Russian).

Lisitsyn, A.P. (1983) The major notions of ocean biogeochemistry (Monin, A.S. and Lisitsyn, A.P. (eds.)). Moscow: Nauka Publishers, pp. 9–32 (in Russian).

Losev, K.S., Gorshkov, V.G., Kondratyev, K.Y., Kotlyakov, V.M., Zapikhanov, M.Ch., Danilov-Danilyan, V.I., Gavrilov, I.T., Revyakin, V.S., and Grakovich, V.F. (1993) Problems of Russian ecology. Moscow: Federal Ecological Foundation of the Russian Federation. 348 pp. (in Russian).

Lovelock, J. (1995) The ages of Gaia: a biography of our living Earth. New York: Norton Co., 255 pp.

Lovelock, J. and Kump, L. (1994) Failure of climate regulation in a geophysical model. *Nature*, **369 (6483)**: 732–734.

Lowe, R. and Pilesbury, R. (1995) Shifts in benthic algal community structure and function following the appearance of zebra mussel. *J. Great Lakes Res.*, **21**: 558–566.

Lukin, E.N. (1968) Hirudinea. *In* Life of animals, Vol. 1 (Zenkevich, L.A. (ed.)). Moscow: Prosveshchenie Publishers, pp. 509–525 (in Russian).

Lukinykh, N.A. (1972) Purification of sewage waters containing synthetic surfactants. Moscow: Stroyizdat Publishers, 92 pp. (in Russian).

Lukyanenko, V.I. (1983) General ichtyotoxicology. Moscow: Legpishchepromizdat Publishers, 320 pp. (in Russian).

Lukyanov, V.S. and Sidorov, S.S. (1989) Method of bioassaying the toxicity of an aqueous medium by the optomotor reaction of fish. Methods of bioassaying the quality of an aqueous medium (Filenko, O.F. (ed.)). Moscow: Moscow University Press, pp. 96–106 (in Russian).

MacBerthouex, P. and Rudd, D.F. (1977) Strategy of pollution control. New York: John Wiley and Sons, 606 pp.

Maki, A. and Bishop, W. (1985) Chemical safety evaluation. *In* Fundamentals of aquatic toxicology (eds. Rand, G. and Petrocelli, S.). New York: Hemisphere Publ. Corporation, pp. 619–635.

Maksimov, V.N. and Zholdakov, I.A. (1985) Use of symplex plans for studies of a combined action of heavy metals on growing corn roots. *Biol. Nauki*, **4**: 107 (in Russian).

Maksimov, V.N., Nagel, H., and Ostroumov, S.A. (1986) Experimental studies of the reaction of *Fagopyrum esculentum* seedlings on the pollution of the aqueous medium by detergents. *In* Problems of ecological monitoring and modeling of ecosystems. Leningrad: Gidrometeoizdat Publishers, Vol. 9, pp. 87–97 (in Russian).

Maksimov, V.N., Nagel, H., and Ostroumov, S.A. (1987) Inhibition of growth of buckwheat seedlings under the action of sodium dodecyl sulfate. *Biol. Nauki*, **12**: 81–84 (in Russian).

Maksimov, V.N., Nagel, H., and Ostroumov, S.A. (1988) Bioassaying of waters containing surfactants sulfonol and DNOC. *Gidrobiol. Zhurn.*, **24** (4): 54–55 (in Russian).

Maksimov, V.N., Nagel, H., Ostroumov, S.A., and Kovaleva, T.N. (1988) Bioassaying of waters polluted by sulfonol. *Vodn. Resursy*, **1**: 165–168 (in Russian).

Malakhov, V.V. and Medvedeva, L.A. (1991) Embryonal development of bivalve mollusks in the norm and at the impact of heavy metals. Moscow: Nauka Publishers, 132 pp. (in Russian).

Malone, T., Conley, D., Fisher, T., Gilbert, P., Harding, L, and Sellner, K. (1996) Scales of nutrient-limited phytoplankton productivity in Chesapeake Bay. *Estuaries*, **19**: 371–385.

Malyarevskaya, A.Y. and Karasina, F.M. (1983) Effects of some surfactants on invertebrates. *Gidrobiol. Zhurn.*, **5**: 84–90 (in Russian).

Manyakhina, L.G. (1990) Effects of synthetic surfactants on the quality of water in Moskva River. Proc. Moscow Center of Hydrometeorology and Observation of the environment (Goskomgidromet), Vol. 2, pp. 105–109 (in Russian).

Marcomini, A., Filipuzzi, F., and Giger, W. (1988) Aromatic surfactants in laundry detergents and hard-surface cleaners: linear alkylbenzenesulphonates and alkylphenol polyethoxylates. *Chemosphere*, **17**: 853–863.

Marfenina, O.E. (1988) Change of the structural and morphoecological indices of fungi in pollution of soils by heavy metals. Ecotoxicology and protection of the environment (Krivolutsky, D.A., Bocharov, B.V., et al. (eds.)). Riga: Inst. of Biology, Latvian Acad. Sci., pp. 103–104 (in Russian).

Matorin, D.N. (1993) Effect of natural factors and anthropogenic pollution on the primary processes of microalgal photosynthesis. DSc thesis. Moscow: Moscow University Press, 45 pp. (in Russian).

Matorin, D.N. (1998) Personal communication.

Matorin, D.N., Vavilin, D.V., and Venediktov, P.S. (1990) On the possibility of using fluorescent methods for studies of crustacean nutrition. *Nauchn. Dokl. Vyssh. Shkoly, Biol. Nauki*, **1**: 147–152 (in Russian).

Matorin, D.N., Vavilin, D.V., Popov, I.V., and Venediktov, P.S. (1989) Method of bioassaying natural waters by recording the delayed fluorescence of microalgae. Methods of bioassaying the quality of an aqueous medium (Filenko, O.F. (ed.)). Moscow: Moscow University Press, pp. 10–20 (in Russian).

Matvienko, A.M. (1977) Protococcophyceae. *In* Life of plants, Vol. 3, (Gollerbakh, M.M. (ed.)). Moscow: Prosveshchenie Publishers, pp. 273–280 (in Russian).

Maximum permissible concentrations of chemical substances in water of aquatic objects of household and social-amenities water use. Hygienic norms (Kurlyandsky, B.A. and Sidorov, K.K. (eds.)) (1998) Moscow: Russian register of potentially hazardous chemical and biological substances, Russian Ministry of Health, 126 pp. (in Russian).

McCutcheon, S. (1997) Personal communication.

McCutcheon, S., Wolfe, N.L., Carreria, L, and Ou, T. (1995) Phytoremediation of hazardous wastes. *In* Innovative technologies for site remediation and hazardous waste management. Proceedings of the National Conference. Pittsburgh, Pennsylvania, July 23–26, pp. 597–604.

McEvoy, J. and Giger, W. (1985) Accumulation of linear alkylbenzensulphonate surfactants in sewage sludges. *Naturwissenschaften*, **72 (8)**: 429.

Medina, V. and McCutcheon, S. (1996) Phytoremediation: modeling removal of TNT. *Remediation*, winter issue, pp. 31–45.

Meekes, H. (1985) Ultrastructure, differentiation and cell wall texture of trichoblasts and root hairs of *Ceratopteris thalictroides* (L.) Brongn. (Parkeriaceae). Aquatic Botany, **21**: 347–362.

Metelev, V.V., Kanaev, A.I., and Dzasokhova, N.G. (1971) Aquatic toxicology. Moscow: Kolos Publishers, 248 pp. (in Russian).

Methods of bioassaying the quality of an aqueous medium (Filenko, O.F. (ed.)). Moscow: Moscow University Press, 124 pp. (in Russian).

Meyer, O., Andersen, P., Hansen, E., and Larsen, J. (1988) Teratogenicity and *in vitro* mutagenicity studies on nonoxynol-9 and -30. *Pharmacol. Toxicol.*, **62**: 236–238.

Mileikovsky, S.A. (1977) Bottom-dwelling invertebrate larvae. *In* Biology of the ocean, Vol. 1 (Vinogradov, M.E. (ed.)). Moscow: Nauka Publishers, pp. 96–106 (in Russian).

Mill, T., Hendry, D., and Richardson, H. (1980) Free-radical oxidants in natural waters. *Science*, **207**: 886–887.

Miller, E.W. and Miller, R.M. (1991) Environmental hazards: toxic waste and hazardous material. Santa Barbara: ABC-CLIO, 286 pp.

Mironov, O.G. (ed.) (1985) Effect of oil and petroleum products on marine organisms and their communities. Problems of chemical pollution of World ocean waters, Vol. 4. Leningrad: Gidrometeoizdat Publishers, 136 pp. (in Russian).

Mironov, O.G. (2000) Biological problems of marine oil pollution. *Gidrobiol. Zhurn.*, **36 (1)**: 82–96 (in Russian).

Mishustina, I.E. (ed.) (1993) Marine microbiology. Vladivostok: Far Eastern National University Press, 192 pp. (in Russian).

Mishustina, I.E. and Baturina, M.V. (1984) Ultramicroorganisms and organic matter of the ocean. Moscow: Nauka Publishers, 96 pp. (in Russian).

Mishustina, I.E., Moskvina, M.I., Rodikova, L.P., and Severina, I.I. (1994) Cyanobacteria of the genus *Synechococcus* in Arctic seas. *Dokl. Akad. Nauk*, **336 (4)**: 562–565 (in Russian).

Mishustina, I.E., Shcheglova, I.K., and Mitskevich, I.N. (1985) Marine microbiology. Vladivostok: Far Eastern National University Press, 184 pp. (in Russian).

Mitin, A.V. (1984) Effect of some factors of the medium on water-clarifying activity of bivalve mollusks. DSc thesis. Moscow: Moscow University Press, 22 pp. (in Russian).

Mitin, A.V. and Voskresensky, K.A. (1982) Clarification of suspensions by zebra mussels during the change of their concentration and temperature. *Nauchn. Dokl.*

Vyssh. Shkoly, Biol. Nauki, **7**: 52–55 (in Russian).

Mochalova, A.S. and Antonova, N.M. (2000) Intensification of natural processes of reservoir self-purification from oil pollution. *Vodn. Resursy*, 27 (**2**): 232–236 (in Russian).

Moiseyenko, T.I. (1999) Methodology and techniques of determining the critical loads (as applied to the surface waters of the Kola Subarctic). *Izv. Akad. Nauk, Ser. Geograph.*, **6**: 68–78 (in Russian).

Moiseyenko, T.I. and Yakovlev, V.A. (1990) Anthropogenic transformations of the aquatic ecosystems of the South Kola Peninsula. Leningrad: Nauka Publishers, 220 pp. (in Russian).

Moiseyev, P.A. (ed.) (1985) Biological resources of the ocean. Moscow: Agropromizdat Publishers, 288 pp. (in Russian).

Monakov, A.V. (1998) Nutrition of freshwater invertebrates. Moscow: Severtsov Institute of Problems of Ecology and Evolution, 322 pp. (in Russian).

Monk, B., Montesinos, C., Leonard, K., and Serrano, R. (1989) Sidedness of yeast plasma membrane vesicles and mechanisms of activation of the ATPase by detergents. *Biochim. Biophys. Acta*, **981** (**2**): 226–234.

Moore, J. and Ramamoorthy, S. (1984) Organic chemicals in natural waters. New York: Springer, 289 pp.

Moore, R. and Weiner, R. (1989) Genus *Hyphomonas* (ex Pongratz 1957) Moore, Weiner and Geberts 1984. *In* Holt, J. et al. (eds.) Bergey's manual of systematic bacteriology, 9th edition, Baltimore: Williams & Wilkins, Vol. 3 (Staley, J., ed.), pp. 1904–1910.

Mosharov, S.A. (1996) Characterization of biogenic sedimentation in the Baltic and Black Seas. DSc thesis. Moscow, 24 pp. (in Russian).

Mozhayev, E.A. (1976) Pollution of reservoirs by surfactants. Moscow: Meditsyna, 96 pp. (in Russian).

Mozhayev, E.A. (1989) Alkyl benzene sulfonates. Moscow: Center of International Projects, State Committee for Science and Technology, 17 pp. (in Russian).

Mudryi, I.V. (1990) Soil hygiene under conditions of irrigation of agricultural lands by sewage waters containing surfactants (a review). *Gigiena Sanit.*, **8**: 27–29 (in Russian).

Mudryi, I.V. (1994) Effect of anionic surfactants in combination with other priority pollutants on the quality of water of Dnieper River, some of its tributaries and a cascade of water storage reservoirs. *Gigiena Sanit.*, **3**: 17–19 (in Russian).

Mudryi, I.V. (1995) On the possible disturbance of the ecological and hygienic equilibrium by surfactants under conditions of a integrated anthropogenic pollution of the environment (a review). *Gigiena Sanit.*, **3**: 35–38 (in Russian).

Murphy, D. and Woodrow, I. (1984) The effect of Triton X-100 and *n*-octyl *b-D*-glucopyranoside on energy transfer in photosynthetic membranes. *Biochem. J.*, **224** (**3**): 989–993.

Nagel, H., Ostroumov, S.A., and Maksimov, V.N. (1987) Inhibition of growth of buckwheat seedlings under the action of sodium dodecyl sulfate. *Biol. Nauki*, **12**: 81–84 (in Russian).

Naumova, R.P., Lisin, G.R., Cherepneva, I.E., and Belousova, T.O. (1981) Surfactants as mutagens and prophage inducers. Materials of the 4th Congress of the All-Union Hydrobiological Society (Kiev, December 1–4). Kiev: Naukova Dumka, Part 1, pp. 135–136 (in Russian).

Nesterova, M.P. (1980) Ecological aspects of using chemical means for elimination of the consequences of oil spills in the sea. *In* Man and biosphere (Fedorov, V.D.

(ed.)). Moscow: Moscow University Press, Vol. 5, pp. 110–118 (in Russian).

Newell, R. (1998) Ecological changes in Chesapeake Bay. are they the result of over-harvesting the American oyster, *Crassostrea virginica*? *In* Understanding the estuary: Advances in Chesapeake Bay Research. Proceedings of a Conference. 29–31 March. Baltimore, Maryland: Chesapeake Research Consortium Publication 129, CBP/TRS 24/88, pp. 536–546.

Newell, R. and Ott, J. (1999) Macrobenthic communities and eutrophication. Macrobenthic communities and eutrophication, Chapter 9. *In* Malone, T.C., Malej, A., Harding, L.W., Jr., Smodlaka, N., and Turner, R.F. (eds). Coastal and estuarine studies. Washington, DC: American Geophysical Union, Vol. 55, pp. 265–293.

Nielsen, T. and Ekelund, N. (1995) Influence of solar ultraviolet radiation on photosynthesis and motility of marine phytoplankton. *FEMS Microbiol. Ecol.*, 18 (4): 281–288.

Niemi, S. (1983) Alternatives to animal experimentations. *Int. J. Study Anim. Probl.*, 4 (3): 241–249.

Nikitin, D.P. (ed.) (1980) Large cattle-breeding complexes and the environment. Moscow: Meditsina Publishers, 256 pp. (in Russian).

Nimmo, D. (1985) Pesticides. *In* Fundamentals of aquatic toxicology (eds. Rand, G. and Petrocelli, S.). New York: Hemisphere Publ. Corporation, pp. 335–373.

Nyberg, H. (1985) The influence of ionic detergents on the phospholipid fatty acid compositions of *Porphyridium purpureum*. *Phytochemistry*, 24 (3): 435–440.

Nyberg, H. (1988) Growth of *Selenastrum capricornutum* in the presence of synthetic surfactants. *Water Res.*, 22 (2): 217–223.

Nyberg, H. and Koskimies-Soininten, K. (1984a) The phospholipid fatty acids of *Porphyridium purpureum* cultured in the presence of Triton X-100 and sodium desoxycholate. *Phytochemistry*, 23 (11): 2489–2495.

Nyberg, H. and Koskimies-Soininten, K. (1984b) The glycolipid fatty acids of *Porphyridium purpureum* cultured in the presence of detergents. *Phytochemistry*, 23 (4): 751–757.

Obroucheva, N.V. (1992) Seedlings root growth in terms of cell division and elongation. *In* Root ecology and its practical applications (eds. Kutschera, L., Hubl, E., Lichtenegger, E., Persson, H., and Sobotik, M.). Klagenfurt: Verein für Wurzelforschung, pp. 13–16.

Officer, C., Smayda, T., and Mann, R. (1982) Benthic filter feeding: a natural eutrofication control. *Mar. Ecol. Prog. Ser.*, 9: 203–210.

Ogilvie, S. and Mitchell, S. (1995) A model of mussel filtration in a shallow New Zealand lake, with reference to eutrophication control. *Arch. Hydrobiol.*, 133 (4): 471–482.

Oksiyuk, O.P. (1999) Ecological standards of water quality for the Shatsk Lakes, *Gidrobiol. Zhurn.*, 35 (5): 74–86 (in Russian).

Okuno, M. and Morisawa, M. (1989) Effects of calcium on motility of rainbow trout sperm flagella demembranated with Triton X-100. *Cell Motil. Cytoskeleton*, 14 (2): 194–200.

Ono, K., Akakawa, O., Onoue, Y., Matsumoto, S., Shihara, H., Takeda, K., Nakanashi, M., Yamaji, T., Uemura, K., and Ogawa, Y. (1998) Synthetic surfactants for protecting cultured fish against toxic phytoplankton. *Aquacult. Res.*, 29 (8): 569-572.

Organization for Economic Cooperation and Development (1984) Terrestrial Plants: Growth Test. OESD Guidelines for Testing of Chemicals, No 208, Paris.

Osanai, S., Yoshida, Y., Fukushima, K., and Yoshikawa, S. (1989) Preparation of optically active amphoteric surfactants and their surface and antimicrobial properties. *Yukagaku*, **38 (8)**: 633–641.

Ostroumov, S.A. (1981) Protection of the environment. Veterinary encyclopedic dictionary. Moscow: Sov. Entsiklopediya, pp. 205–206 (in Russian).

Ostroumov, S.A. (1984) Problems of the protection of ecosystems: a conceptual analysis. Moscow: Chelovek i Priroda, Vol. 5, pp. 3–15 (in Russian).

Ostroumov, S.A. (1986a) Introduction to biochemical ecology. Moscow: Moscow University Press, 176 pp. (in Russian).

Ostroumov, S.A. (1986b) Pollution of the biosphere. *In* Biological encyclopedic dictionary. Moscow: Sov. Entsyklopediya, p. 416 (in Russian).

Ostroumov, S.A. (1989) Chemical pollution of the environment and development of carcinogenesis. Problems of the environment and natural resources. Moscow: VINITI Press, Vol. 8, pp. 12–19 (in Russian).

Ostroumov, S.A. (1990a) Bioassaying of solutions of xenobiotics using water lettuce. Ecological and technological aspects of decontamination of industrial wastes of polymer productions. Proc. of a conference, February 15–17, Donetsk: NIITEKHIM Minkhimnefteprom SSSR, pp. 12–13 (in Russian).

Ostroumov, S.A. (1990b) Some aspects of assessing the biological activity of xenobiotics. *Vestn. MGU, Ser. Biol.*, **6 (2)**: 27–34 (in Russian).

Ostroumov, S.A. (1991a) Reaction of test organisms to the pollution of the aqueous medium by a quaternary ammonium compound. *Vodn. Resursy*, **2**: 112–116 (in Russian).

Ostroumov, S.A. (1991b) Biological activity of waters containing surfactants. *Khim. Tekhnol. Vody*, **13 (3)**: 270–283 (in Russian).

Ostroumov, S.A. (1994) Some aspects of ecotoxicology and biochemical ecology of surfactants. Proceedings of the 6th International Congress of Ecology (August 21–26, Manchester), p.127.

Ostroumov, S.A. (1998) Biological filtering and ecological machinery for self-purification and bioremediation in aquatic ecosystems: towards a holistic view. *Rivista di Biologia / Biology Forum*, **91**: 247–258.

Ostroumov, S.A. (1999) Triton X-100 (inhibition of *Lepidium sativum*). *Toksikol. Vestn.*, **4**: 41 (in Russian).

Ostroumov, S.A. (1999a) Integrity-oriented approach to ecological biomachinery for self-purification and bioremediation in aquatic ecosystem. Limnology and oceanography: navigating into the next century. Waco, Texas: ASLO, p. 134.

Ostroumov, S.A. (1999b) The ability of mussels to filter and purify the sea water is inhibited by surfactants. Limnology and oceanography: navigating into the next century. Waco, Texas: ASLO, p. 134.

Ostroumov, S.A. (2000) Effects of some xenobiotics on marine and freshwater bivalves. *Ecol. Studies, Hazards, Solutions*, **3**: 22–23.

Ostroumov, S.A. (2000a) Tetradecyl trimethyl ammonium bromide (action on *L. stagnalis*). *Toksikol. Vestn.*, **1**: 42–43 (in Russian).

Ostroumov, S.A. (2000b) Criteria of the ecological hazard of anthropogenic impacts on the biota: search for a system. *Dokl. Akad. Nauk*, **371 (6)**: 844–846 (in Russian).

Ostroumov, S.A. (2000c) The concept of aquatic biota as a labile and vulnerable link of the water self-purification system. *Dokl. Akad. Nauk*, **372 (2)**: 279–282 (in Russian).

Ostroumov, S.A. (2000d) Biological effects of surfactants in relation to anthropogenic effects on the biosphere. Moscow: MAX Press, 116 pp. (in Russian).

Ostroumov, S.A. (2000e) Some approaches to the assessment of carbon transfer to the lower layers of the aqueous mass and benthic sediments of the aquatic ecosystems. Aquatic Ecosystems and Organisms, 2. Moscow: MAX Press, pp. 57–58 (in Russian).

Ostroumov, S.A. (2001) The synecological approach to the problem of eutrophication. *Doklady Biological Sciences*, **381**: 559–562 (translated from *Doklady Akad. Nauk*, **381**: 709–712 (2001).

Ostroumov, S.A. (2002) Inhibitory analysis of top-down control: new keys to studying eutrophication, algal blooms, and water self-purification. *Hydrobiologia*, **469**: 117–129.

Ostroumov, S.A. (2003a) Studying effects of some surfactants and detergents on filter- feeding bivalves. *Hydrobiologia*, **500**: 341–344.

Ostroumov, S.A. (2003b) Anthropogenic effects on the biota: towards a new system of principles and criteria for analysis of ecological hazards. *Rivista di Biologia / Biology Forum*, **96(1)**: 159–169.

Ostroumov, S.A. (2004) Suspension-feeders as factors influencing water quality in aquatic ecosystems. *In* The comparative roles of suspension-feeders in ecosystems, Dame, R.F. and Olenin, S. (eds.), Springer-Verlag, Dordrecht, pp. 147–164.

Ostroumov, S.A. (2005a) Some aspects of water filtering activity of filter-feeders. *Hydrobiologia*, **542**: 275–286.

Ostroumov, S.A. (2005b) On some aspects of maintaining water quality and its self-purification. *Vodn. Resursy*, **32**: 337–347 (in Russian).

Ostroumov, S.A. and Donkin, P. (1997) Impact of pollution of an aqueous medium by surfactants on the biological mechanisms of removing a suspension of phytoplankton particles from a water column with possible consequences for the optical characteristics of the aquatic ecosystem. Physical problems of ecology. Moscow: Moscow University Press, Vol. 2, pp. 71–72 (in Russian).

Ostroumov, S.A. and Fedorov, V.D. (1999) Basic components of ecosystems self-purification and the possibility of its disturbance as a result of its chemical pollution. *Vestn. MGU, Ser. Biol.*, **1**: 24–32 (in Russian).

Ostroumov, S.A. and Golovko, A.E. (1992) Bioassaying of the toxicity of a surfactant (sulfonol) using rice seedlings as a test object. *Gidrobiol. Zhurn.*, **28 (3)**: 72–75 (in Russian).

Ostroumov, S.A. and Khoroshilov, V.S. (1992) Bioassaying of waters polluted by surfactants. *Izv. Akad. Nauk, Ser. Biol.*, **3**: 452–458 (in Russian).

Ostroumov, S.A. and Kolesnikov, M.P. (2000) Biocatalysis of the transfer of matter in the mesocosm is inhibited by a contaminant: impact of a surfactant on *Lymnaea stagnalis*. *Dokl. Akad. Nauk*, **373 (2)**: 278–280 (in Russian).

Ostroumov, S.A. and Kolotilova, N.N. (1998) Cetyl trimethyl ammonium bromide (CTAB, hexadecyl trimethyl ammonium bromide). *Toksikol. Vestnik*, **5**: 30 (in Russian).

Ostroumov, S.A. and Kolotilova, N.N. (2000) Growth of cyanobacteria in the presence of surfactant-containing preparations. Aquatic ecosystems and organisms, 2. Moscow: MAX Press, p. 60 (in Russian).

Ostroumov, S. and Maertz-Wente, M. (1991) Effects of the nonionic surfactant on marine diatoms. Papers presented at the 201st National Meeting of American Chemical Society, Atlanta, GA, April 14–19, Vol. 31, No 1, pp. 18–19.

Ostroumov, S.A. and Maksimov, V.N. (1988a) Degradation of algae under pollution of the aqueous environment with the surfactant ethonium. *Ekologiya*, **6**: 165–168

(in Russian).

Ostroumov, S.A. and Maksimov, V.N. (1988b) Disturbance of the ontogenesis of *Camelina sativa* and *Triticum aestivum* by the action of a nonionogenic surfactant. Ecotoxicology and protection of the environment. Riga: Inst. of Biology, pp. 54–55 (in Russian).

Ostroumov, S.A. and Semykina, N.A. (1990) Bioassaying of CHMA polymer aqueous solutions on plants. Ecological and technological aspects of decontamination of industrial emissions of polymer productions. Cherkassy: NIITEKhIM, pp. 13–14 (in Russian).

Ostroumov, S.A. and Semykina, N.A. (1991) Reaction of macrophyte seedlings to pollution of an aqueous medium by high-molecular surfactants. *Ekologiya*, **4**: 83–85 (in Russian).

Ostroumov, S.A. and Semykina, N.A. (1993) Reaction of *Fagopyrum esculentum* Moench to pollution of an aqueous medium by a polymer surfactant. *Ekologiya*, **6**: 50–55 (in Russian).

Ostroumov, S.A. and Tretyakov, A.N. (1990). Effect of the pollution of the medium by cationic surfactant on algae and seedlings of *Fagopyrum esculentum*. *Ekologiya*, **2**: 43–46 (in Russian).

Ostroumov, S.A. and Vasternak, K. (1991) Reaction of photoorganotrophically grown green flagellates to the pollution of the aqueous medium with "Kristall" detergent. *Vestn. MGU, Ser. Biol.*, **2**: 67–69 (in Russian).

Ostroumov, S.A. and Vorobiev, L.N. (1976) Membrane potential as a possible polyfunctional regulator of membrane protein activity. *Biol. Nauki*, **7**: 22 (in Russian).

Ostroumov, S.A. and Vorobiev, L.N. (1978) Membrane potential and surface charge densities as possible generalized regulators of membrane protein activities. *J. Theor. Biol.*, **75**: 289–297.

Ostroumov, S.A., et al. (1994) Ecotoxicology and biological activity of surfactants. Third European Conference on Ecotoxicology (Zurich, August 28–31, 1994), Abstract No. 6.26, p. 141.

Ostroumov, S.A., Borisova, E.V., Leonova, L.I., and Maksimov, V.N. (1990a) Effect of sulfonol on the algal culture *Dunaliella asymmetrica* and on seedlings of *Fagopyrum esculentum*. *Gidrobiol. Zhurn.*, **26 (2)**: 96–98 (in Russian).

Ostroumov, S.A., Dodson, S., Hamilton, D., Peterson, S., and Wetzel, R.G. (2003) Medium-term and long-term priorities in ecological studies. *Rivista di Biologia / Biology Forum*, **96**: 327–332.

Ostroumov, S.A., Donkin, P., and Staff, F. (1997a) Inhibition of the ability of mussels *Mytilus edulis* to filter and purify sea water by an anionic surfactant. *Vestn. MGU, Ser. Biol.*, **3**: 30–35 (in Russian).

Ostroumov, S.A., Donkin, P., and Staff, F. (1997b) Biofiltration of water and its disturbance in pollution of the medium. Ecological problems of large administrative units of megacities. Moscow: Prima Press, pp. 117–118 (in Russian).

Ostroumov, S.A., Golovko, A.E., and Khoroshilov, V.S. (1990b) Biodiagnostics and bioassaying of polluted waters and xenobiotics: search for nontraditional test objects and methods. Ecological and technological aspects of decontamination of industrial emissions of polymer productions,. Cherkassy: NIITEKhIM, pp. 14–15 (in Russian).

Ostroumov, S.A., Golovko, A.E., and Khoroshilov, V.S. (1990c) Bioassaying of surfactants and surfactant-containing preparations. Methodology of ecological regulation (All-Union conference, Kharkov, April 16–20). Kharkov: VNIIOVO, Vol. 2, pp. 139 (in Russian).

Ostroumov, S.A., Jasaitis, A.A., and Samuilov, V.D. (1979) Electrochemical proton gradient across the membranes of photophosphorylating bacteria. *Biomembranes*, **10**: 209–233.

Ostroumov, S.A., Kaplan, A.Y., Kovaleva, T.N., and Maksimov, V.N. (1988) Study of some aspects of the ecotoxicology of anionic surfactant sulfonol on plants and other objects. Ecotoxicology and protection of the environment. Riga: Inst. of Biology, pp. 134–136 (in Russian).

Ostroumov, S.A., Kolotilova, N.N., Piskunova, N.F., Kartasheva, N.V., Lyamin, M.Y., and Kraevsky, V.M. (1999a) Impact of surfactants from the class of quaternary ammonium compounds on unicellular cyanobacteria, green algae, and rotifers. Aquatic ecosystems and organisms. Moscow: Dialog-MGU, p. 45 (in Russian).

Ostroumov, S.A., Kraevsky, V.M., and Lyamin, M.Y. (1999b) Tetradecyl trimethyl ammonium bromide (TDTMA). *Toksikol. Vestnik*, **1**: 35–36 (in Russian).

Ostroumov, S.A., Samuilov, V.D., and Yasaitis, A.A. (1979) Electrochemical gradient of hydrogen ions on membranes of photosynthetic bacteria. *Usp. Sovr. Biol.*, **87 (3)**: 155–169 (in Russian).

Otstavnova, N.A. and Kurmakayev, V.A. (1997) On the state of the environment of Moscow in 1996. *Ekol. Vestnik Moskvy*, Nos 6–8, pp. 16–47 (in Russian).

Oviatt, C.A. (1981) Effects of different mixing schedules on phytoplankton, zooplankton, and nutrients in marine microcosms. *Mar. Ecol. Progr. Ser.*, **4 (1)**: 57–67.

Paal, L.L., Karu, Y.Y., Melder, H.A., and Repin, B.N. (1994) Reference manual on purification of natural and sewage waters. Moscow: Vysshaya Shkola Publishers, 336 pp. (in Russian).

Paine, R.T. (1966) Food web complexity and species diversity. *Am. Naturalist*, **100**: 65–75.

Painter, H.A. (1992) Anionic surfactants. *In* The handbook of environmental chemistry, Vol. 3, Part F. Berlin: Springer, pp. 1–88.

Palaski, M. and Booth, H. (1995) Zebra mussel pseudofaeces production, degradation, and their potential for removal of PCBs from freshwater. Abstr. Pap. Present. Annu. Meet. Mich. Acad. Ferris State Univ., Ann Arbor, Mich., March 10–11, Mich. Acad., Vol. 27, No 3, p. 381.

Palenik, B., Zafirion, O., and Morel, F.M.M. (1987) Hydrogen peroxide production by a marine phytoplankter. *Limnol. Oceanogr.*, **32 (6)**: 1365.

Pantani, C., Spreti, N., Maggiti, M., and Germani, R. (1995) Acute toxicity of some synthetic cationic and zwitterionic surfactants to freshwater amphipod *Echinogammarus tibaldii*. *Bull. Environm. Contam. Toxicol.*, **55 (2)**: 179–186.

Parshikova, T.V. (1996) Effect of surfactants on the growth, propagation, and functional activity of algae on cultures and natural populations. Ecol. and physiol. studies of algae and their significance for the assessment of the condition of natural waters. A short conference report, Borok, December 3–5. Yaroslavl: pp. 161–163 (in Russian).

Parshikova, T.V. and Negrutsky, S.F. (1988) Effect of surfactants on algae. *Gidrobiol. Zhurn.*, **24 (6)**: 46–57 (in Russian).

Parshikova, T.V., Veselovsky, V.V., Veselova, T.V., and Dmitrieva, A.G. (1994) Effect of surfactants on the functioning of the photosynthetic apparatus of chlorella. *Algologiya*, **4 (1)**: 38–46 (in Russian).

Parsons, T., Takahashi, M. and Hargrave, B. (1988) Biological oceanographic processes. Pergamon Press: Oxford, 330 pp.

Pashchenko, N.I. and Kasumyan, A.O. (1984) Degenerative and regenerative processes in the olfactory lining of the grass carp *Ctenopharyngodon idettea* (Val.) (Cyprinidae) after the action of detergent Triton X-100. *Vopr. Ikhtiol.*, **24 (1)**: 128–139 (in Russian).

Passet, B.V., Golubyatnikova, A.A., Enina, N.V., Nekrasov, S.V., and Mordvinova, E.T. (1985) On the relation between the structure and antimicrobial activity of anionic surfactants. *Khim.-Farm. Zhurn.*, **19 (11)**: 1356–1361 (in Russian).

Patin, S.A. (1977) Chemical pollution and its impact on hydrobionts. Biology of the ocean. Vol. 2, Biological productivity of the ocean (Vinogradov, M.E. (ed.)). Moscow: Nauka Publishers, pp. 322–331 (in Russian).

Patin, S.A. (1979) Impact of pollution on the biological resources and productivity of the World Ocean. Moscow: Legpishchepromizdat Publishers, 304 pp. (in Russian).

Patin, S.A. (1985) Ecological and toxicological aspects of aquaculture. *In* Biological bases of aquaculture in the seas of the European part of the USSR. Moscow: Nauka Publishers, pp. 65–72 (in Russian).

Patin, S.A. (1988a) Aquatic toxicology and optimization of bioproduction processes in aquaculture. Moscow: VNIRO Publishers, p. 5 (in Russian).

Patin, S.A. (1988b) Ecological aspects of aquatic toxicology. *In* Ecotoxicology and protection of the environment (Krivolutsky, D.A., Bocharov, B.V., et al., eds.) Riga: Inst. of Biology, Latvian Acad. Sci., pp. 137–139 (in Russian).

Patin, S.A. (1988c) Theoretical and applied aspects of aquatic toxicology. *In* Ecological chemistry of an aqueous medium (Skurlatov, Yu.S., ed.). Moscow: Inst. Chem. Phys., USSR Acad. Sci., Vol. 2, pp. 96–107 (in Russian).

Patin, S.A. (1997) Ecological problems of the development of oil and gas resources of the marine shelf. Moscow: All-Russian Research Inst. of Marine Fisheries and Oceanography (VNIRO), 350 pp. (in Russian).

Pavlov, D.S. and Kasumyan, A.O. (1990) Sensor bases of the food behavior in fish. *Vopr. Ikhtiologii*, **30 (5)**: 720–732 (in Russian).

Pickup, J. (1990) Detergents and the environment: an industry view. *Chemistry and Industry*, March 19, pp. 174–177.

Pinnaduwage, P., Schmitt, L. and Huang, L. (1989) Use of a quaternary ammonium detergent in liposome-mediated DNA transfection of mouse L-cells. *Biochim. Biophys. Acta*, **985**: 225–232.

Piskunova, N.F. and Ostroumov, S.A. (1999) Effect of cationic surfactant tetradecyl trimethyl ammonium bromide on freshwater green algae. *Toksikol. Vestnik*, **3**: 27–29 (in Russian).

Pittinger, C., Woltering, D., and Masters, J. (1989) Bioavailability of sediment-sorbed and aqueous surfactants to *Chironomus riparius* (midge). *Environm. Toxicol. Chem.*, **8 (11)**: 1023–1033.

Plekhanov, S.E., Bratkovskaya, L.B., Svetlova, E.N., and Elias, V.V. (1997) Physiological reactions of charophytes on the components of sewage waters (microelements, sulfates, phenols) of sulfate-cellulose production. Moscow: Int. Biotechnol. Center at the Moscow State University. Deposited in VINITI on 27 March, 1997, No 982–B97, 32 pp. (in Russian).

Plymouth Marine Laboratory (1997) Annual Report 1996–1997. Plymouth: PML, 72 pp.

Pokarzhevsky, A.D. (1988) Ecotoxicology and geochemical ecology of animals. Ecotoxicology and protection of the environment. (Krivolutsky, D.A., Bocharov, B.V., et al. (eds.)). Riga: Inst. of Biology, Latvian Acad. Sci., pp. 143–144 (in

Russian).

Pokarzhevsky, A.D. (1992) Ecological regulation, ecotoxicology, and soil biology. Ecological regulation: problems and methods. Moscow: Nauka Publishers (in Russian).

Pokarzhevsky, A.D. and Semenova, N.L. (1988) Introduction to biochemical ecology (book review). *Ekologiya*, **2**: 89–90 (in Russian).

Polikarpov, G.G. and Egorov, V.N. (1989) Marine dynamic radiochemoecology. Moscow: Energoatomizdat Publishers, 176 pp. (in Russian).

Pollution of Black and Azov Seas: A review (1975). Sevastopol: State Ocanographic Institute, 180 pp. (in Russian).

Poremba, K., Gunkel, W., Lang, S., and Wagner, F. (1991) Marine biosurfactants. III. Toxicity testing with marine microorganisms and comparison with synthetic surfactants. *Z. Naturforsch.*, **45**: 210–216.

Potapova, N.A. and Galagan, N.P. (1983) Effect of sodium alkylbenzene sulfonate on Kiev water reservoir bacterioplankton. *Gidrobiol. Zhurn.*, **19 (1)**: 44 (in Russian).

Prager, M., Kanar, M., Farmer, J., and Vanderzee, J. (1985) Effect of dimethyldioctadecylammonium bromide-induced macrophages on malignant cell proliferation. *Cancer Lett.*, **27 (2)**: 225–232.

Pritchard, P.H. and Bourquin, A. (1985) Microbial toxicity testing. *In* Fundamentals of aquatic toxicology, eds. Rand, G. and Petrocelli, S. New York: Hemisphere Publ. Corporation, pp. 177–220.

Pupyrev, E.I. (ed.) (1992) Problems of Moscow ecology. Moscow: Moscow Division of Gidrometeoizdat Publishers, 198 pp. (in Russian).

Purmal, A.P. (1988) Physico-chemical bases of processes in aqueous media. Ecological chemistry of an aqueous medium. Vol. 1, pp. 23–37 (in Russian).

Ramade, F. (1987) Ecotoxicology. Chichester: Wiley, 262 pp.

Rand, G. (1985) Behavior. *In* Fundamentals of Aquatic Toxicology (Rand, G. and Petrocelli, S., eds.). New York: Hemisphere Publ. Corporation, pp. 221–263.

Rand, G. and Petrocelli, S. (eds.) (1985) Fundamentals of aquatic toxicology. New York: Hemisphere Publ. Corporation, 666 pp.

Randall, R. and Henry, S. (1989) The stress proteins response as a measure of pollutant stress in the infaunal clam *Macoma nasata*. Ocean '89: Int. Conf. Address. Meth. Understand. Global Ocean, Seattle, Wash., Sept. 18–21, Vol. 2, New York (N.Y.), p. 701.

Rebandel, H. and Dryl, S. (1980) Dependence of toxic effects of detergents in *Paramecium caudatum* on ionic composition of external medium. *Acta Protozool.*, **19**: 261–268.

Rebhun, M. and Manka, J. (1971) Classification of organics in secondary effluents. *Envir. Sci. Technol.*, **7**: 606–609.

Recommendations on maximum permissible concentrations of pollutants for waters of fish-economy reservoirs. Compilers: Patin, S.A., Lesnikov, L.A., Filenko, O.F., Kiryushina, L.P., Balabanova, T.S., Yarushek, N.E., Dmitrieva, A.G., Mosiyenko, T.K., Stroganov, N.S., Korol, V.M., Golubkova, E.G., Isakova, E.F., Kolosova, L.V., Danilchenko, O.P., Buzinova, N.S., Pugintsev, A.I., et al. Moscow: VNIRO Publishers, 88 pp. (in Russian).

Renzoni, A. (1971) The influence of some detergents on the larval life of marine bivalve larvae. *Rev. Int. Oceanogr. Med.*, **24**: 50–52.

Review of the state of the environment in the USSR (1990) Moscow: Gidrometeoizdat Publishers, 115 pp. (in Russian).

Röderer, G. (1987) Toxic effects of tetraethyl lead and its derivatives on the chryso-phyte *Poteriochromonas malhamensis*. VIII. Comparative studies with surf-actants. *Arch. Environ. Contam. Toxicol.*, **16 (3):** 291–301.

Romanenko, V.D. and Romanenko, A.V. (1992) At the junction of sciences. Book review. Introduction to problems of biochemical ecology: biotechnology, agri-culture, protection of the environment. *Gidrobiol. Zhurn.,* **28 (2):** 82–83 (in Russian).

Romankevich, E.A. and Lyutsarev, S.V. (1984) Biogeochemical carbon cycle in the ocean. Biogeochemistry of pericontinental regions of the ocean. *In* Book of abstracts of the All-Union conference. Moscow: Nauka Publishers, p. 60 (in Russian).

Roper, D. and Hickey, C. (1995) Effect of food and silt on filtration, respiration and condition of the freshwater mussel *Hyridella menziesi* (Unionacea: Hyriidae): implications for bioaccumulation. *Hydrobiologia,* **312 (1):** 17–25.

Rosenbaum, W. (1991) Environmental politics and policy. Washington, DC: Con-gressional Quarterly Inc., 336 pp.

Rowan, A. (1983) The LD_{50} – the beginning of the end. *Int. J. Study Anim. Probl.,* **4 (1):** 4–24.

Rozenberg, G.S., Mozgovoi, D.P., and Gelashvili, D.B. (1999) Ecology. Samara: Samara Scientific Center, Russian Academy of Sciences, 396 pp. (in Russian).

Rumbold, D. and Snedaker, S. (1997) Evaluation of bioassays to monitor surface microlayer toxicity in tropical marine waters. *Arch. Environ. Toxicol.,* **32:** 135–140.

Ryzhikova, O.P. and Ryabukhina, E.V. (1998) The role of detergents in the change of the filtration activity of bivalve mollusks. Deposited in VINITI 16 October 1998, No 3035–B98 (in Russian).

Sadchikov, A.P. (1997) Production and transformation of organic matter by different groups of phyto- and bacterioplankton (by the example of Moscow Region reservoirs). DSc thesis. Moscow: Moscow University Press, 54 pp. (in Russian).

Safonova, T.A. (1977) Euglenophyta. *In* Life of plants, Vol. 3 (Gollerbakh, M.M. (ed.)). Moscow: Prosveshchenie Publishers, pp. 259–265 (in Russian).

Saitoh, Y., Yokosawa, H., and Ishii, S. (1989) Sodium dodecyl sulfate-induced con-formational and enzymatic changes of multicatalytic proteinase. *Biochim. Bio-phys. Res. Commun.,* **162 (1):** 334–339.

Sakunthala, B., Shanmukhappa, H., and Neelakantan, B. (1990) Comparative toxi-city of four detergents on a prawn *Metapenaeus dobsoni. Fish. Technol.,* **27 (2):** 109–111.

Sala, M. and Güde, H. (1993) Development of microbial ectoenzyme activities in littoral sediments of Lake Constance. Int. Ver. Theor. Angew. Limnol., Stuttgart, p. 633.

Salanki, J. (1995) Biomonitoring of a natural medium. *Zhurn. Obshch. Biol.,* **46 (6):** 743–752 (in Russian).

Sanden, P. and Hakansson, B. (1996) Long-term trends in Secchi depth in the Baltic Sea. *Limnol. Oceanogr.,* **41 (2):** 346–351.

Sanotsky, I.V. (1984) The concept of prophylactic toxicology and the principles of justification of sanitary standards. Prophylactic toxicology. Moscow: Center of International Projects, State Comm. for Science and Technology, Vol. 1, pp. 68–82 (in Russian).

Savarese, M., Patterson, M., Chernykh, V., and Fialkov, V. (1997) Trophic effects of sponge feeding within Lake Baikal's littoral zone. 1. *In situ* pumping rates.

Limnol. Oceanogr., **42**: 171–178.

Schauerte, W., Lay, J., Klein, W., and Korte, F. (1982) Influence of 2,4,6-trichloro-phenol and pentachlorophenol on the biota of aquatic systems. *Chemosphere*, **11**: 71–79.

Schnoor, J., Light, L., McCutcheon, S., Wolfe, N.L., and Carreira, L. (1995) Phyto-remediation of organic and nutrient contaminants. *Env. Sci. Techn.*, **29 (7)**: 318A–323A.

Schuurmann, G. (1991) Acute aquatic toxicity of alkyl phenol ethoxylates. *Ecotoxi-col. Environm. Safety*, **21**: 227–233.

Self-purification and bioindication of polluted waters (1980) Moscow: Nauka Publishers (in Russian).

Sericano, J., Wade, T., Jackson, T., Brooks, J., Tripp, B., Harrington, J., Mee, L., Readmann, J., Villeneuve, J.-P., and Goldberg, E. (1995) Trace organic conta-mination in the Americas: an overview of the US national status and trends and the international "Mussel Watch" programmes. *Mar. Pollut. Bull.*, **31**: 214–225.

Seymour, R. and Geyer, R. (1992) Fates and effects of oil spills. *Annu. Rev. Energy Environ.*, **17**: 261–283.

Shamshurin, A.A. and Krimer, M.Z. (1976) Physico-chemical properties of pesti-cides. Moscow: Khimiya Publishers, 328 pp. (in Russian).

Shcherbakov, Y.A. and Kotlyar, S.G. (1983) Role of bioindication in assessing the extent of reservoir pollution in field studies and laboratory modeling. *In* Theoreti-cal issues of bioassaying (Lukyanenko, V.I., (ed.)). Volgograd, Institute for the Biology of Internal Waters, USSR Acad. Sci., pp. 82–90 (in Russian).

Sheedy, B., Lazorchak, J., Grunwald, D., Pickering, Q., Pilli, A., Hall, D., and Webb, R. (1991) Effects of pollution on freshwater organisms. *Research Journal WPCF*, **63 (4)**: 619–696.

Shelakova, V.V. (1975) Effects of surfactants on pathogenic enterobacteria in an aqueous medium. *Gigiena Sanit.*, **3**: 25–28 (in Russian).

Shevchuk, I.A., Pisko, R.T., and Barilyak, I.R. (1975) Studies of the teratogenic and embryotoxic properties of a new surfactant, ethonium. *In* The physiological role of surfactants. Chernovtsy, pp. 116–117 (in Russian).

Shtamm, E.V. (1988) Redox state of an aqueous medium and issues of reproduction of fish resources. *In* Ecological chemistry of an aqueous medium (Skurlatov, Yu.I. (ed.)). Moscow: Inst. of Chem. Phys., USSR Acad. Sci., Vol. 1, pp. 278–294 (in Russian).

Shtamm, E.V. and Batovskaya, L.O. (1988) Biological and abiotic factors of forming the redox state of a natural aqueous medium. *In* Ecological chemistry of an aqueous medium (Skurlatov, Yu.I. (ed.)). Moscow: Inst. of Chem. Phys., USSR Acad. Sci., Vol. 2, pp. 125–137 (in Russian).

Shtannikov, E.V. and Antonova, A.N. (1978) Review of book by E.A. Mozhayev. Pollution of reservoirs by surfactants. *Gigiena Sanit.*, **3**: 118–119 (in Russian).

Shtannikov, E.V. and Podzemelnikov, E.V. (1978) Efficiency of water pipeline works with respect to organophosphorus pesticides. *Gigiena Sanit.*, **3**: 18–23 (in Russian).

Shtina, E.A. (1985) Soil algae as pioneers of invading the technogenic substrates and as indicators of the state of disturbed lands. *Zhurn. Obshch. Biol.*, **46 (4)**: 435–443 (in Russian).

Shtina, E.A. (1988) Algae as indicators in the assessment of soil pollution. *In* Eco-toxicology and protection of the environment (Krivolutsky, D.A., Bocharov, B.V., et al. (eds.). Riga: Inst. of Biology, Latvian Acad. Sci., pp. 216–217

(in Russian).

Sidorenko, G.I. and Mozhayev, E.A. (1994) Issues of water hygiene abroad. *Gigiena Sanit.*, **3**: 12–17 (in Russian).

Sieburth, J.M. (1979) Sea microbes. New York: Oxford University Press, 491 pp.

Simakov, Y.G. (1986) Ontogenetic and toxicological aspects of protecting hydrobionts from anthropogenic impacts. DSc thesis. Moscow: Moscow University Press, 53 pp. (in Russian).

Sinelnikov, V.E. (1980) Mechanism of reservoir self-purification. Moscow: Stroyizdat Publishers, 111 pp. (in Russian).

Singer, M., George, S., and Tjeerdema, R. (1995) Relationship of some physical properties of oil dispersants and their toxicity to marine organisms. *Arch. Environm. Contam. Toxicol.*, **29 (1)**: 33–38.

Singer, M., George, S., Jacobson, S., Lee, I., Weetman, L., Tjeerdema, R., and Sowby, M. (1995) Acute toxicity of the oil dispersant Corexit 9554 to marine organisms. *Ecotoxicol. Environm. Safety*, **32 (1)**: 81–86.

Singer, M.M., Smalheer, D.L., Tjeerdema, R.S., and Martin, M. (1990) Toxicity of an oil dispersant to the early life stages of four marine species. *Environm. Toxicol. Chem.*, **9**: 1389–1397.

Sirenko, L.A. (1972) Physiological bases of the propagation of blue-green algae in a reservoir. Kicv: Naukova Dumka, 403 pp. (in Russian).

Sirenko, L.A. (1991) Express methods for studies of ecological modifications of phytocenoses. Ecological modifications and criteria of ecological normalization. Proc. of Int. Symposium. Leningrad: Russian Academy of Sciences, pp. 151–163 (in Russian).

Sirenko, L.A. (1992) Biochemical ecology of aquatic ecosystems and its problems (book review). *Gidrobiol. Zhurn.*, **28 (5)**: 108–109 (in Russian).

Sivak, A., Goyer, M., Permak, J., and Thayer, P. (1982) Environment and human health aspects of commercially important surfactants. *In* Solution behavior of surfactants. New York: Plenum Press, Vol. 1, 739 pp.

Skadovsky, S.N. (1955) Ecological physiology of aquatic organisms. Moscow: Sov. Nauka, 337 pp. (in Russian).

Skalskaya, I.A. and Flerov, B.A. (1999) Assessment of the condition of Upper Volga River (territory of the Yaroslavl Region) with respect to zooperiphyton. *Ekologiya*, **6**: 442–448 (in Russian).

Skarlato, O.A. (ed.) (1976) Methods of biological assaying of freshwaters. Leningrad: Nauka Publishers, 165 pp. (in Russian).

Skurlatov, Y.I. (1988) Basics of monitoring the quality of natural waters. *In* Ecological chemistry of an aquatic medium. Moscow: Center of International Projects, Vol. 1, pp. 230–255 (in Russian).

Skurlatov, Y.I., Duka, G.G., and Ernestova, L.S. (1983) Processes of toxication and mechanisms of natural water self-purification under conditions of anthropogenic impact. *Izv. Akad. Nauk Moldavian SSR, Ser. Biol. Khim. Nauk*, **5**: 3–20 (in Russian).

Slegosh, E.I. (1978) Technogenic phytopathology. Biological methods of assessing the environment. Moscow: Nauka Publishers, pp. 208–232 (in Russian).

Slepyan, E.I. (1988) Ecological toxicology and toxicological problems in ecology. Ecotoxicology and protection of the environment (Krivolutsky, D.A., Bocharov, B.V., et al. (eds.)). Riga: Inst. of Biology, Latvian Acad. Sci., pp. 162–163 (in Russian).

Smaal, A.C. and Widdows, J. (1994) Biomonitoring of coastal waters and estuaries.

Kramer, K.J.M. (ed.). Boca Raton: CRC Press, pp. 247–267.

Smirnov, N.A. (1989) On the trophic function of phytoplankton. Problems of man–biosphere interaction. Moscow: Moscow University Press, pp. 96–99 (in Russian).

Smirnov, N.A. (1994) The ecological structure of White Sea phytoplankton. DSc thesis. Moscow: Moscow University Press, 50 pp. (in Russian).

Smirnov, N.A., Perueva, E.G., Polyakova, T.V., and Fedorov, V.D. (1997) Assessment of the trophic function of White Sea phytovorous zooplankton. *Izv. RAN, Ser. Biol.*, **1**: 75–80 (in Russian).

Smirnova, A.N. (1976). The role of particular grouping of hydrobionts in the sanitary assessment of reservoirs by example of Sev. Donets River. Methods of biological assaying of freshwaters. Leningrad: USSR Academy of Sciences, pp. 147–148 (in Russian).

Smirnova, N.N. (1980) Ecological and chemical peculiarities of the root system of river-bank aquatic plants. // *Gidrobiol. Zhurn.*, **16 (3)**: 60–72 (in Russian).

Sokolov, M.S. (1987) Introduction to biochemical ecology (review of the book by S.A. Ostroumov). *Agrokhimiya*, **7**: 135–136 (in Russian).

Solovyeva, A.A. and Pastukhova, E.V. (1981) Structural and functional changes of phytoplankton and benthos communities under the action of agricultural sewage. Materials of the 4th Congress of the All-Union Hydrobiological Society, December, 1–4). Kiev: Naukova Dumka, Part 1, pp. 155–156 (in Russian).

Sorokin, Y.I. (1970) On the aggregatedness of marine bacterioplankton. *Dokl. Akad. Nauk SSSR*, **192 (4)**: 905–906 (in Russian).

Sorokin, Y.I. (1973) Biogeochemical activity and trophic role of bacteria in marine reservoirs. *Zhurn. Obshch. Biol.*, **34 (3)**: 396–406 (in Russian).

Sorokin, Y.I. (1977) Coral reef communities. Biology of the ocean. Vol. 2, Biological productivity of the ocean (Vinogradov, M.E. (ed.)). Moscow: Nauka Publishers, pp. 133–155 (in Russian).

South, G.R. and Whittick, A. (1987) Introduction to Phycology. Oxford: Blackwell Publishing.

Spellman, F. (1996) Stream ecology and self-purification. Lancaster: Technomic, 134 pp.

Stagg, R. and Shuttleworth, T. (1987) Sites of interactions of surfactants with beta-adrenergic responses in trout (*Salmo gairdneri*) gills. *J. Comp. Physiol. B*, **157 (4)**: 429–434.

Stavskaya, S.S. (1981) Biological degradation of ionic surfactants. Kiev: Naukova Dumka, 116 pp. (in Russian).

Stavskaya, S.S. (1988) Introduction to biochemical ecology (review of the book by S.A. Ostroumov). *Fiziol. Biokhim. Kult. Rasten.*, **20 (1)**: 99–100 (in Russian).

Stavskaya, S.S. (1990) Interaction of surfactants with organisms in an aqueous medium. *Khim. Tekhnol. Vody*, **12 (3)**: 265–272 (in Russian).

Stavskaya, S.S., Krivets, I.A., Grigoryeva, T.Y., Samoilenko, T.S., and Nasto-yashchaya, N.I. (1989) Microbial purification of industrial and storm sewage waters from anionic surfactants. *Khim. Tekhnol. Vody*, **11 (3)**: 272–274 (in Russian).

Stavskaya, S.S., Udod, V.M., Taranova, L.A., and Krivets, I.A. (1988) Microbial purification of water from surfactants. Kiev: Naukova Dumka, 184 pp. (in Russian).

Steffann, S., Joly, J., Loubinoux, B., and Dizengremel, P. (1988) Effects of an alkyl-ammonium bromide on oxidative and phosphorylative properties of mitochondria isolated from *Aaricus bisporus*. *Pesticide Biochem. Physiol.*, **32 (1)**: 38–45.

Steinberg, C. and Geller, W. (1993) Biodiversity and interactions within pelagic nutrient cycling and productivity. Biodiversity and ecosystem function. Shulze, E.D. and Mooney, H.A. (eds.). Berlin: Springer, pp. 43–65.

Steinberg, C., Bernhardt, H., and Klapper, H. (1995) Handbuch angewandte Limnologie. München: Kessler Verlag.

Steinman, A. and McIntire, C. (1987) Effects of irradiance on the community structure and biomass of algal assemblages in laboratory streams. *Can. J. Fish. Aquatic Sci.*, **44**: 1640–1648.

Stepanov, A.M. (1988) The concept of ecotoxicology. Ecotoxicology and protection of the environment. (Krivolutsky, D.A., Bocharov, B.V., et al. (eds.)). Riga: Inst. of Biology, Latvian Acad. Sci., pp. 169–171 (in Russian).

Stephanson, M., Martin, M., and Tjeerdema, R.S. (1995) Long-term trends in DDT, polychlorinated biphenyls, and chlordane in California mussels. *Arch. Environm. Contam. Toxicol.*, **28 (4)**: 443–450.

Stoeckman, A. and Garton, D. (1995) Seasonal energy budget for zebra mussels in western Lake Erie: a method for estimating impacts on phytoplankton communities. *Am. Zool.*, **35 (5)**: 89.

Stom, D.I., Timofeeva, S.S., Belykh, L.I., Butorov, V.V., Kashina, N.F., Kojuva, O.M., and Dneprovskaya, N.M. (1980) The role of plants in self-purification of waters from phenol compounds. Self-purification and bioindication of polluted waters. Moscow: Nauka Publishers, pp. 101–108 (in Russian).

Strayer, D., Caraco, N., Cole, J., Findlay, S., and Pace, M. (1999) Transformation of freshwater ecosystems by bivalves. *BioScience*, **49 (1)**: 19–27.

Stroganov, N.S. (1972) Boundaries of sanitary hydrobiology and aquatic toxicology. Theory and practice of biological self-purification of polluted waters. Moscow: Nauka Publishers, pp. 12–19 (in Russian).

Stroganov, N.S. (1976a) Toxic pollution of reservoirs and degradation of aquatic ecosystems. *In* Results of science and technology. General ecology. Biocenology. Hydrobiology. Vol. 3, Aquatic toxicology. Moscow: VINITI, pp. 5–47 (in Russian).

Stroganov, N.S. (1976b) Comparative sensitivity of hydrobionts to toxicants. *In* Results of science and technology. General ecology. Biocenology. Hydrobiology. Vol. 3, Aquatic toxicology. Moscow: VINITI, pp. 151–176 (in Russian).

Stroganov, N.S. (1979) Reaction of hydrobionts on organotin compounds. Moscow: Moscow University Press, 182 pp. (in Russian).

Stroganov, N.S. (1981) Principles of the assessment of normal and pathological state of reservoirs in chemical pollution. Theor. problems of aquatic toxicology. Materials of the 3rd Soviet–American symposium (Borok). Leningrad: Nauka Publishers, pp. 16–29 (in Russian).

Stroganov, N.S. (1982) A short glossary of terms on aquatic toxicology. Yaroslavl: Yaroslavl State University, 44 pp. (in Russian).

Stroganov, N.S., Filenko, O.F., Lebedeva, G.D., Putintsev, A.I., Buzinova, N.S., Dmitrieva, A.G., Isakova, E.F., Kolosova, L.V., Korol, V.M., Krivenko, M.S., and Parina, O.V. (1983a) Basic principles of bioassaying sewage and the assessment of natural reservoirs' water quality. *In* Theoretical issues of bioassaying (Lukyanenko, V.I. (ed.)). Volgograd: Institute for the Biology of Inner Waters, USSR Acad. Sci., pp. 21–29 (in Russian).

Stroganov, N.S., Dmitrieva, A.G., and Korol, V.M. (1983b) Algae and macrophytes as objects of bioassaying. *In* Theoretical issues of bioassaying (Lukyanenko, V.I. (ed.)). Volgograd: Institute for the Biology of Inner Waters, USSR Acad. Sci., pp. 153–158 (in Russian).

Stugren, B. (1987) Introduction to Biochemical Ecology by S.A. Ostroumov (review). *Studia Univ. Babes-Bolyai, Biologia*, No 2, pp. 96–97.

Stuijfzand, S.C., Kraak, M.H.S., Wink, Y.A., and Davids, C. (1995) Short-term effects of nickel on the filtration rate of the zebra mussel *Dreissena polymorpha. Bull. Environm. Contam. Toxicol.*, **54 (3)**: 376–381.

Stupina, V.V. (1989) Significance of algae for solving environment protection problems. *In* Algae (Vasser, S.P. et al. (eds.)). Kiev: Naukova Dumka, pp. 167–169 (in Russian).

Surfactants (1984) *Chemical Week*, **135 (24)**: SAS3–SAS22.

Surfactants and raw materials for their production. Ecological aspects. Shebekino: Synthesis of surfactants, 140 pp. (in Russian).

Sushchenya, L.M. (1975) Quantitative regularities of crustacean nutrition. Minsk: Nauka i Tekhnika, 208 pp. (in Russian).

Sushchenya, L.M., Semenchenko, V.P., Semenyuk, G.A., and Trubetskova, I.L. (1990) Production of planktonic crustaceans and environmental factors. Minsk: Nauka i Technika, 58 pp. (in Russian).

Suzuki, J., Komatsuzawa, H., Sugai, M., Ohta, K., Kozai, K., Nagasaka, N., and Suginaka, H. (1997) Effects of various types of Triton X on the susceptibilities of methicillin-resistant staphylococci to oxacillin. *FEMS Microbiol. Lett.*, **153**: 327–331.

Swisher, R. (1987) Surfactant biodegradation. New York: Marcel Dekker Inc., 1085 pp.

Symonides, E. (1987) Introduction to *Biochemical Ecology*. Review of the book written by S.A. Ostroumov. *Wiadomosci Ekologiczne*, **33 (2)**: 199–201.

Talakin, Y.N., Savchenko, M.V., Nijaradze, M.Z., and Ivanova, L.A. (1985) Effect of powders of synthetic detergents on the immunobiological system and lipid exchange of experimental animals. *Gigiena Sanit.*, **2**: 79–80 (in Russian).

Telitchenko, M.M. (1982) Development and problems of sanitary hydrobiology in the USSR. *Gidrobiol. Zhurn.*, **18 (6)**: 22 (in Russian).

Telitchenko, M.M. and Doskoch, Y.E. (1971) Physiologically active compounds of biogenic origin. Moscow: Moscow University Press, 144 pp. (in Russian).

Telitchenko, M.M. and Ostroumov, S.A. (1990) Introduction to problems of biochemical ecology. Moscow: Nauka Publishers, 288 pp. (in Russian).

The IUCN invertebrate red data book (1983) Gland: IUCN, 632 pp.

Theory and practice of biological self-purification of polluted waters (Telitchenko, M.M. (ed.)). Moscow: Nauka Publishers, 240 pp. (in Russian).

Thiele, B., Gunther, K., and Schwuger, M.J. (1997) Alkylphenol ethoxylates: trace analysis and environmental behavior. *Chem. Rev.*, **97**: 3247–3272.

Thomas, O. and White, G.F. (1989) Metabolic pathway for the biodegradation of sodium dodecyl sulphate by *Pseudomonas* sp. C12B. *Biotechnol. Appl. Biochem.*, **11**: 318–327.

Tomas, C.R. (ed.) (1997) Identifying marine phytoplankton. Academic Press: San Diego, New York, Boston etc., 858 pp.

Toth, M., Gimes, G., and Hertelendy, F. (1987) Triton X-100 promotes the accumulation of phosphatidic acid and inhibits the synthesis of phosphatidylcholine in human decidua. *Biochim. Biophys. Acta*, **921 (3)**: 417–425.

Trägner, D. and Csordas, A. (1987) Biphasic interaction of Triton detergents with the erythrocyte membrane. *Biochem. J.*, **244 (3)**: 605–609.

Traina, S., McAvoy, D., and Versteeg, D. (1996) Association of linear alkylbenzene-sulfonates with dissolved humic substances and its effect on bioavailability. *Environm. Sci. Technol.*, **30 (4)**: 1300–1309.

Tretyakova, A.N. (1987) Algological assessment of the use of mineral fertilizers. Topical problems of modern algology. Kiev: Naukova Dumka, pp. 176–177 (in Russian).

Tretyakova, A.N. (1988) Use of algae as bioindicators in agriculture. Ecotoxicology and protection of the environment. (Krivolutsky, D.A., Bocharov, B.V., et al. (eds.)). Riga: Inst. of Biology, Latvian Acad. Sci., pp. 186–187 (in Russian).

Trikulenko, V.I. (1978) Biological action of some new detergents and the level of their harmfulness in their intake by the reservoirs. *Gigiena Sanit.*, **3**: 14–18 (in Russian).

Tsomides, H., Hughes, J., Thomas, J.M., and Ward, C. (1995) Effect of surfactant addition on phenanthren biodegradation in sediments. *Environm. Toxicol. Chem.*, **14**: 953–959.

Tsyban, A.V. and Mosharov, S.A. (1995) Biogenic sedimentation and its role in the transfer and deposition of pollutants in marine ecosystems. *Meteorol. Gidrologiya*, **11**: 63–71 (in Russian).

Tukaj, Z. (1994) The toxicity of three dispersed diesel fuel oils and dispersants towards some *Scenedesmus* (Chlorococcales) species. *Oceanologia*, **36 (2)**: 175–186.

Unified criteria for the quality of waters (1982) Moscow: Council for Mutual Economic Assistance, 70 pp. (in Russian).

Unified methods of water quality studies (1975) Part 3: Methods of biological analysis of waters. Moscow: Council for Mutual Economic Assistance, 176 pp. (in Russian).

US EPA (1982) Seed germination/root elongation toxicity tests. EG-12. Office of Toxic Substances, Washington, DC.

US Food and Drug Administration (1987) Seed germination and elongation. FDA environment technical assistance document 4.06. US Department of Health and Human Services, Washington, DC.

Varfolomeyev, S.D., Spivak, S.I., and Zavialova, N.V. (1997) Rates and dynamics of bioremediation. Wild, J.R., Varfolomeyev, S.D., and Scozzafava, A. (eds.). Perspectives in bioremediation: technologies for environmental improvement. NATO ASI Series. 3. High Technology, Vol. 19. Kluwer Academic Publishers: Dordrecht, pp. 39–56.

Vasser, S.P., Kondratyeva, N.V., Masyuk, N.P., et al. (1989) Algae. Kiev: Naukova Dumka, 608 pp. (in Russian).

Vasternak, K. and Ostroumov, S.A. (1990) Impact of polluting an aquatic medium with synthetic washing agent Bio-S on euglena. *Gidrobiol. Zhurn.*, **26 (6)**: 78–79 (in Russian).

Vasilyev, L.A. and Vasilyev, A.L. (1993) Use of natural biocenoses of reservoirs in purification of natural waters. *Vodosnabzh. Santekhn.*, **11–12**: 20–21 (in Russian).

Vavilin, V.A. (1983) Nonlinear models of biological purification and self-purification processes in rivers. Moscow: Nauka Publishers, 158 pp. (in Russian).

Vavilin, V.A. (1986) Biomass turnover time and degradation of organic matter in biological purification systems. Moscow: Nauka Publishers, 143 pp. (in Russian).

Vavilin, V.A. and Vasilyev, V.B. (1979) Mathematical modeling of sewage biological purification by activated sludge. Moscow: Nauka Publishers, 119 pp. (in Russian).

Vavilin, V.A., Vasilyev, V.B., and Rytov, S.V. (1992) Modeling of organic matter degradation by a microbial community. Moscow: Nauka Publishers, 202 pp. (in Russian).

Veldre, I.A., Itra, A.R., and Paalme, L.P. (1977) Experimental study of the effect of detergents on the stability of benz(a)pyrene. *Gigiena Sanit.*, **3**: 89–90 (in Russian).

Venitsianov, E.V. (1992) On the state of Russia's water resources, primary problems of water management, and the concept of increasing the water-resource potential of Russia. *Biol. Nauki*, **8**: 13–16 (in Russian).

Vernadsky, V.I. (1926) Biosphere. Leningrad: Nauchn. Khim.-Tekhn. Izd., 146 pp. (in Russian).

Vernadsky, V.I. (1944) Several words on the noosphere. *Usp. Sovr. Biol.*, **18 (2)**: 113–120 (in Russian).

Versteeg, D., Stalmans, M., Dyer, S., and Janssen, C. (1997a) *Ceriodaphnia* and *Daphnia*: a comparison of their sensitivity to xenobiotics and utility as a test species. *Chemosphere*, **34**: 869–892.

Versteeg, D., Stanton, D., Pence, M., and Cowan, C. (1997b) Effects of surfactants on the rotifer, *Brachionus calyciflorus*, in a chronic toxicity test and in the development of QSARs. *Environm. Toxicol. Chem.*, **16 (5)**: 1051–1058.

Veselova, T.V., Veselovsky, V.A., and Chernavsky, D.S. (1993) Stress in plants. Moscow: Moscow University Press, 145 pp. (in Russian).

Vetrov, V.A. and Chugay, V.V. (1988) Invertebrates as indicators of pollution of background freshwater ecosystems by heavy metals. Problems of ecological monitoring and modeling of ecosystems. Leningrad: Gidrometeoizdat Publishers, V. 11, pp. 61–75 (in Russian).

Vig, N. and Kraft, M. (eds.) (1994) Environmental policy in the 1990s. Washington, DC: Congressional Quarterly Inc., 442 pp.

Vinberg, G.G. (ed.) (1973) Biological processes and self-purification at a polluted area of a river. Minsk: Belorussian University Press, 192 pp. (in Russian).

Vinberg, G.G. (ed.) (1980) Benthos of the Uchinskoe reservoir. Moscow: Nauka Publishers, 252 pp. (in Russian).

Vinberg, G.G., Alimov, A.F., Balushkina, E.V., Nikulina, V.N., Finogenova, V.P., and Tsalolikhin, S.Y. (1977) Experience of using various systems of biological indication of polluted waters. Scientific bases for monitoring the quality of waters by hydrobiological parameters. Proc. Soviet-American Seminar. Leningrad: Gidrometeoizdat Publishers, pp. 124–131 (in Russian).

Vinogradov, M.E. (ed.) (1977) Biology of the ocean. Vol. 2. Biological productivity of the ocean. Moscow: Nauka Publishers, 400 pp. (in Russian).

Vinogradov, M.E. and Shushkina, E.A. (1987) Functioning of the ocean epipelageal planktonic communities. Moscow: Nauka Publishers, 240 pp. (in Russian).

Vives-Rego, J., Vaque, D., and Martinez, J. (1986) Effect of heavy metals and surfactants on glucose metabolism, thymidine incorporation and exoproteolytic activity in sea water. *Water Res.*, **20 (11)**: 1411–1415.

Vlasenko, A.G., Krainyukova, A.N., Nesmeyanov, Y.M., and Katrichenko, G.N. (1983) Instrumental express method of assessing the toxicity of sewage by fish escapement reaction. *In* Theoretical issues of bioassaying (Lukyanenko, V.I., (ed.)). Volgograd, Institute for the Biology of Internal Waters, USSR Acad. Sci., pp. 100–103 (in Russian).

Volkov, I.V., Zalicheva, I.N., Moiseeva, V.P., Samylin, A.F., and Kharin, V.N. (1997) Regional aspects of aqueous toxicology. *Vodn. Resursy*, **5**: 556–562 (in Russian).

Voskresensky, K.A. (1948) The filter feeder belt as a biohydrological marine system. Proc. State Oceanographic Inst., Issue 6 (18), pp. 55–120 (in Russian).

Wallace, K. and Niemi, G. (1988) Structure–activity relationships of species selectivity in acute chemical toxicity between fish and rodents. *Environm. Toxicol. Chem.*, **7 (3)**: 201–212.

Wang, W. (1985) Use of millet root elongation for toxicity tests of phenolic compounds. *Environm. Intern.*, **11**: 95–98.

Wang, W. (1987) Root elongation method for toxicity testing of organic and inorganic pollutants. *Environm. Toxicol. Chem.*, **6**: 409–414.

Wang, W. and Williams, J. (1990) The use of phytotoxicity tests (common duckweed, cabbage, and millet) for determining effluent toxicity. *Environm. Monitor. Assessment*, **14**: 45–58.

Watanabe, R., Ogasawara, N., Tanaka, H., and Uchiyama, T. (1988) Effect of fungal lytic enzymes and nonionic detergents on the actions of some fungicides against *Pyricularia oryzae*. *Agric. Biol. Chem.*, **52 (4)**: 895–901.

Waterby, J. and Ostroumov, S.A. (1994) Action of a non-ionogenic surfactant on cyanobacteria. *Mikrobiologiya*, **63 (2)**: 258–262 (in Russian).

Weiner, J. and Rudy, B. (1988) Effects of detergent on the binding of solubilized sodium channels to immobilized wheat germ agglutinin: structural implications. *Biochim. Biophys. Acta*, **944 (3)**: 521–526.

Weiner, R. and Ostroumov, S.A. (1998) Triton X-100. *Toksikol. Vestnik*, **4**: 42–43 (in Russian).

Weiner, R., Devine, R., Powell, D., Dagasan, L., and Moore, R. (1985) *Hyphomonas oceanitis* sp. nov., *Hyphomonas hirschiana* sp. nov., and *Hyphomonas jannaschiana* sp. nov. *Int. J. System. Bacteriol.*, **35**: 237–243.

Wetzel, R.F., Hatcher, P., and Bianchi, T. (1995) Natural photolysis by ultraviolet irradiance of recalcitrant dissolved organic matter to simple substrates for rapid bacterial metabolism. *Limnol. Oceanogr.*, **40**: 1369–1380.

Widdows, J. and Page, D.S. (1993) Effects of tributyltin and dibutyltin on the physiological energetics of the mussel *Mytilus edulis*. *Mar. Environ. Res.*, **35**: 233–249.

Wilson, J. and Fraser, F. (1977) Handbook of teratology. General principles. New York: Plenum Press, 476 pp.

Witteberg, C. and Triplett, E. (1985) A detergent-activated tyrosinase from *Xenopus laevis*. *J. Biol. Chem.*, **260 (23)**: 12535–12541.

Wong, S. (1985) Algal assay evaluation of trace contaminants in surface water using the nonionic surfactant, Triton X-100. *Aquatic Toxicol.*, **6**: 115–131.

World Resources 1990–1991 (1990) New York, Oxford: Oxford University Press, 383 pp.

World Resources 1994–1995 (1994) New York, Oxford: Oxford University Press, 404 pp.

Wotton, R., Malmqvist, B., Muotka, T., and Larsson, K. (1998) Fecal pellets from a dense aggregation of suspension feeders in a stream: an example of ecosystem engineering. *Limnol. Oceanogr.*, **43 (4)**: 719–725.

Wraige, E., Donkin, P., Rowland, S., and Lowe, D. (1994) The effect of a model low MW UCM (unresolved complex mixture of oil hydrocarbons) on the feeding rate of mussels. Paper presented at SETAC–Europe (UK Branch) 5th Annual

Meeting, University of Sheffield, September 13–14.

Yablokov, A.V. and Ostroumov, S.A. (1979) Protection of the animal world: problems and prospects. Moscow: Znanie Publishers, 64 pp. (in Russian).

Yablokov, A.V. and Ostroumov, S.A. (1983) Protection of the living nature: problems and prospects. Moscow: Lesnaya Promyshlennost Publishers, 272 pp. (in Russian).

Yablokov, A.V. and Ostroumov, S.A. (1985) Levels of protection of the living nature. Moscow: Nauka Publishers, 176 pp. (in Russian).

Yablokov, A.V. and Ostroumov, S.A. (1989a) Conservacion de la naturaleza viva. Problemas y perspectivas. Moscow: Vneshtorgizdat Press, Editorial Cientifico-Tecnica, 238 pp.

Yablokov, A.V. and Ostroumov, S.A. (1989b) Opazvane na zhivata priroda. Sofia: Zemizdat, 192 pp. (in Bulgarian).

Yablokov, A.V. and Ostroumov, S.A. (1991) Conservation of living nature and resources: problems, trends and prospects. Springer Verlag: New York, etc., 271 pp.

Yakovlev, V.A. (2000) The trophic structure of zoobenthos as an indicator of the state of aqueous ecosystems and quality of water. *Vodn. Resursy*, **27 (2)**: 237–244 (in Russian).

Yamada, H. and Hirari, S. (1986) Initiation of cleavage in starfish eggs by the injection of Triton-treated spermatozoa. *Gamete Research*, **13 (13)**: 135–141.

Yamane, A., Okada M., and Sudo, R. (1984) The growth inhibition of planktonic algae due to surfactants used in washing agents. *Water Res.*, **18 (9)**: 1101–1105.

Yamaoka, K., Kato, M., and Kamihara, T. (1989) Effects of alkylammonium ions on the activity of dissimilatory nitrate reductase in the cells of *Pseudomonas denitrificans*. *Agric. Biol. Chem.*, **53 (5)**: 1451–1452.

Yamasu, T. and Mizofuchi, S. (1989) Effects of synthetic, neutral detergent and red clay on short-term measurement of O_2 production in an Okinawan reef coral. *Galaxea*, **8 (1)**: 127–142.

Yanusheva, N.Y., Voloshchenko, O.I., Chernichenko, I.A., and Balenko, N.V. (1982) On the blastomagenic action of some surfactants – components of detergents. *Gigiena Sanit.*, **7**: 9–12 (in Russian).

Yurmalietis, R.V. (1987) Fauna and ecology of Ostracoda in Latvian reservoirs. PhD thesis, Moscow: Moscow University Press, 18 pp. (in Russian).

Yusfina, E. and Leontyeva, Z. (1975) Physiological processes in the animal organism at the impact of synthetic surfactants. *In* The physiological role of surfactants. Chernovtsy, pp. 120–122 (in Russian).

Zaika, V.E. (ed.) (1992) Perennial changes of Black Sea zoobenthos. Kiev: Naukova Dumka, 248 pp. (in Russian).

Zaika, V.E., Valovaya, N.A., Povchun, A.S., and Revkov, N.K. (1990) Black Sea mytilides. Kiev: Naukova Dumka, 208 pp. (in Russian).

Zaitsev, Y.P. and Golovenko, V.K. Effect of DDT on the Black Sea mollusk *Mytilus edulis*. *Dokl. Akad. Nauk Ukrainian SSR*, **10**: 72–75 (in Russian).

Zak, G.L. (1960) Self-purification of reservoirs. Moscow: Izd. Kommun. Khoz. RSFSR, 160 pp. (in Russian).

Zakharov, V. (ed.) (1999) Priorities for Russia's environmental policy. Moscow: Center for Russian Environmental Policy, 96 pp. (in Russian).

Zamolodchikov, D.G. (1993) Assessment of ecologically admissible levels of anthropogenic impact on freshwater ecosystems. *In* Problems of ecological monitoring and modeling of ecosystems. Leningrad: Gidrometeoizdat Publishers, Vol.

15, pp. 214–233 (in Russian).

Zatsepin, V.I. and Filatova, Z.A. (1968) Class of bivalves ((Bivalvia). *In* Life of animals, Vol. 2 (Zenkevich, L.A., ed.). Moscow: Prosveshchenie Publishers, pp. 95–111 (in Russian).

Zatsepin, V.I. and Rittikh, L.A. (1975) Lists of mass and characteristic invertebrates – macrofauna of benthic communities of Northern, Southern, and Far-eastern seas of the USSR. Moscow: Moscow University Press, 88 pp. (in Russian).

Zatsepin, V.I., Rittikh, L.A., and Krasnova, G.L. (1978) Lists of mass and characteristic forms of macrofauna of benthic biocenoses of the continental reservoirs of the European part of the USSR. Moscow: Moscow University Press, 78 pp. (in Russian).

Zavarzin, G.A. (1984) Bacteria and the composition of the atmosphere. Moscow: Nauka Publishers, 192 pp. (in Russian).

Zeifert, D.V., Petrov, S.S., and Rudakov, K.M. (1991) Ecological modifications of the phytocenoses of higher aqueous plants in the mid-river Belaya under the action of anthropogenic pollution. Ecological modifications and the criteria of ecological normalization. Proc. of Int. Symposium. Leningrad: Gidrometeoizdat Publishers, pp. 198–212 (in Russian).

Zevina, G.B. and Lebedev, E.M. (1985) Ecology of marine fouling. Ecological bases of the protection against biodamage. Moscow: Nauka Publishers, pp. 64–77 (in Russian).

Zezina, O.N. (1997) Contemporary brachiopods within the natural benthic biofilter strata of Russia's seas. Moscow: Paleontological Institute, Russian Acad. Sci., 84 pp. (in Russian).

Zhadin, V.I. (1964) Bottom-dwelling biocenoses of Oka River and their changes in 35 years. Pollution and self-purification of Oka River. Moscow, Leningrad: Nauka Publishers, pp. 226–288 (in Russian).

Zhmur, N.S. (1997) Monitoring of the process and control of the result of purification of sewage waters in installations with aeration tanks. Moscow: Luch Publishers, 172 pp. (in Russian).

Zhukinsky, V.N., Oksinyuk, O.P., Oleinik, G.N., and Kosheleva, S.I. (1980) Criteria for the integrated assessment of the quality of surface freshwaters. Self-purification and bioindication of polluted waters. Moscow: Nauka Publishers, pp. 57–63 (in Russian).

Zhang, Y. and Prepas, E. (1996) Regulation of the dominance of planctonic diatoms and cyanobacteria in four eutrophic hardwater lakes by nutrients, water column stability, and temperature. *Can. J. Fish Aquat. Sci.*, **53**: 621–633.

Zinchenko, T.D. and Rosenberg, G.S. (eds.) (1996) Ecological status of the Chapaevka River basin under anthropogenic impact (biological indexation): ecological safety and stable development of the Samara Region. Issue 3, Institute of Volga Basin Ecology, Togliatti, 342 pp. (in Russian).

Zlochevskaya, I.V., Absalyamov, S.U., Galimova, L.M., and Reshetnikova, I.A. (1981) On the mechanism of the fungicidal action of quaternary ammonium salts. *Nauchn. Dokl. Vyssh. Shkoly, Biol. Nauki*, **3**: 80–84 (in Russian).

Recent Data on Bioeffects of Surfactants: Addendum to the English Edition

In addition to the text of the book that was written some time ago, the author would like to mention some more recent relevant information and references.

More information became available about the production and consumption of surfactants.

According to some evaluations, in 1995, the estimated global use of the major classes of surfactants was: linear alkyl benzene sulfonates (LAS), 1.77 million metric tons (MMT); alcohol ethoxylates (AE), 0.35 MMT; alkylphenol ethoxylates (APE), 0.32 MMT; alcohol sulfates (AS), 0.3 MMT (Britton 1998). However, those evaluations probably did not include the use of surfactants in Russia and some adjacent countries.

Since 2000, several books were published that covered the issues as diverse as: environmental impact of surfactants and detergents (distribution, behavior, fate, and effects of surfactants and their degradation products in the environment; biodegradation assessments of detergent/surfactants; toxicology and ecotoxicology of detergent chemicals; predicting environmental concentrations for use in detergent risk assessments; life cycle assessment: a novel approach to the environmental profile of detergent consumer products; biodegradability and toxicity of surfactants; the biodegradability of detergent ingredients in an environmental context; environmental risk assessment of surfactants, and quantitative structure–activity relationships for aquatic toxicity. Those issues were covered, e.g., in the *Handbook of Detergents: Environmental Impact* (2004) edited by Uri Zoller. Environmental aspects of surfactants were also considered in the book by Knepper T., Barcelo D., de Voogt P., entitled *Surfactants: Properties, Production and Environmental Aspects*, 2003 (occurrence of surfactants in the environment; toxicity), and in the book entitled *Emulsions, Foams, and Suspensions: Fundamentals and Applications* by Laurier L. Schramm (2005). The same scientist, L. Schramm, also edited the book entitled *Surfactants: Fundamentals and Applications in the Petroleum Industry* (2000). Additional information on the books about surfactants may be found at: http://www.esia.it/asp/Bibliografie/Ing_Ind_200111/surfactants.htm.

A useful review of literature is available on the Internet (Britton 1998). Britton concludes his extensive review with the words: "The overall conclusion is that, despite the huge volumes of surfactants entering the environment, they do not represent a serious threat to it." This conclusion is a typical opinion about surfactants, which our data and analysis (see the main part of the book) do not support.

In recent literature, new data on ecotoxicology of surfactants were published. Some of the data are mentioned below.

Anionic Surfactants

The data on ecotoxicology of surfactants include assessments of biological effects of linear alkyl benzene sulfonates (LAS). Thus, LAS produced some impact on phosphatase activity in testis of the teleostean fish, *Heteropneustes fossilis* (Bloch) (Trivedi et al. 2001). In the same species of fish, LAS produced a haematotoxic effect (Rani et al. 2002a). LAS changed the erythrocyte sedimentation rate of *Heteropneustes fossilis* (Rani et al. 2002b). Toxicity of LAS to early life stages of the marine fish, seabream (*Sparus aurata*) was found (Hampel and Blasco 2002, Hampel et al. 2002). LAS may produce genotoxic effects, which was demonstrated using *Tetrahymena pyriformis* (Wu et al. 1992). Integration of data on aquatic fate and ecological responses to LAS in model stream ecosystems was done by Belanger et al. (2002).

LAS produced toxic effects on metabolic activity, growth rate, and microcolony formation of *Nitrosomonas* and *Nitrosospira* strains (Brandt et al. 2001).

Dodecyl alkyl sulfates were also actively studied. Ecotoxicological data were obtained also when investigating effects of dodecyl alkyl sulfate, e.g., on protistan communities (McCormick et al. 1997).

SDS (0.5 mg/l) inhibited water filtration by oysters *Crassostrea gigas* (Ostroumov 2003).

Toxicity of anionic detergents to *Saccharomyces cerevisiae* was also shown using microarray analysis (Sirisattha et al. 2004).

Nonionic Surfactants

Triton X-100 inhibited the filtration rate of marine mussels *M. edulis* and freshwater bivalves *Unio tumidus* (Ostroumov 2005). It was shown that Triton X-100 (TX100) at a concentration of 0.5–5 mg/l induced a sharp slowdown in water filtration by *M. edulis* (Ostroumov 2002). At concentrations of 1 and 5 mg/l, TX100 caused a decrease in the filtration rate by freshwater mussels *Unio tumidus* (Ostroumov 2002).

Cationic Surfactants

Tetradecyl trimethyl ammonium bromide (TDTMA), 2 mg/l, inhibited the feeding rate of mollusks *Lymnaea stagnalis* by 65.5% (when feeding on phytomass of leaves of macrophytes) (Ostroumov 2000a).

TDTMA, 1 mg/l, inhibited the filtration rate of juveniles of mussels *Mytilus galloprovincialis* Lam., when grazing (feeding on) algae *Monochrysis lutheri*; the removal of algae *Monochrysis lutheri* was decreased (Ostroumov 2000b).

Recently new data were obtained that demonstrated that TDTMA inhibited water filtration by natural hybrids *Mytilus edulis* × *M. galloprovincialis* (Ostroumov and Widdows 2004).

Surfactant TDTMA (0.5 mg/l) caused a pronounced decrease in the filtration rate by oysters *Crassostrea gigas* (Ostroumov et al. 2003).

Effects of TDTMA on freshwater bivalves were also studied. TDTMA at 1–2 mg/l constrained the filtering activity of freshwater bivalve *Unio pictorum* (Ostroumov 2002).

Not only bivalves but also other filter feeders were used in assessment of ecological hazards from cationic surfactants. TDTMA (0.5 mg/l) inhibited feeding by culture of rotifers *Brachionus calyciflorus* in the turbidostat (Ostroumov et al. 2003).

In the presence of TDTMA (2 mg/l), the feeding rate of mollusks *Lymnaea stagnalis* was decreased by 27.9–70.9%. Under the effect of the cationic surfactant, TDTMA (2 mg/l), the production of pellets per 1 g of the wet weight of mollusks was decreased by 41.7% (Ostroumov and Kolesnikov 2003).

Amphoteric Surfactants

Surface-active chemicals with both alkaline and acidic properties are known as amphoteric surfactants. They are used in all-purpose and industrial cleaning agents, and in personal care products (e.g., hair shampoos and conditioners, liquid soaps, hand dishwashing agents, and cleansing lotions). The production and consumption of amphoteric surfactants is expected to increase in future because of the request for milder surfactants. An Internet website presents a review of data on two groups of amphoteric surfactants, namely betaines (environmental fate including aerobic and anaerobic biodegradability, effects on the aquatic environment, effects on human health) and imidazoline derivatives (environmental fate, effects on the aquatic environment, effects on human health) (see "Amphoteric surfactants," the Website listed in the references). EC_{50} values were determined for the green algae *Scenedesmus subspicatus* affected by cocoamidopropyl betaine. For this species of algae, the EC_{50} was between 0.55 and 48 mg/l, the geometric mean was 3.1 mg/l. For *Daphnia magna* and cocoamidopropyl betaine, NOEC (48 h) was 1.6 mg/l. For zebrafish (*Brachydanio rerio*) and cocoamidopropyl betaine, NOEC (96 h) was 1.7 mg/l. Betains and amphoteric imidazoline derivatives (cocoamphodiacetate, alkylamphopropionates, alkylamphoacetates) are not included in Annex 1 of the list of dangerous substances of Council Directive 67/548/EEC.

Environmental aspects of amphoteric surfactants were summarized at a Website (Amphoteric surfactants, http://www.mst.dk/udgiv/publications/2001/87-7944-596-9/html/kap06_eng.htm. It covers information on betaines (environmental fate, effects on the aquatic environment, effects on human health) and on imidazoline derivatives.

Some information on amphoteric surfactants as viewed by the U.S. EPA is presented at the website http://www.epa.gov/fedrgstr/EPA-PEST/2005/February/Day-10/p2620.htm (also in the list of references).

Modern *In Vitro* Tests for Surfactants

Using cell lines and assays of cell proliferation and viability, valuable data on cyto-toxicity of surfactants were obtained. The cytotoxicity of two surfactants, benzalko-nium chloride (BAC) and polyoxyethylene-20-stearyl ether (Brij®78, PSE) was independently studied in four laboratories in the E.U. by using an immortalized human corneal epithelial (HCE) cell line. After one-hour exposure, the average EC_{50} value in BAC-treated cells in the presence of serum was 0.0650 ± 0.0284 (mean $\pm$ SD) mM. The average EC_{50} value for PSE was 0.0581 ± 0.300 mM (Huhtala et al. 2003).

Mixtures of Chemicals

Some papers analyze the environmental hazards from surfactants that are compo-nents of mixtures. Among the mixtures, important examples are pesticides, dispers-ants, laundry detergents, and surfactant-based fire-suppressant foams.

Pesticides

There are indications that surfactants that are components of the pesticide glyphosate pose some hazard to environment and may produce negative effects of aquatic organ-isms. Among the publications on surfactants that are components of glyphosate is the study of the polyethoxylated tallowamine surfactant (POEA) commonly used in glyphosate-based herbicides (Howe et al. 2004). It was toxic to the North American amphibian species *Rana clamitans* and *R. pipiens*. The tadpoles (*R. pipiens*) chronically exposed to environmentally relevant concentrations of POEA or glypho-sate formulations containing POEA showed decreased snout-vent length at metamor-phosis and increased time to metamorphosis, tail damage, and gonadal abnormalities. These effects may be caused, in some part, by disruption of hormone signaling (Howe et al. 2004). Data on environmental hazards from glyphosate (Glyphosate factsheet, 2004) are presented at: https://secure.virtuality.net/panukcom/subs.htm. Another source of the data is the website cited as *"Glyphosate herbicides, Roundup, 2005"* (see the list of references).

Discussion of toxicity of the herbicide Roundup (which contains surfactants) to amphibians is presented at: http://forums.gardenweb.com/forums/load/natives/msg 0812580527444.html.

Additional discussion of new information on toxicity of surfactants that are components of herbicides including Roundup is on the website: http://www.

environmentalhealthnews.org/archives.jsp?sm=fr18%3Bcontaminationagent12%3
B5Surfactants11%3Bsurfactants.

Oil dispersants

There is additional evidence on the surfactant-based toxicity of oil dispersants to aquatic invertebrates, such as developing nauplii of *Artemia* (Cotou et al., 2001). National Academies Press published information on *"Oil Spill Dispersants: Efficacy and Effects"* (2005) at http://www.nap.edu/books/030909562X/html/289.html.

Detergents

Warne and Schifko studied the components of laundry detergents that were toxic to a freshwater cladoceran (Warne and Schifko 1999).

Detergents Lotos-Extra (25 and 50 mg/l) and Tide-Lemon (33 and 43 mg/l) inhibited the filtration by *Mytilus galloprovincialis*; no increase in mortality was observed at this concentration (Ostroumov 2000c). In another study, four detergents (6.7–50 mg/l) hindered the filtration by *M. galloprovincialis* (Ostroumov 2002). In the same study, it was shown that three detergents (1–30 mg/l) slowed down the clearance rate during water filtration by *Crassostrea gigas*.

Three synthetic detergent mixtures inhibited water filtration by *Mytilus galloprovincialis* and *Crassostrea gigas* (Ostroumov 2003).

Two liquid detergents, Biospul and Kashtan, arrested the elongation of plant seedlings (Ostroumov and Khoroshilov 2001).

Surfactant-based fire-suppressant foams

Laboratory studies were conducted to determine the acute toxicity of surfactant-based fire-suppressant foams (FireFoam 103B, FireFoam 104, Fire Quench, ForExpan S, and Pyrocap B-136), and anionic surfactants (linear alkylbenzene sulfonate [LAS] and sodium dodecyl sulfate [SDS]) to juvenile rainbow trout *Oncorhynchus mykiss* (Buhl and Hamilton 2000). Toxicity (96-h concentration lethal to 50% of test organisms [96-h LC_{50}]) for the ammonia-based fire retardants was as follows: Phos-Chek 259F,168 mg/l; Fire-Trol LCA-F, 942 mg/l; Fire-Trol LCM-R, 1,141 mg/l. The toxicity for the foams was as follows: FireFoam 103B, 12.2 mg/l; Fire-Foam 104, 13.0 mg/l, ForExpan S, 21.8 mg/l; Fire Quench, 39.0 mg/l; Pyrocap B-136, 156 mg/l. Except for Pyrocap B-136, the foams were more toxic than the fire retardants. Linear alkylbenzene sulfonate (5.0 mg/l) was about five times more toxic than SDS (24.9 mg/l). Based on estimated anionic surfactant concentrations at the 96-h LC_{50} of the foams and reference surfactants, LAS was intermediate in toxicity and SDS was less toxic to rainbow trout when compared with the foams. Comparisons of recommended application concentrations to the test results indicate that accidental inputs of these chemicals into streams require substantial dilutions

(100–1,750-fold) to reach concentrations nonlethal to rainbow trout (Buhl and Hamilton 2000).

Rawlings et al. (2004) studied mixture toxicity of anionic surfactants to the fathead minnow (*Pimephales promelas*). *Pimephales promelas* was used in acute and chronic tests with the anionic surfactants sodium dodecylbenzene sulfonate (LAS), sodium dodecyl sulfate (AS) and tetradecyl diethylene sulfate (AES) singularly and in a mixture. The acute mixture exposure at equitoxic concentrations of LAS, AS and AES resulted in LC_{50} and LBB_{50} values of 0.039 mmol/l and 0.17 mmol/kg. EC_{20} and EBB_{20} values in the chronic mixture study were 0.012 mmol/l and 0.082 mmol/kg. Both acute and chronic mixture data compared well with single compound data. The laboratory studies conform to concentration addition models for water column and tissue-based exposures. Rawlings et al. concluded that it is likely that tissue-based assessments will prove most predictive for environmental risk assessments.

Other mixtures and formulations

The shampoo AHC (Avon Herbal Care) at concentrations of 5–60 mg/l arrested the filtration by *M. galloprovincialis* (Ostroumov 2002).

Detailed data on surfactants as components of many agricultural formulations are presented at the website cited on the list of references (*Surfactants and Other Additives in Agricultural Formulations*, 2001).

Biodegradation of Surfactants and Some Technologies to Treat Polluted Water

The relevant information on environmental role of surfactants includes the data on biodegradation of surfactants. The issues of biodegradation were considered by many authors. Among them findings on LAS degradation by immobilized *Pseudomonas aeruginosa* under low intensity ultrasound (Lijun et al. 2005).

Biodegradation of LAS by activated sludge was studied by Temmink and Klapwijk (2004).

New modifications of treatment of wastewater that contains anionic surfactants using electrospray were studied by Mantzavinos et al. (2001).

Prospects for using phytotechnologies to treat waters polluted with surfactants were analyzed and several factors were studied that affect LAS removal in subsurface flow constructed wetlands (Huang et al. 2004).

References

Amphoteric surfactants (2001) http://www.mst.dk/udgiv/publications/2001/87-7944-596-9/html/kap06_eng.htm.

Amphoteric surfactants as viewed by U.S. EPA (2005) http://www.epa.gov/fedrgstr
/EPA-PEST/2005/February/Day-10/p2620.htm.

Books about surfactants (2001) http://www.esia.it/asp/Bibliografie/Ing_Ind_2001
11/surfactants.htm.

Belanger, S., Bowling, J.W., Lee, D.M., LeBlanc, E.M., Kerr, K.M., McAvoy, D.C.,
Christman, S.C., and Davidson, D.H. (2002) Integration of aquatic fate and
ecological responses to linear alkyl benzene sulfonate (LAS) in model stream
ecosystems. *Ecotoxicol. Environ. Saf.*, **52(2)**: 150–171.

Brandt, K.K., Hesselsoe, M., Roslev, P., Henriksen, K., and Sorensen, J. (2001) Toxic
effects of linear alkylbenzene sulfonate on metabolic activity, growth rate, and
microcolony formation of *Nitrosomonas* and *Nitrosospira* strains. *Appl. Environ.
Microbiol.*, **67(6)**: 2489–2498.

Britton, L.N. (1998) Surfactants and the Environment. http://www.cler.com/new/
britton.html.

Buhl, K. J., and Hamilton S.J. (2000) Acute toxicity of fire-control chemicals,
nitrogenous chemicals, and surfactants to rainbow trout. *Trans. Am. Fisheries
Society*, **129**: 408–418.

http://www.npwrc.usgs.gov/resource/fish/fchem/fchem.htm (Version 06JUL2000).

Cotou, E., Castritsi-Catharios, I., and Moraitou-Apostolopoulou, M. (2001) Surfactant-based oil dispersant toxicity to developing nauplii of *Artemia*: effects on
ATPase enzymatic system, *Chemosphere*, **42(8)**: 959–964.

Glyphosate factsheet (2004): https://secure.virtuality.net/panukcom/subs.htm.

Glyphosate herbicides, Roundup (2005) http://maritimes.indymedia.org/news/
2005/09/10959.php.

Hampel, M. and Blasco, J. (2002) Toxicity of linear alkylbenzene sulfonate and one
long-chain degradation intermediate, sulfophenyl carboxylic acid on early
life-stages of seabream (*Sparus aurata*). *Ecotoxicol. Environ. Saf.* **51(1)**: 53–59.

Hampel, M., Moreno-Garrido, I., and Blasco, J. (2002) Can early life-stages of the
marine fish *Sparus aurata* be useful for the evaluation of the toxicity of linear
alkylbenzene sulphonates homologues (LAS C10-C14) and commercial LAS?
ScientificWorldJournal, **2**: 1689–1698.

Howe, C. M., Berrill, M., Pauli, B.D., Helbing, C.C., Werry, K., and Veldhoen, N.
(2004) Toxicity of glyphosate-based pesticides to four North American frog
species. *Environm. Toxicol. Chem.*, **23(8)**: 1928–1938.

Huang, Y., LaTorre, A., Barcelo, D., Garcia, J., Aguirre, P., Mujeriego, R., and
Bayona, J.M. (2004) Factors affecting linear alkylbenzene sulfonates removal in
subsurface flow constructed wetlands. *Environ. Sci. Technol.*, **38(9)**: 2657–2663.

Huhtala, A., Alajuuma, P., Burgalassi, S., Chetoni, P., Diehl, H., Engelke, M.,
Marselos, M., Monti, D., Pappas, P., Saettone, M. F., Salminen, L., Sotiro-
poulou, M., Tähti, H., Uusitalo, H., and Zorn-Kruppa, M. (2003) A collaborative
evaluation of the cytotoxicity of two surfactants by using the human corneal
epithelial cell line and the WST-1 test. *J. Ocular Pharmacol. Therapeutics*, **19**:
11–21.

Knepper, T., Barcelo, D., and de Voogt, P. (2003) Surfactants: Properties, production and environmental aspects (ISBN: 0-444-50935-6), Elsevier, 960 pp.

Lijun, X., Bochu, W., Zhimin, L., Chuanren, D., Qinghong, W., and Liu, L. (2005)
Linear alkyl benzene sulphonate (LAS) degradation by immobilized *Pseudomo-
nas aeruginosa* under low intensity ultrasound. *Colloids and Surfaces B:
Biointerfaces*, **40(1)**: 25–29.

Mantzavinos, D., Burrows, D.M., Willey, R., Lo Biundo, G., Zhang, S.F., Livingston,

A.G., and Metcalfe, I.S. (2001) Chemical treatment of an anionic surfactant wastewater: Electrospray-MS studies of intermediates and effect on aerobic biodegradability. *Water Res.*, **35(14)**: 3337–3344.

McCormick, P.V., Belanger, S.E., and Cairns, J. (1997) Evaluating the hazard of dodecyl alkyl sulphate to natural ecosystems using indigenous protistan communities. *Ecotoxicology*, **6**: 67–85.

Oil Spill Dispersants: Efficacy and Effects (2005) http://www.nap.edu/books/030909562X/html/289.html.

Orme, S. and Kegley, S. (2000–2004) [Alkanolamide surfactants] PAN Pesticide Database, Pesticide Action Network, North America (San Francisco, CA): http: www.pesticideinfo.org.

Ostroumov, S.A. (2000a) Tetradecyltrimethylammonium bromide (TDTMA). *Toxicological Bulletin* (in Russian, Toxicologicheskiy Vestnik = Toxicological Review, Moscow, ISSN 0869-7922) **1**: 42–43.

Ostroumov, S.A. (2000b) Tetradecyltrimethylammonium bromide (TDTMA). *Toxicological Bulletin* (in Russian, Toxicologicheskiy Vestnik = Toxicological Review, Moscow, ISSN 0869-7922), **3**: 34–35.

Ostroumov, S.A. (2000c) Detergents Lotos-Extra and Tide-Lemon. *Toxicological Bulletin* (in Russian, Toxicologicheskiy Vestnik = Toxicological Review, Moscow, ISSN 0869-7922), **4**: 35–37.

Ostroumov, S.A. (2002) Inhibitory analysis of top-down control: new keys to studying eutrophication, algal blooms, and water self-purification. *Hydrobiologia*, **469**: 117–129.

Ostroumov, S.A. (2003) Studying effects of some surfactants and detergents on filter-feeding bivalves. *Hydrobiologia*, **500**: 341–344.

Ostroumov, S.A. (2005) On some aspects of maintaining water quality and its self-purification. *Vodn. Resursy*, **32**: 337–347 (in Russian).

Ostroumov, S.A. and Khoroshilov, V.S. (2001) Liquid detergents Biospul and Kashtan. *Toxicological Bulletin* (in Russian, Toxicologicheskiy Vestnik = Toxicological Review, Moscow, ISSN 0869-7922), **6**: 41–43.

Ostroumov, S.A. and Kolesnikov, M.P. (2003) Mollusks in biogeochemical flows (C, N, P, Si, Al) and water self-purification: effects of surfactants. *Vestnik MGU* (Bulletin of Moscow University), Ser. 16, Biology (ISSN 0201-7385; ISSN 0137-0952), **1**: 15–24.

Ostroumov, S.A. and Widdows, J. (2004) Effects of cationic surfactant on mussels: inhibition of water filtration. *Vestnik MGU* (Bulletin of Moscow University), Ser. 16, Biology (ISSN 0201-7385; ISSN 0137-0952), **4**: 38–41.

Ostroumov, S.A., Walz, N., and Rusche, R. (2003) Effect of a cationic amphiphilic compound on rotifers. *Doklady Biological Sciences*, **390**: 252–255 (translated from *Doklady Akademii Nauk*, ISSN 0012-4966; distributed by Springer, orderdept@springer-sbm.com).

Rani, R., Trivedi, S.P., Singh, P., and Singh, R.K. (2002a) Haematotoxic effect of linear alkyl benzene sulphonate (LAS) in fish, *Heteropneustes fossilis. J. Environ. Biol.*, **23(1)**: 101–103.

Rani, R, Trivedi, S.P., Singh, P., and Singh, R.K. (2002b) Effect of linear alkyl benzene sulphonate on erythrocyte sedimentation rate of teleost fish, *Heteropneustes fossilis. J. Environ. Biol.*, **23(2)**: 213–214.

Rawlings, J., Belanger, S., Price, B., Smith, C., and Versteeg, D. (2004) Body burdens and mixture toxicity assessment of anionic surfactants to the fathead minnow (*Pimephales promelas*). http://abstracts.co.allenpress.com/pweb/setac

2004/document/?ID=41405.

Schramm, L.L. (2005) Emulsions, foams, and suspensions: Fundamentals and applications (ISBN: 3-527-30743-5), Wiley, 464 pp.

Schramm, Laurier L. (ed.) (2000) Surfactants: Fundamentals and applications in the petroleum industry. Cambridge University Press, Cambridge, 621 pp.

Sirisattha, S., Momose, Y., Kitagawa, E., and Iwahashi, H. (2004) Toxicity of anionic detergents determined by *Saccharomyces cerevisiae* microarray analysis. *Water Res.*, **38(1)**: 61–70.

Surfactants and other additives in agricultural formulations (2001) http://www.marketresearch.com/map/prod/722276.html

Surfactants in herbicides: http://www.environmentalhealthnews.org/archives.jsp?sm=fr18%3Bcontaminationagent12%3B5Surfactants11%3Bsurfactants; and http://forums.gardenweb.com/forums/load/natives/msg0812580527444.html.

Temmink, H. and Klapwijk, B. (2004) Fate of linear alkylbenzene sulfonate (LAS) in activated sludge plants. *Water Res.*, **38(4)**: 903–912.

Trivedi, S.P., Kumar, M., Mishra, A., Banerjee, I., and Soni, A. (2001) Impact of linear alkyl benzene sulphonate (LAS) on phosphates activity in testis of the teleostean fish, *Heteropneustes fossilis* (Bloch). *J. Environ. Biol.*, **22(4)**: 263–266.

Warne, M.S. and Schifko, A.D. (1999) Toxicity of laundry detergent components to a freshwater cladoceran and their contribution to detergent toxicity. *Ecotoxicol. Environ. Saf.*, **44(2)**: 196–206.

Wu, Y. and Shen, Y. (1992) Genotoxic effects of linear alkyl benzene sulfonate, sodium pentachlorophenate and dichromate on *Tetrahymena pyriformis*. *J. Protozool.*, **9(4)**: 454–456.

Zoller, U. (ed.) (2004) Handbook of detergents: Environmental impact. New York: Marcel Dekker, 1120 pp.

Subject Index

T - #0115 - 101024 - C0 - 234/156/16 [18] - CB - 9780849325267 - Gloss Lamination